CORVETTE
SHOP MANUAL
1966-82
Covers all small-block
and big-block
Chevrolet Corvettes

CON

The contents list enables the reader to find any section by means of the black tabs on the edge of the page. The tab on the first page of each section is in line with the section name in the contents.

Full maintenance, repair, troubleshooting and tune-up information for home mechanic and professional alike

MOTORBOOKS

First published in 1987 by Motorbooks, an imprint of MBI Publishing Company, Galtier Plaza, Suite 200, 380 Jackson Street, St. Paul, MN 55101-3885 USA

Motorbooks titles are also available at discounts in bulk quantity for industrial or sales-promotional use. For details write to Special Sales Manager at MBI Publishing Company, Galtier Plaza, Suite 200, 380 Jackson Street, St. Paul, MN 55101-3885 USA.

ISBN-13: 978-0-87938-236-0
ISBN-10: 0-87938-236-8

This manual is intended to provide the car owner and the professional mechanic with information necessary to perform the required service operations. The information, illustrations, and specifications in this manual are those available at the time of original publication. No responsibility can be assumed for design or specification changes made to the cars by the manufacturer which in any way differ from those contained in the manual.

This manual is published independently of General Motors, Chevrolet Division.

Portions of materials contained herein have been reprinted with permission of General Motors Corporation Service Technology Group.

Printed in the United States of America

TUNE-UP .. MAINTENANCE .. LUBRICATION 1

1

INDEX

COMPRESSION TEST

Remove air cleaner and block throttle and choke in wide open position. Hook up starter remote control cable and insert compression gauge firmly in spark plug port. Whenever the engine is cranked remotely at the starter, with a jumper cable or other means, the distributor primary lead must be disconnected from the negative post on the coil and the ignition switch must be in the "ON" position. Failure to do this will result in a damaged grounding circuit in the ignition switch.

Crank engine through at least four compression strokes to obtain highest possible reading. Check and record compression of each cylinder. If one or more cylinders read low or uneven, inject about a tablespoon of engine oil on top of pistons in low reading cylinders (through spark plug port). Crank engine several times and recheck compression. If compression comes up but does not necessarily reach normal, rings are worn. If compression does not improve, valves are burnt, sticking or not seating properly. If two adjacent cylinders indicate low compression and injecting oil does not increase compression, the cause may be a head gasket leak between the cylinders. Engine coolant and/or oil in cylinders could result from this defect.

Carburetor adjustments

The adjustments described apply to all carburetors used, except as noted. All adjustments are made with the engine at normal operating temperature.

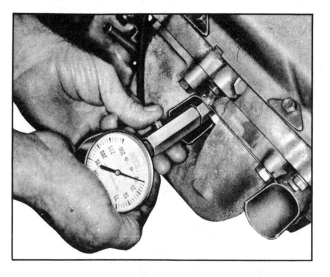

Checking compression

Idle speed and mixture — 66-70

Remove distributor vacuum line at distributor and plug hose. Start engine and set the throttle stop screw for recommended idle speed. The choke valve must be wide open and the fast idle inoperative. Adjust one idle mixture screw at a time for smoothest, fastest idle speed. On

A.I.R. cars, turn one adjusting screw at a time until engine speed drops approx. 30 rpm and starts to roll (lean mixture), then turn screw out exactly 1/4 turn for final setting. Readjust throttle stop screw for recommended idle speed. Basic setting for idle mixture screws is 2 turns open from fully closed for 1966-68, 3 turns for 1969, 1 turn for Holley 2300, Air conditioner to be ON, except Mark IV and all 1972.

When adjusting the idle speed be sure that the idle compensator is closed. Close it manually if necessary. After idle speed is adjusted, check by pressing down on the compensator. If speed drops, readjust idle speed.

Note: Idle speed adjustments on cars with automatic transmissions must be made with transmission in Drive and idle stator switch, if so equipped, closed. Be sure parking brake is on.

Idle speed 1971

The 1971 models use a Combined Emission Control Valve (C.E.C.) to adjust idle speed and adjust ignition according to demand.

Disconnect the vacuum hose at the distributor and plug, remove gas tank cap, and be sure A/C is off. Manually extend C.E.C. valve plunger to contact the throttle lever at the limit of its travel. Adjust plunger length to obtain specified rpm. Adjust idle mixture screw as previously described, if necessary.

Idle speed 1972-1976

The idle stop solenoid requires two idle speed settings. The curb idle speed is normal engine idle speed, the low idle speed is set for conditions when the solenoid is de-energized, as when the ignition is turned off. This prevents engine run-on. Idle speed adjustments are made with the engine at normal operating temperature, air cleaner on, choke open, air conditioning off, and fuel tank

hose from vapor canister disconnected (on later models). Set parking brake and block driving wheels. Disconnect electrical connector at idle stop solenoid. With automatic transmission in Drive, or manual transmission in neutral, turn low idle screw to obtain low idle speed of 500 rpm, reconnect electrical connector to solenoid and crack throttle slightly, to extend solenoid plunger. Then turn solenoid plunger screw in or out to set curb idle speed of approximately 600 rpm for automatic transmission models in Drive, or approximately 850 rpm for manual transmission models in neutral. Shut off engine and reconnect fuel tank hose to vapor canister.

Idle speed 1977-1979

See emission label on vehicle. Set engine for adjustments. Set ignition timing. For carburetors without solenoid and with air conditioner off, turn idle speed screw to set curb idle speed to specifications. For carburetors with solenoid, energize solenoid, disconnect air conditioner at compressor, turn air conditioner on, set A/T in drive, M/T in neutral and turn solenoid screw to adjust speed to specified RPM.

Idle mixture 1972-1976

The idle mixture is factory preset and the screws are capped with plastic limiter caps. These caps allow about one full turn for adjustment. If more is required remove the caps or break off the tabs with needle nose pliers.

Adjusting the idle mixture is done with the engine running at normal operating temperature, air cleaner on, choke open, air conditioning off, and fuel tank hose from canister disconnected. Set the parking brake and block the driving wheels. Put automatic transmission in Drive; manual transmission in neutral. Turn in or out one idle mixture screw at a time to obtain smoothest, fastest idle speed; then adjust the other screw to make engine run even smoother and faster. A basic starting point for adjusting screws is four turns out from fully in position with needle just touching seat. Readjust idle speed if necessary. Reconnect fuel tank hose from canister.

Idle mixture 1977-1979

Idle mixture screws have been preset at the factory and capped. Do not remove the caps during normal engine maintenance. Idle mixture should be adjusted only in the case of major carburetor overhaul, throttle body replacement or high idle CO level as determined by inspections.

Idle speed and mixture — Holley 2300

All adjustments are same as previously described except as follows: On models equipped with idle stop solenoid,

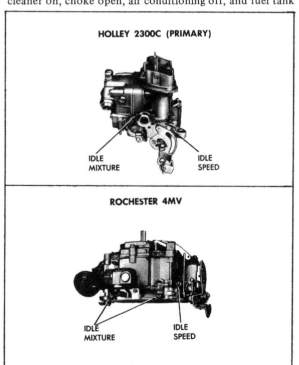

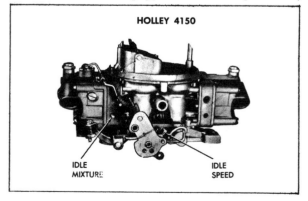

Idle speed and mixture screws

adjust idle stop solenoid screw to give 1000 rpm, then adjust idle mixture adjusting screw for highest steady rpm; readjust idle stop solenoid screw to specified rpm. Turn idle mixture screw in (leaner mixture) until engine speed drops 20 rpm, then turn out 1/4 turn. Disconnect lead at idle stop solenoid (throttle lever will rest against regular stopscrew). Adjust this stopscrew for idle speed of 500 rpm. Do not change setting of idle stop solenoid stopscrew or idle mixture screw.

Fast idle 1966-1976

With the transmission in neutral, position the cam follower on the high step (2nd step, 1971-72) of fast idle cam. Adjust fast idle screw of Rochester carburetors to obtain recommended fast idle speed. Bend fast idle lever on Holleys. On 1970-72 models, disconnect transmission controlled spark solenoid.

Fast idle 1977-1979

Use choke valve measuring gage J-26701. Rotate degree scale of tool until zero is opposite pointer. With choke completely closed place magnet squarely on top of choke valve. Rotate bubble until it is centered. Rotate scale so that number of degreees specified is opposite pointer. Place cam follower on second step of cam next to high step. Close choke by pushing upward on choke coil lever. To adjust, bend tang on fast idle cam until bubble is centered. Remove gage.

Dashpot adjustment

With slow idle speed correctly adjusted, fully open choke and make sure fast idle cam follower is off steps of cam. With dashpot fully compressed, adjust for 1/16" clearance between dashpot plunger and throttle lever.

Automatic choke

Remove the air cleaner and check to see that choke valve and rod move freely. Disconnect choke rod at choke lever.

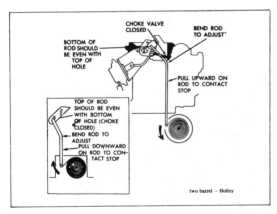

Automatic choke adjustment

Check choke adjustment by holding choke valve closed and position rod so that it contacts stop. If necessary, adjust rod length by bending rod at offset. Bend must be such that rod enters choke lever hole freely and squarely. Connect rod at choke lever and install air cleaner.

AIR INJECTION REACTOR (A.I.R.)
Description and operation

The A.I.R. system is used to burn the unburned portion of the exhaust gases to reduce its hydrocarbon and carbon monoxide content. The system forces compressed air into the exhaust manifold where it mixes with the hot exhaust gases. The hot exhaust gases contain unburned particles that complete their combustion when the additional air is supplied.

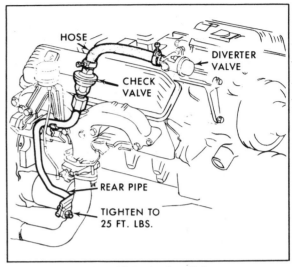

1975-76 A.I.R. pipe installation

The system consists of: an air pump, diverter valve, check valve(s), AIR pipe assemblies and connecting hoses and fittings. Carburetors and distributors for AIR engines are made to be used with the system and should not be replaced with components intended for use with engines that do not have the system.

The air pump is a two-vane pump which compresses fresh filtered air and injects it into the exhaust manifold. The pump consists of: a housing, centrifugal filter, set of vanes that rotate about the centerline of pump housing bore, the rotor, and the seals for the vanes. The centrifugal filter is replaced by first removing the drive belt and pump pulley; then pulling filter off with pliers. Care should be taken to prevent fragments from entering the air intake hole. Note: A new filter may squeal when first

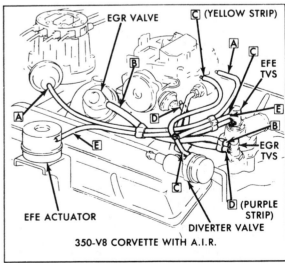

Emission hose routing

put into operation. Note: Great care should be taken in working on the air compressor as the aluminum used is quite soft and thin. The air pump is operating satisfactorily when the air flow from it increases as engine speed increases.

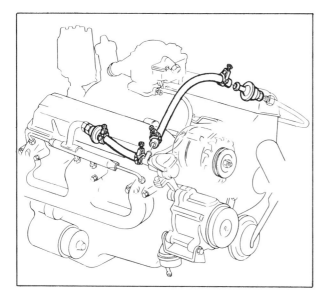

Small V8 A.I.R. hoses (big block similar)

The diverter valve is intended to keep the output of the air pump from the exhaust pipe during periods of deceleration, when the fuel mixture is unusually rich in fuel. If air from the pump were allowed to mix with the mixture, a backfire could occur. This unit cannot be serviced and must be replaced if it doesn't work.

The check valve is a one-way valve that allows air to flow in only one direction, from the pump to the exhaust manifold. This keeps any of the exhaust gases from entering the pump. The valve may be checked after removal from system by blowing through it by mouth. You should be able to blow through in only one direction. If not, the valve must be replaced.

The air hoses should be replaced only with hoses which are designed for AIR system use, as no other type hoses can withstand the high temperature.

Check and adjust dwell

Start engine then check ignition dwell. With engine running at idle, raise the adjustment screw window and insert an Allen wrench in the socket of the adjusting screw. Turn the adjusting screw as required until a dwell reading of 30° is obtained. A 2° variation is allowable for wear. Close access cover fully to prevent the entry of dirt into the distributor. If a dwell meter is not available, turn adjusting screw clockwise until engine starts to misfire, then turn screw one-half turn in the opposite direction to complete adjustment.

Slowly accelerate engine to 1500 rpm and note dwell reading. Return engine to idle and note dwell reading. If dwell variation exceeds specifications, check for worn distributor shaft, worn distributor shaft bushing or loose breaker plate.

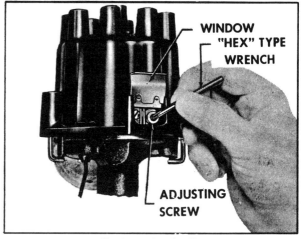

Setting point dwell

IGNITION SYSTEM

Remove distributor cap, clean cap and inspect for cracks, carbon tracks and burned or corroded terminals. Replace cap where necessary. Clean rotor and inspect for damage or deterioration. Replace rotor where necessary. Replace brittle, oil soaked or damaged spark plug wires. Install all wires to proper spark plug. Proper positioning of spark plug wires in supports is important to prevent cross-firing. Tighten all ignition system connections. Replace or repair any wires that are frayed, loose or damaged.

Ignition timing

Disconnect the distributor spark advance hose and plug the vacuum source opening. Start engine and run at idle speed (See tune up chart in Specifications section).

Ignition timing marks

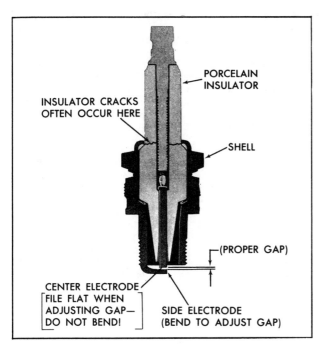

Spark plug detail

Aim timing light at timing tab. The markings on the tabs are in 2° increments (the greatest number of markings on the "A" side of the "Q"). The "O" marking is TDC and all BTDC settings fall on the "A" (advance) side of the "O". Adjust the timing by loosening the distributor clamp and rotating the distributor body as required, then tighten the clamp, and recheck timing. Stop engine and remove timing light and reconnect the spark advance hose.

Spark plugs

Inspect each plug individually for badly worn electrodes, glazed, broken or blistered porcelains and replace plugs where necessary. Clean serviceable spark plugs thoroughly, using an abrasive-type cleaner such as sand blast. File the center electrode flat. Inspect each spark plug for make and heat range. All plugs must be of the same make and number. Adjust spark plug gaps to .035 in. using a round feeler gauge. If available, test plugs with a spark plug tester. Inspect spark plug hole threads and clean before installing plugs. Install spark plugs with new gaskets and torque to specifications. Connect spark plug wiring.

Transistorized distributor   H.E.I. system

There are no moving parts in the ignition pulse amplifier, and the distributor shaft and bushings have permanent type lubrication, therefore no periodic maintenance is required for the magnetic pulse ignition system.

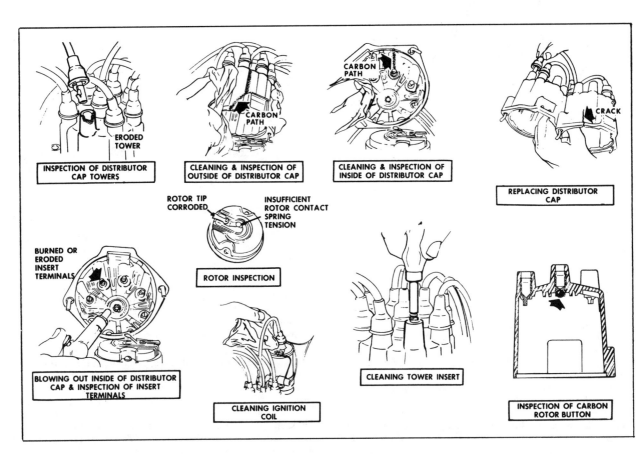

Cleaning and inspecting distributor cap, rotor, and coil

(Breaker Point) distributor

Check the distributor centrifugal advance mechanism by turning the distributor rotor in a clockwise direction as far as possible, then releasing the rotor to see if the springs return it to its retarded position. If the rotor does not return readily, the distributor must be disassembled and the cause of the trouble corrected.

Check to see that the vacuum spark control operates freely by turning the movable breaker plate counter-clockwise to see if the spring returns to its retarded position. Any stiffness in the operation of the spark control will affect the ignition timing. Correct any interference or binding condition noted.

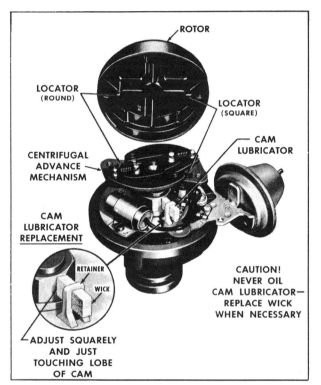

Distributor

Examine distributor points and clean or replace if necessary. Contact points with an overall gray color and only slight roughness or pitting need not be replaced. Dirty points should be cleaned with a clean point file. Use only a few strokes of a clean, fine-cut contact file. The file should not be used on other metals and should not be allowed to become greasy or dirty. Never use emery cloth or sandpaper to clean contact points since particles will embed and cause arcing and rapid burning of points. Do not attempt to remove all roughness nor dress the point surfaces down smooth. Merely remove scale or dirt. Clean cam lobe with cleaning solvent, and rotate cam lubricator wick end for end (or 180° as applicable). Replace points that are burned or badly pitted.

Where prematurely burned or badly pitted points are encountered, the ignition system and engine should be checked to determine the cause of trouble so that it can be eliminated. Unless the condition causing point burning

or pitting is corrected, new points will provide no better service than the old points. Refer to Section 6Y for an anlysis of point burning or pitting.

Check point alignment then adjust distributor contact point gap to .019" (new points) or .016" (used points). Breaker arm rubbing block must be on high point of lobe during adjustment. If contact points have been in service, they should be cleaned with a point file before adjusting with a feeler gauge.

Check distributor point spring tension (contact point pressure) with a spring gauge hooked to breaker lever at the contact and pull exerted at 90 degrees to the breaker lever. The points should be closed (cam follower between lobes) and the reading taken just as the points separate. Spring tension should be 19-23 ounces. If not within limits, replace. Excessive point pressure will cause excessive wear on the points, cam and rubber block. Weak point pressure permits bouncing or chattering, resulting in arcing and burning of the points and an ignition miss at high speed.

Install rotor and distributor cap. Press all wires firmly into cap towers.

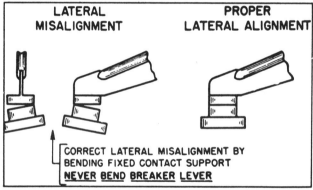

Point alignment

Battery and battery cables

Measure the specific gravity of the electrolyte in each cell. If it is below 1.230 (corrected to 80°F.) recharge with a slot rate charger, or if desired, further check battery. Connect a voltmeter across the battery terminals and measure the terminal voltage of the battery during cranking (disconnect the coil primary lead at the negative terminal during this check to prevent engine from firing). If the terminal voltage is less then 9.0 volts at room temperature, approximately 80° +20°F., the battery should be further checked. Inspect for signs of corrosion on battery, cables and surrounding area, loose or broken carriers, cracked or pluged cases, dirt and acid, electrolyte leakage and low electrolyte level. Fill cells to proper level with distilled water or water passed through a "demineralizer."

The top of the battery should be clean and the battery hold-down bolts properly tightened. Particular care should be taken to see that the top of the battery is kept clean of acid film and dirt. When cleaning batteries, wash first with a dilute ammonia or soda solution to neutralize any acid present and then flush off with clean water. Keep vent plugs tight so that the neutralizing solution does not enter the cell. The hold-down bolts should be kept tight enough to prevent the battery from shaking around in its holder, but they should not be tightened to the point where the battery case will be placed under a severe strain.

To insure good contact, the battery cables should be tight on the battery posts. Oil battery terminal felt washer. If the battery posts or cable terminals are corroded, the cables should be cleaned separately with a soda solution and wire brush. After cleaning and before installing clamps, apply a thin coating of petrolatum to the posts and cable clamps to help retard corrosion.

If the battery has remained undercharged, check for loose or defective fan belt, defective alternator, high resistance in the charging circuit, oxidized regulator contact points, or a low voltage setting. If the battery has been using too much water, the voltage output is too high.

Valve adjustment, hydraulic
Engine should be at normal operating temperature. Remove valve cover and start engine. While the engine is idling back off valve rocker arm nut until it starts to clatter. Turn nut down slowly until clatter stops; zero backlash. Turn nut down 1/4 turn more and wait 10 seconds. Turn nut additional 1/4 turns, waiting 10 seconds each time, until nut has been turned down one full turn from zero backlash. Repeat operation on other valves and replace valve cover. Check for leaks around cover, replace gasket if leaks appear.

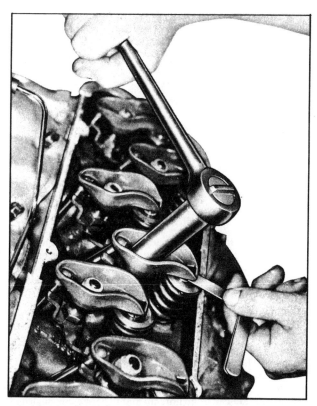

Valve adjustment (mechanical)

Valve adjustment, mechanical
Engine should be at normal operating temperature. Remove valve covers. Use a socket wrench on self-locking rocker arm stud nut and adjust as needed for proper clearance. See Specifications section. Be sure valves are fully closed. Check by removing distributor cap and noting toward which spark plug wire the rotor is pointing. Both valves of that cylinder are fully closed.

Adjustments may also be made with engine idling by inserting feeler gauge leaf between rocker arm and end of valve, while adjusting nut.

Manifold heat valve
Check manifold heat control valve for smooth operation. If shaft is sticking, free it up with manifold heat control solvent or its equivalent. Tap shaft end to end to help free it up.

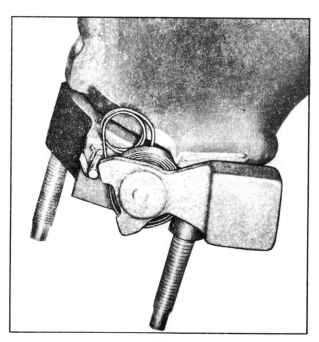

Manifold heat control valve

Manifold
Tighten intake manifold bolts. A slight leak at the intake manifold destroys engine performance and economy.

Cooling system
Inspect cooling system for leaks, weak hoses, loose hose clamps and correct coolant level, and service as required.

ADDITIONAL CHECKS AND ADJUSTMENTS
Testing crankcase ventilation valve
Connect tachometer and vacuum gauge as for idle speed and mixture adjustment. Set parking brake, start engine and adjust idle speed and mixture. Remove vent valve (PCV) from rocker cover grommet (with hose attached), block opening of valve and read engine rpm change. A decrease of less than 50 rpm indicates a plugged ventilation valve — replace the valve.

Installing PCV valve

Crankcase ventilation

Ventilation valve may be checked as outlined under "Additional Checks and Adjustments." Inspect for deteriorated or plugged hoses. Inspect all hose connections. On engines with closed element air cleaners, inspect crankcase ventilation filter and replace if necessary. On engines with open element air cleaners, remove flame arrestor and wash in solvent then dry with compressed air.

Crankcase ventilation filter

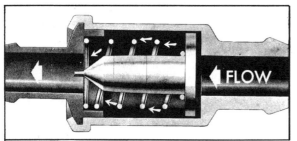

Crankcase ventilation valve

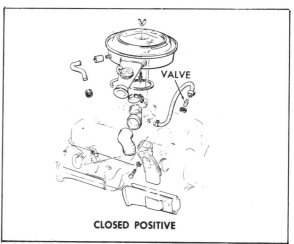

Crankcase ventilation systems

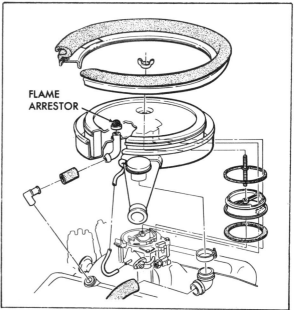

Air cleaner flame arrestor

Brakes

Check the brake fluid regularly, for as the brake pads wear the level will drop rapidly. It should be replenished only with the recommended fluid. Check disc brake assemblies to see if they are wet; it would indicate a leaking cylinder.

Disc brakes do not need periodic adjustments; they are self adjusting. The pads should be replaced when the friction material gets down to 1/16". This is when the groove in the center of the pad is gone. Check by removing wheel and looking directly into caliper.

Parking brake

To adjust the parking brake, jack up the car and remove wheel and knock-off wheel adapter, if used. Through the hole in disc insert a large screwdriver and engage adjustment ratchet. Pull handle downward to tighten brake, continue until disc will not move. Then back off 10 notches. Do this to both wheels. Pull up parking brake handle, inside car, four notches and tighten cable qualizer to provide light drag with wheel mounted. Fully released there should be no drag on wheels.

Parking brake, 1977-79

Raise the vehicle and remove the rear wheels. Loosen the equalizer check nuts until the levers move freely to the "off" position with slack in the cables. Turn the disc until the adjusting screw is visible through the hole in the disc. Insert a screwdriver and tighten the adjusting screw by moving the screwdriver handle upward. Adjust both sides. Tighten until the disc will not move, then back off six to eight notches. Install the wheels and place the brake handle in the applied position-13 notches. Tighten the check nuts until an 80 pound pull is required to pull the handle into the fourteenth notch. Torque the check nuts to 70 in. lbs. With the hand brake off, there should be no drag on either of the rear wheels.

Clutch pedal play

Check clutch action by holding pedal ½" from floor and move shift lever between first and reverse several times, with engine running. If shift is not smooth adjust clutch. Free play with pedal released is approx. 1-1/4" to 2" and 2" to 2-1/2" for heavy duty.

Clutch adjustment

At clutch lever near firewall remove clutch return spring. To decrease clutch pedal free play remove clutch pedal return spring and loosen lower nut on clutch pedal rod; take up play with upper nut. Continue until proper play is obtained, then securely tighten top nut and replace spring. To increase pedal play work nuts in opposite sequence.

Clutch adjustment, 1975 and later

Disconnect the clutch return spring at the cross shaft. Push the clutch lever until the pedal is against the rubber stop under the dash. Loosen the two shaft locknuts and push the shaft until the throwout bearing just touches the pressure plate spring. Tighten the top locknut toward the swivel until the distance between it and the swivel is 0.4". Tighten the bottom locknut against the swivel. The pedal free travel should now be 1½".

Accelerator linkage

Disconnect control linkage at carburetor throttle lever. Hold carburetor throttle lever in wide position. Pull control linkage to wide open position. (On vehicles equipped with automatic transmission, pull through detent.) Adjust control linkage to freely enter hole in carburetor throttle lever. Connect control linkage at throttle lever.

Throttle linkage adjustment, powerglide

Remove air cleaner, disconnect accelerator linkage at carburetor. Disconnect accelerator return and trans. rod return springs. Pull upper rod forward until transmission is through detent. Open carburetor wide open, at which point ball stud must contact end of slot in upper rod. Adjust swivel on end of rod if necessary.

Throttle linkage adjustment, Turbo Hydra-matic 350

Disengage the snap lock and position the carburetor to wide open throttle. Push the snap lock downward until the top is flush with the rest of the cable.

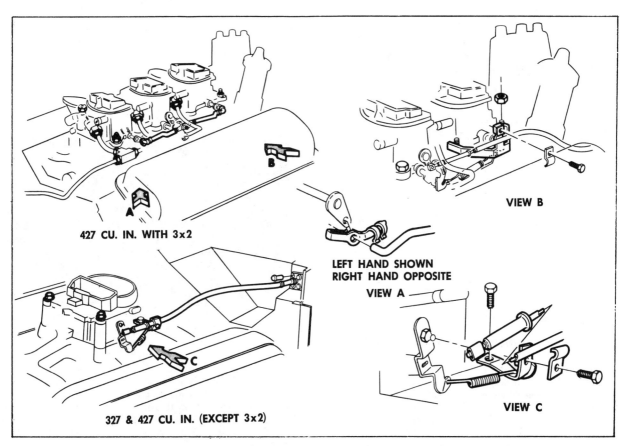

427 CU. IN. WITH 3x2

327 & 427 CU. IN. (EXCEPT 3x2)

VIEW B

LEFT HAND SHOWN
RIGHT HAND OPPOSITE
VIEW A

VIEW C

Throttle linkage

Throttle linkage adjustment, Turbo Hydra-matic 400
Pull detent switch driver to rear until hole in switch body
lines up with hole in driver. Insert a 3/16" pin through
holes to depth of 1/8", and loosen mounting bolts. Open
throttle fully and move switch forward until lever touches
accelerator lever. Tighten mounting bolt and remove pin.

EXHAUST GAS RECIRCULATION (EGR) SYSTEM
Description and operation
To reduce the formation of nitrous oxide (NOx) during
combustion, a small amount of inert gas from the exhaust system is introduced into the combustion process.
The EGR valve controls the amount of inert gas added in
relation to engine vacuum. A thermal switch opens vacuum to EGR valve when engine temperature reaches
about 100°. The EGR valve is designed to open the most
during periods of low vacuum, as when the car is
accelerating.

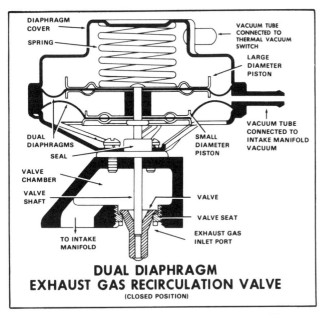

Dual Diaphragm EGR valve (similar to single)

EGR Valve Check
A rough idling engine may be caused by a malfunction of
the valve. Check by pinching vacuum hose to carburetor
with engine idling. If idling smooths out, the valve should
be removed for cleaning or replacement if something
appears to be broken.

Thermal Vacuum Switch Check
The switch should be open whenever the engine coolant
is above 100°, and closed when it is below. If switch is
defective it should be replaced.

TRANSMISSION CONTROLLED SPARK SYSTEM (TCS)
Description and operation
TCS is used only on cars with manual transmissions. It
eliminates ignition vacuum advance when the car is
operating in reverse, neutral or low forward gears. Corvettes have a dual override switch that cuts out the TCS
when coolant temperature is below 93° or above 232°F.
Is is also used on all models with automatic transmissions.

Transmission controlled spark system consists of a two-
position idle stop solenoid, vacuum advance solenoid,
time relay, temperature switch, and the transmission
switch. These individual components cannot be serviced
and must be replaced if defective.

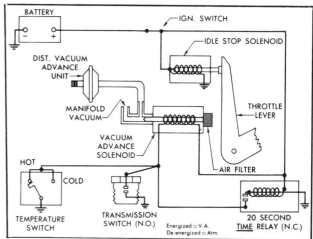

TCS system

The idle stop solenoid when de-energized (plunger retracted) allows the throttle plates to close beyond the idle
position; thereby shutting off air supply so engine will
shut down without misfiring. The vacuum advance solenoid supplies vacuum to the distributor when it is energized, and shuts it off when the solenoid is de-energized.
The time relay contains a bi-metal strip that requires
about 20 seconds to heat up after the ignition switch is
turned on. Engine must be started within that time, else
the switch has to be allowed to cool before it can go
through the 20 second starting cycle, if not, vacuum advance will be denied the distributor.

Transmission Switch Location [Manual]

The temperature switch is located on the right cylinder
head between #6 and #8 exhaust ports. It is a three
position switch that is in neutral position when engine
coolant temperature is in 'normal' range. The transmission switch is located on the outside of the transmission
case near the 2-3 or 3-4 shifter shaft. It is a spring loaded,
mechanically operated switch. The switch is closed in all
high forward gears, and open in all other gear positions.

LUBRICATION
Engine oil

The car should be standing on level ground and the oil level checked with the dipstick. Withdraw the dipstick, wipe it with a clean rag, replace and withdraw again. The mark made by the oil on the lower end of the dipstick will indicate the oil level. If necessary, oil should be added through the filler cap.

Never let the oil level fall so low that it does not show at all on the dipstick. If in doubt, it is better to have a bit too much oil than too little. Never mix oils of different brands, the additives may not be compatable.

Engine oil drain and replacement

Place a pan under the oil pan drain plug and remove plug. Be sure pan is of a large enough capacity to hold the oil. Move pan under filter and remove filter by turning it counterclockwise. Clean gasket surface of cylinder block. Coat gasket of new filter with engine oil. Thread filter into adapter. Tighten securely by hand. Do not over-

tighten filter. Remove drip pan.

Remove drain pan. Inspect oil pan drain plug gasket and replace if broken, cracked, or distorted. Install drain plug and tighten. Fill crankcase to required level with recommended oil. Operate engine at fast idle and check for oil leakage.

Crankcase capacity, 327 & 350 engines.....................4 qts.
427 & 454 engines.....................5 qts.
When changing oil filter, add......................................1 qt

Transmission, automatic

Check fluid level with engine idling, transmission in neutral and engine at normal operating temperature. Add fluid as needed to bring level to mark. Do not overfill.

Every 12,000 miles or sooner, depending on service, remove fluid from sump and add new fluid. Operate transmission and check fluid level. Every 24,000 miles the transmission sump strainer of the Turbo Hydra-Matic

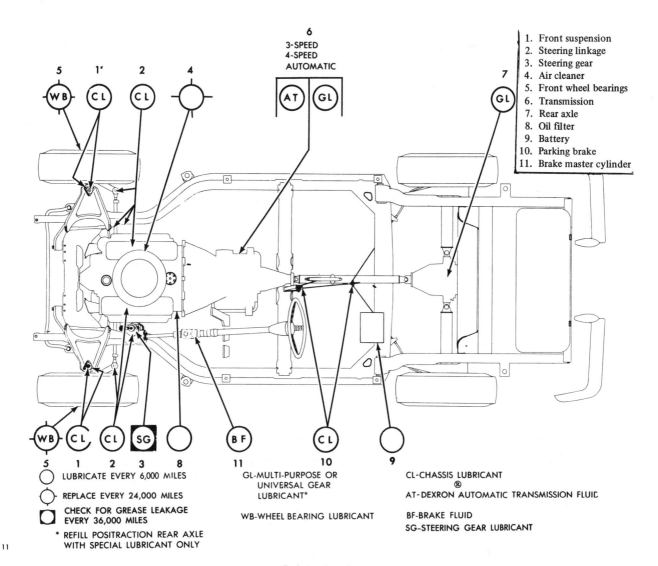

1. Front suspension
2. Steering linkage
3. Steering gear
4. Air cleaner
5. Front wheel bearings
6. Transmission
7. Rear axle
8. Oil filter
9. Battery
10. Parking brake
11. Brake master cylinder

6
3-SPEED
4-SPEED
AUTOMATIC

○ LUBRICATE EVERY 6,000 MILES

◑ REPLACE EVERY 24,000 MILES

▣ CHECK FOR GREASE LEAKAGE EVERY 36,000 MILES

* REFILL POSITRACTION REAR AXLE WITH SPECIAL LUBRICANT ONLY

GL-MULTI-PURPOSE OR UNIVERSAL GEAR LUBRICANT*

WB-WHEEL BEARING LUBRICANT

CL-CHASSIS LUBRICANT ®
AT-DEXRON AUTOMATIC TRANSMISSION FLUID

BF-BRAKE FLUID
SG-STEERING GEAR LUBRICANT

Lubrication diagram

transmission should be replaced.

Refill capacity, Powerglide..2 qts
Turbo Hydra-Matic.........................7½ pts

Distributor
Change cam lubricator end for end at 12,000 mile intervals. Replace at 24,000 mile intervals.

Transmission, manual
Raise car on lift, clean dirt and grease from area around the filler plug. Plug is located on side of transmission case. Remove plug and place finger tip inside hole. The oil should be just about level with the bottom edge of the hole. Add oil as needed, using a plastic syringe.

Differential
With the car standing level, clean dirt and grease from area around filler plug. Remove plug and place finger tip inside hole. The oil should be just about level with the bottom edge of the hole. Add oil, with a plastic syringe, as needed.

Steering gear
Check fluid level by removing lowest, outboard side cover retaining screw. If lubricant does not run out add enough through same hole until it does.

Power steering
Check fluid level with engine at normal operating temperature. Bring to full mark on stick, as required.

RECOMMENDED LUBRICANTS
Engine oil
Recommended SAE viscosity numbers and temperature range anticipated before next oil change.

Temperature	single grade	multi-grade
Min. below 5°F. .... SAE 10W		.... 10W-30
Min. between 5° and 32°F. .... SAE 20W		.... 20W-40
Between 32° and 95° .... SAE 30		.... 20W-40
Over 95°F. .... SAE 40		.... 20W-40

Automatic transmission ....Auto transmission type 'A'
Manual transmission ....multi-purpose SAE 90
extreme cold ....SAE 80
Differential, standard ....Multi-purpose SAE 90
positraction...............................Positraction lubricant
Steering box .... EP Chassis lubricant
Power steering pump ....Auto transmission type 'A'

MAINTENANCE SCHEDULE
The time or mileage intervals indicated in this section are intended as a guide for establishing regular maintenance and lubrication periods. Sustained heavy duty or high speed driving, or driving under adverse conditions may require more frequent servicing.

Every 300 miles or 2 weeks, whichever comes first, perform the following service items.
1. Check tire pressure
2. Check battery water level
3. Check oil level in engine

Every 3000 miles or 3 months, whichever comes first, add the following service items.
4. change engine oil
5. check transmission and rear end oil level
6. check distributor point gap
7. check or replace spark plugs
8. check engine idle
9. check thickness of disc brake pads
10. check brake system for leaks
11. check fan belt tension

Every 6,000 miles or 6 months, whichever comes first, add the following service items.
12. lubricate distributor and coat cam face
13. check steering gear box for oil
14. change oil filter
15. check ignition timing
16. check valve clearances
17. adjust clutch pedal, if necessary
18. check and tighten suspension parts
19. rotate wheels, check balance
20. check wires for loose or broken connections

Every 12,000 miles or 12 months, whichever comes first, add the following service items.
21. change brake fluid
22. check fuel pump operation
23. check specific gravity of battery electrolyte
24. check alternator, regulator and starter function
25. retighten cylinder head and manifold nuts and bolts
26. check engine mountings
27. replace distributor breaker points
28. replace air cleaner element
29. check crankcase ventilation system
30. check exhaust emission system
31. check brake vacuum — servo rubber parts
32. check and tighten steering mechanism
33. check wheel alignment
34. retighten transmission and rear end nuts and bolts

Every 24,000 miles or 24 months, whichever comes first, add the following service items.
35. lubricate handbrake linkage, wiper motor linkage, window regulator
36. change engine cooling water
37. check compression
38. change gasoline filter
39. overhaul carburetor
40. overhaul brake vacuum-servo unit
41. check shock absorbers
42. check exhaust system
43. check condition of valve timing belt

Every 30,000 miles or 30 months, whichever comes first, add the following service items.
44. change transmission and rear end oil
45. change wheel bearing grease and check bearing for wear
46. change grease in drive shaft splines & U-joints
47. overhaul master cylinder and wheel cylinders
48. check headlight aiming
49. change valve timing belt.

TROUBLESHOOTING 2

INDEX

INTRODUCTION

When using this troubleshooting section search out the probable cause in the left hand column first. Eliminate those components and causes which do not apply to your particular model. The center column will give the most likely cause of the condition. The right-hand column will provide the solution or cure.

In attempting to eliminate a condition, it is recommended that the simplest correction be attended to first, if the exact cause is not immediately apparent. Should the condition appear to be due to more than one component, make adjustment to one component at a time. This could eliminate much unnecessary work and expense.

FUEL SYSTEM

Condition	Possible Cause	Correction
POOR IDLING	(a) Idle air bleed carbonized or of incorrect size.	(a) Disassemble carburetor. Then, use compressed air to clear idle bleed after soaking it in a suitable solvent.
	(b) Idle discharge holes plugged or gummed.	(b) Disassemble carburetor. Then, use compressed air to clear idle discharge holes after soaking main and throttle bodies in a suitable solvent.
	(c) Throttle body carbonized or worn throttle shaft.	(c) Disassemble carburetor. Check throttle valve shaft for wear. If excessive wear is apparent, replace throttle body assembly.
	(d) Damaged or worn idle mixture needle.	(d) Replace throttle body assembly.
	(e) Low grade fuel or incorrect float level.	(e) Test fuel level in carburetor. Adjust as necessary to obtain correct float level.
	(f) Loose main body to throttle body screws.	(f) Tighten main body to throttle body screws securely to prevent air leaks.
	(g) Worn or corroded needle valve and seat.	(g) Clean and inspect needle valve and seat. If found to be in questionable condition, replace assembly. Then, test fuel pump pressure. Refer to Specifications for correct fuel pump pressure.
	(h) Incorrect valve lash.	(h) Adjust valves.
	(i) Engine miss (ignition.)	(i) Check ignition system.
POOR ACCELERATION	(a) Accelerator pump piston (or plunger) leather too hard, worn, or loose on stem.	(a) Disassemble carburetor. Replace accelerator pump assembly if leather is hard, cracked or worn. Test follow-up spring for compression.
	(b) Faulty accelerator pump discharge ball.	(b) Disassemble carburetor. Use compressed air to clean discharge nozzle and channels after soaking main body in a suitable solvent. Test fuel pump capacity.
	(c) Faulty accelerator pump inlet check ball.	(c) Disassemble carburetor. Check accelerator pump inlet, check ball for poor seat or release. If part is faulty, replace.

Condition	Possible Cause	Correction
	(d) Incorrect fuel or float level.	(d) Test fuel or float level in carburetor. Adjust as necessary to obtain correct float level.
	(e) Worn accelerator pump and throttle linkage.	(e) Disassemble carburetor. Replace worn accelerator pump and throttle linkage and measure for correct position.
	(f) Manifold heat valve sticking.	(f) Free up manifold heat control valve, using recommended solvent.
	(g) Incorrect pump setting.	(g) Reset pump.
CARBURETOR FLOODS OR LEAKS	(a) Cracked body.	(a) Disassemble carburetor. Replace cracked body. Make sure main to throttle body screws are tight.
	(b) Faulty body gaskets.	(b) Disassemble carburetor. Replace defective gaskets and test for leakage. Be sure screws are tightened securely.
	(c) High float level.	(c) Test fuel level in carburetor. Make necessary adjustment to obtain correct float level.
	(d) Worn needle valve and seat.	(d) Clean and inspect needle valve and seat. If found to be in a questionable condition, replace complete assembly and test fuel pump pressure. Refer to specifications for correct fuel pump pressure.
	(e) Excessive fuel pump pressure.	(e) Test fuel pump pressure. If pressure is in excess of recommended pressure (refer to Specifications), replace fuel pump.
POOR PERFORMANCE MIXTURE TOO RICH	(a) Restricted air cleaner.	(a) Remove and clean air cleaner or replace element.
	(b) Leaking float.	(b) Disassemble carburetor. Replace leaking float. Test float level and correct as necessary, to proper level.
	(c) High float level.	(c) Adjust float level as necessary to secure proper level.
	(d) Excessive fuel pump pressure.	(d) Test fuel pump pressure. Refer to specifications for recommended pressure. If pressure is in excess of recommended pressure, replace fuel pump assembly.
	(e) Worn metering jet.	(e) Disassemble carburetor. Replace worn metering jet, using a new jet of the correct size and type.
CARBURETOR MIXTURES LEAN	(a) Air leak bypassing carburetor.	(a) Repair.

ENGINE RUNS EXCESSIVELY RICH AFTER COLD START

CHOKE SYSTEM RICH	(a) Choke thermostat adjustment richer than specified.	(a) Correct.
	(b) Choke vacuum diaphragm inoperative or misadjusted.	(b) Correct or replace.
	(c) Choke vacuum passage blocked or leaking.	(c) Correct.
CARBURETOR RICH	(a) Incorrect gasket or gasket installation between carburetor and intake manifold.	(a) Replace or correct.

Condition	Possible Cause	Correction
EXCESSIVE STALLS AFTER COLD START		
	(a) Choke System Lean.	(a) Check items under "Poor Starting—Choke Valve Fails to Close."
	(b) Choke vacuum diaphragm adjustment lean.	(b) Adjust to specifications.
ENGINE OUTPUT LOW	(a) Fast idle speed low.	(a) Adjust to specification.
	(b) Fast idle cam position adjustment incorrect.	(b) Adjust to Specifications.
	(c) Engine lubrication oil of incorrect viscosity.	(c) Recommend No. 5W-20.
CARBURETOR LEAN	(a) Curb idle set very lean. (CAS Carbs.)	(a) Adjust to CAS Specifications.
	(b) Air leak bypassing the carburetor.	(b) Repair.

POOR COLD ENGINE STARTING

Condition	Possible Cause	Correction
CHOKE VALVE FAILS TO CLOSE	(a) Choke thermostat adjustment leaner than specified.	(a) Adjust.
	(b) Choke thermostat corroded such that it has cracked and distorted lean.	(b) Replace assembly.
	(c) Choke linkage, shaft or related parts corroded, bent or dirty such that the system is not entirely free to move from the open to the closed position.	(c) Repair, clean or replace.
	(d) Choke valve improperly seated.	(d) Reseat valve.
	(e) Air cleaner gasket interferes with choke valve or linkage.	(e) Reinstall gasket properly.
LOW ENGINE OUTPUT (10°F or lower)	(a) Engine lubricating oil or incorrect viscosity.	(a) Recommended 5W-20.
	(b) Valve lash incorrect.	(b) Readjust.
	(c) Choke thermostat adjustment incorrect, rich.	(c) Adjust to correct setting.

ENGINE RUNS LEAN, FIRST HALF MILE

Condition	Possible Cause	Correction
CHOKE LEAN	(a) Check items under (Poor Starting).	(a) See "Choke Valve Fails to Close."
	(b) Diaphragm adjustment lean.	(b) Readjust to specification.

ENGINE RUNS LEAN AFTER HALF MILE

Condition	Possible Cause	Correction
ENGINE HEAT INSUFFICIENT	(a) Heat valve stuck open.	(a) Free with solvent.
	(b) Heat valve thermostat distorted.	(b) Replace thermostat.
	(c) Heat valve failed within exhaust. See engine section for proper diagnosis.	(c) Replace heat valve.
	(d) Water temperature sub-normal.	(d) Check thermostat.

FUEL PUMP

Condition	Possible Cause	Correction
FUEL PUMP LEAKS—FUEL	(a) Worn, ruptured or torn diaphragm.	(a) Install new pump.
	(b) Loose diaphragm mounting plates.	(b) Install new pump.
	(c) Loose inlet or outlet line fittings.	(c) Tighten line fittings.
FUEL PUMP LEAKS—OIL	(a) Cracked or deteriorated pull rod oil seal.	(a) Install new pump.
	(b) Loose rocker arm pivot pin.	(b) Install new pump.

15

Condition	Possible Cause	Correction
INSUFFICIENT FUEL DELIVERY	(c) Loose pump mounting bolts.	(c) Tighten mounting bolts securely.
	(d) Defective pump to block gasket.	(d) Install new gasket.
	(a) Vent in tank restricted. (This will also cause collapsed fuel tank.)	(a) Unplug vent and inspect tank for leaks.
	(b) Leaks in fuel line or fittings.	(b) Tighten line fittings.
	(c) Dirt or restriction in fuel tank.	(c) Install new fuel filter and clean out tank.
	(d) Worn, ruptured, or torn diaphragm.	(d) Install new pump.
	(e) Frozen gas lines.	(e) Thaw lines and drain tank.
	(f) Improperly seating valves.	(f) Install new fuel pump.
	(g) Push rod worn	(g) Replace push rod.
	(h) Vapor lock.	(h) Install heat shield where lines or pump are near exhaust.
	(i) Low pressure.	(i) Install new fuel pump.
	(j) Incorrect fuel pump.	(j) Install correct fuel pump.
	(k) Restricted fuel filter.	(k) Install new filter.
FUEL PUMP NOISE	(a) Loose mounting bolts.	(a) Tighten mounting bolts.
	(b) Scored or worn rocker arm.	(b) Install new fuel pump.
	(c) Weak or broken rocker arm spring.	(c) Install new spring.
	(d) Stiff inlet hose	(d) Install 3 in. longer hose

IGNITION SYSTEM

Condition	Possible Cause	Correction
BURNED OR PITTED DISTRIBUTOR CONTACTS	(a) Dirt or oil on contacts.	(a) If oil is on contact face, determine cause and correct condition. Clean distributor cam of dirt and grease, apply a light film of distributor cam lubricant to cam lobes; wipe off excess. See "Distributor Lubrication." Replace contact set and adjust as necessary.
	(b) Alternator voltage regulator setting too high.	(b) Test alternator voltage regulator setting, adjust as necessary. Replace contact set and adjust as necessary.
	(c) Contacts misaligned or gap too small.	(c) Align and adjust contacts.
	(d) Faulty coil.	(d) Test and replace coil if necessary. Replace and adjust contacts.
	(e) Ballast resistor not in circuit.	(e) Inspect conditions, and correctly connect the coil.
	(f) Wrong condenser or faulty condenser.	(f) Test condenser and replace if necessary. Replace and adjust contacts.
	(g) Faulty ignition switch.	(g) Replace ignition switch.
	(h) Bushings worn.	(h) Replace housing.
	(i) Touching contacts with the hands during installation.	(i) Replace and adjust contacts.
IGNITION COIL FAILURE	(a) Coil damaged by excessive heat from engine.	(a) Replace coil. Inspect condition of the distributor contacts.
	(b) Coil tower carbon-tracked.	(b) Replace the coil.
	(c) Oil leak at tower.	(c) Replace the coil.

COOLING SYSTEM

Condition	Possible Cause	Correction
EXTERNAL LEAKAGE	(a) Loose hose clamp.	(a) Replace the hose clamp.
	(b) Hose leaking.	(b) Replace the hose.
	(c) Leaking radiator.	(c) Repair or replace the radiator as necessary.
	(d) Water pump leaking through vent hole.	(d) Replace the water pump.
	(e) Loose core hole plug.	(e) Install new core hole plug.
	(f) Damaged gasket, or dry gasket, if engine has been stored.	(f) Replace gaskets as necessary.
	(g) Cylinder head bolts loose, or tightened unevenly.	(g) Replace the cylinder head gasket and torque head in correct sequence. correct sequence.
	(h) Leak at heater connection.	(h) Clean the heater connections and replace the hoses and clamps if necessary.
	(i) Leak at water temperature sending unit.	(i) Tighten the water temperature sending unit.
	(j) Leak at water pump attaching bolt.	(j) Tighten the water pump attaching bolts
	(k) Leak at exhaust manifold stud.	(k) Seal and re-drive the stud.
	(l) Cracked thermostat housing.	(l) Replace the thermostat housing.
	(m) Dented radiator inlet or outlet tube.	(m) Straighten the radiator inlet or outlet tube as necessary.
	(n) Leaking heater core.	(n) Repair or replace the heater core.
	(o) Cracked or porous water pump housing.	(o) Replace the water pump assembly.
	(p) Warped or cracked cylinder head.	(p) Replace the cylinder head.
	(q) Cracked cylinder block.	(q) Replace the cylinder block.
	(r) Sand holes or porous condition in block or head.	(r) Replace the cylinder block or cylinder head as necessary.
	(s) Faulty pressure cap.	(s) Replace pressure cap.
	(t) Loose or stripped oil cooler fittings.	(t) Tighten or replace as necessary.
INTERNAL LEAKAGE	(a) Faulty head gasket.	(a) Install a new head gasket.
	(b) Refer to causes (f), (g), (p), (q), (r) and (t) listed under External Leakage.	(b) Refer to corrections (f), (g), (p), (q), (r) and (t) listed under External Leakage.
	(c) Crack in head into valve compartment.	(c) Pressure test cooling system, replace the cylinder head.
	(d) Cracked valve port.	(d) Pressure test cooling system, replace the cylinder head.
	(e) Crack in block into push rod compartment.	(e) Pressure test cooling system, replace the cylinder block.
	(f) Cracked cylinder wall.	(f) Pressure test cooling system, replace the cylinder block.
	(g) Leaking oil cooler.	(g) Repair or replace the oil cooler.
POOR CIRCULATION	(a) Low coolant level.	(a) Fill radiator to correct level.
	(b) Collapsed radiator hose. (A bottom hose with faulty spring may collapse only at medium or high engine speeds.)	(b) Replace the hose and spring.
	(c) Fan belt lose, glazed, or oil soaked.	(c) Tighten or replace the fan belt as necessary.
	(d) Air leak through bottom hose.	(d) Reposition hose clamps or replace the hose. Check radiator outlets for dents or out-of-rounds.

Condition	Possible Cause	Correction
	(e) Faulty thermostat.	(e) Replace the thermostat.
	(f) Water pump impeller broken or loose on shaft.	(f) Replace the water pump.
	(g) Restricted radiator core water passages.	(g) Flush the radiator thoroughly or rod out if necessary.
	(h) Restricted engine water jacket.	(h) Flush the engine cooling system thoroughly.
OVERHEATING (refer to Causes and Corrections listed under "Poor Circulation")	(a) Blocked radiator air passages.	(a) Clean out the radiator air passages.
	(b) Incorrect ignition timing.	(b) Time the engine ignition system.
	(d) Incorrect valve timing.	(d) Correct the engine valve timing.
	(e) Inaccurate temperature gauge.	(e) Replace the temperature gauge.
	(f) Restricted overflow tube.	(f) Remove restriction from overflow tube.
	(g) Faulty radiator pressure cap or seat.	(g) Replace the radiator cap. Clean or replace seat.
	(h) Frozen heat control valve.	(h) Free up manifold heat control valve.
	(i) Dragging brakes.	(i) Adjust the brakes.
	(j) Excessive engine idling.	(j) Increase idle R.P.M. or stop engine.
	(k) Frozen coolant.	(k) Thaw out cooling system, add anti-freeze as required.
	(l) Faulty fan drive unit.	(l) Replace the fan drive unit.
	(m) Faulty temperature sending unit.	(m) Replace the sending unit.
	(n) Faulty vacuum By-Pass valve.	(n) Replace valve.
OVERFLOW LOSS (Also refer to Causes and Corrections listed under "Poor Circulation and Overheating")	(a) Overfilling.	(a) Adjust coolant to the correct level.
	(b) Coolant foaming due to insufficient corrosion inhibitor.	(b) Flush the radiator and add antifreeze as required.
	(c) Blown head gasket.	(c) Replace the head gasket.
	(d) Broken or shifted lower hose spring.	(d) Replace lower hose.
CORROSION	(a) Use of water containing large concentration of lime and minerals.	(a) Use only clean soft water with anti-freeze.
	(b) Insufficient corrosion inhibitor.	(b) Use antifreeze as required.
	(c) Use of antifreeze for extended length of time.	(c) Drain cooling system and replace with new antifreeze.
TEMPERATURE TOO LOW—SLOW ENGINE WARM-UP	(a) Faulty thermostat.	(a) Replace the thermostat.
	(b) Inaccurate temperature gauge.	(b) Replace the temperature gauge.
	(c) Faulty temperature sending unit.	(c) Replace the sending unit.
	(d) Faulty heater controls.	(d) Adjust heater controls
WATER PUMP NOISY	(a) Seal noisy.	(a) Add Water Pump Lube.
	(b) Bearing corroded.	(b) Replace water pump.

ACCESSORY DRIVE BELTS

Condition	Possible Cause	Correction
INSUFFICIENT ACCESSORY OUTPUT DUE TO BELT SLIPPAGE	(a) Belt too loose.	(a) Adjust belt tension.
	(b) Belt excessively glazed or worn.	(b) Replace and tighten as specified.

18

Condition	Possible Cause	Correction
BELT SQUEAL WHEN ACCELERATING ENGINE	(a) Belts too loose.	(a) Adjust belt tension.
	(b) Belts glazed.	(b) Replace belts.
BELT SQUEAK AT IDLE	(a) Belt too loose.	(a) Adjust belt tension.
	(b) Dirt and paint imbedded in belt.	(b) Replace belt.
	(c) Non-uniform belt.	(c) Replace belt.
	(d) Misaligned pulleys.	(d) Align accessories (file brackets or use spacers as required).
	(e) Non-uniform groove or eccentric pulley.	(e) Replace pulley.

ENGINE

Condition	Possible Cause	Correction
ENGINE WILL NOT START	(a) Weak battery.	(a) Test battery specific gravity. Recharge or replace as necessary.
	(b) Corroded or loose battery connections.	(b) Clean and tighten battery connections. Apply a coat of petroleum to terminals.
	(c) Faulty starter.	(c) Refer to "Starting Motor".
	(d) Moisture on ignition wires and distributor cap.	(d) Wipe wires and cap clean and dry.
	(e) Faulty ignition cables.	(e) Replace any cracked or shorted cables.
	(f) Faulty coil or condenser.	(f) Test and replace if necessary.
	(g) Dirty or corroded distributor contacts.	(g) Clean or replace as necessary.
	(h) Incorrect spark plug gap.	(h) Set gap
	(i) Incorrect ignition timing.	(i) Refer to "Ignition Timing."
	(j) Dirt or water in fuel line or carburetor.	(j) Clean lines and carburetor.
	(k) Carburetor flooded.	(k) Adjust float level—check seats.
	(l) Incorrect carburetor float setting.	(l) Adjust float level—check seats.
	(m) Faulty fuel pump.	(m) Install new fuel pump.
	(n) Carburetor percolating. No fuel in the carburetor.	(n) Measure float level. Adjust bowl vent. Inspect operation of manifold control valve.
ENGINE STALLS	(a) Idle speed set too low.	(a) Adjust carburetor.
	(b) Incorrect choke adjustment.	(b) Adjust choke.
	(c) Idle mixture too lean or too rich.	(c) Adjust carburetor.
	(d) Incorrect carburetor float setting.	(d) Adjust float setting.
	(e) Leak in intake manifold.	(e) Inspect intake manifold gasket and replace if necessary.
	(f) Dirty, burned or incorrectly gapped distributor contacts.	(f) Replace contacts and adjust.
	(g) Worn or burned distributor rotor.	(g) Install new rotor.
	(h) Incorrect ignition wiring.	(h) Install correct wiring.
	(i) Faculty coil or condenser.	(i) Test and replace if necessary.
	(j) Incorrect tappet lash.	(j) Adjust to specifications.
ENGINE LOSS OF POWER	(a) Incorrect ignition timing.	(a) Refer to "Ignition Timing."
	(b) Worn or burned distributor rotor.	(b) Install new rotor.
	(c) Worn distributor shaft or cam.	(c) Remove and repair distributor.
	(d) Dirty or incorrectly gapped spark plugs.	(d) Clean plugs and set gap

19

Condition	Possible Cause	Correction
	(e) Dirt or water in fuel line, carburetor or filter.	(e) Clean lines, carburetor and replace filter.
	(f) Incorrect carburetor float setting.	(f) Adjust float level.
	(g) Faulty fuel pump.	(g) Install new pump.
	(h) Incorrect valve timing.	(h) Refer to "Checking Valve Timing."
	(i) Blown cylinder head gasket.	(i) Install new head gasket.
	(j) Low compression.	(j) Test compression of each cylinder.
	(k) Burned, warped or pitted valves.	(k) Install new valves.
	(l) Plugged or restricted exhaust system.	(l) Install new parts as necessary.
	(m) Faulty ignition cables.	(m) Replace any cracked or shorted cables.
	(n) Faulty coil or condenser.	(n) Test and replace as necessary.
ENGINE MISSES ON ACCELERATION	(a) Dirty, burned, or incorrectly gapped distributor contacts.	(a) Replace contacts and adjust.
	(b) Dirty, or gap too wide in spark plugs.	(b) Clean spark plugs and set gap
	(c) Incorrect ignition timing.	(c) Refer to "Ignition Timing."
	(d) Dirt in carburetor.	(d) Clean carburetor.
	(e) Acceleration pump in carburetor.	(e) Install new pump.
	(f) Burned, warped or pitted valves.	(f) Install new valves.
	(g) Faulty coil or condenser.	(g) Test and replace if necessary.
ENGINE MISSES AT HIGH SPEED	(a) Dirty or incorrectly gapped distributor contacts.	(a) Clean or replace as necessary.
	(b) Dirty or gap set too wide in spark plug.	(b) Clean spark plugs and set gap
	(c) Worn distributor shaft or cam.	(c) Remove and repair distributor.
	(d) Worn or burned distributor rotor.	(d) Install new rotor.
	(e) Faulty coil or condenser.	(e) Test and replace if necessary.
	(f) Incorrect ignition timing.	(f) Refer to "Ignition Timing."
	(g) Dirty jets in carburetor.	(g) Clean jets.
	(h) Dirt or water in fuel line, carburetor or filter.	(h) Clean lines, carburetor and replace filter.
NOISY VALVES	(a) High or low oil level in crankcase.	(a) Check for correct oil level.
	(b) Thin or diluted oil.	(b) Change oil.
	(c) Low oil pressure.	(c) Check engine oil level.
	(d) Dirt in tappets.	(d) Clean tappets.
	(e) Bent push rods.	(e) Install new push rods.
	(f) Worn rocker arms.	(f) Inspect oil supply to rockers.
	(g) Worn tappets.	(g) Install new tappets.
	(h) Worn valve guides.	(h) Ream and install new valves with O/S stems.
	(i) Excessive run-out of valve seats or valve faces.	(i) Grind valve seats and valves.
	(j) Incorrect tappet lash.	(j) Adjust to specifications.
CONNECTING ROD NOISE	(a) Insufficient oil supply.	(a) Check engine oil level.
	(b) Low oil pressure.	(b) Check engine oil level. Inspect oil pump relief valve and spring.
	(c) Thin or diluted oil.	(c) Change oil to correct viscosity.
	(d) Excessive bearing clearance.	(d) Measure bearings for correct clearance.
	(e) Connecting rod journals out-of-round.	(e) Replace crankshaft or regrind journals.
	(f) Misaligned connecting rods.	(f) Replace bent connecting rods.
MAIN BEARING NOISE	(a) Insufficient oil supply.	(a) Check engine oil level.

20

Condition	Possible Cause	Correction
	(b) Low oil pressure.	(b) Check engine oil level. Inspect oil pump relief valve and spring.
	(c) Thin or diluted oil.	(c) Change oil to correct viscosity.
	(d) Excessive bearing clearance.	(d) Measure bearings for correct clearance.
	(e) Excessive end play	(e) Check thrust bearing for wear on flanges.
	(f) Crankshaft journal out-of-round or worn.	(f) Replace crankshaft or regrind journals.
	(g) Loose flywheel or torque converter.	(g) Tighten to correct torque.
OIL PUMPING AT RINGS	(a) Worn, scuffed, or broken rings.	(a) Hone cylinder bores and install new rings.
	(b) Carbon in oil rings slots.	(b) Install new rings.
	(c) Rings fitted too tight in grooves.	(c) Remove the rings. Check grooves. If groove is not proper width, replace piston.
OIL PRESSURE DROP	(a) Low oil level.	(a) Check engine oil level.
	(b) Faulty oil pressure sending unit.	(b) Install new sending unit.
	(c) Clogged oil filter.	(c) Install new oil filter.
	(d) Worn parts in oil pump.	(d) Replace worn parts or pump.
	(e) Thin or diluted oil.	(e) Change oil to correct viscosity.
	(f) Excessive bearing clearance.	(f) Measure bearings for correct clearance.
	(g) Oil pump relief valve stuck.	(g) Remove valve and inspect, clean, and reinstall.
	(h) Oil pump suction tube loose, bent or cracked.	(h) Remove oil pan and install new tube if necessary.
OIL PUMPING AT RINGS	(a) Worn, scuffed, or broken rings.	(a) Hone cylinder bores and install new rings.
	(b) Carbon in oil ring slot.	(b) Install new rings.
	(c) Rings fitted too tight in grooves.	(c) Remove the rings. Check grooves. If groove is not proper width, replace piston.
OIL PRESSURE DROP	(a) Low oil level.	(a) Check engine oil level.
	(b) Faulty oil pressure sending unit.	(b) Install new sending unit.
	(c) Clogged oil filter.	(c) Install new oil filter.
	(d) Worn parts in oil pump.	(d) Replace worn parts or pump.
	(e) Thin or diluted oil.	(e) Change oil to correct viscosity.
	(f) Excessive bearing clearance.	(f) Measure bearings for correct clearance.
	(g) Oil pump relief valve stuck.	(g) Remove valve and inspect, clean, and reinstall.
	(h) Oil pump suction tube loose, bent or cracked.	(h) Remove oil pan and install new tube if necessary.

CLUTCH

Condition	Possible Cause	Correction
CLUTCH CHATTER	(a) Worn or damaged disc assembly.	(a) Replace disc assembly.
	(b) Grease or oil on disc facings.	(b) Replace disc assembly and correct cause of contamination.
	(c) Improperly adjusted cover assembly.	(c) Replace cover assembly.
	(d) Broken or loose engine mounts	(d) Replace or tighten mounts
	(e) Misaligned clutch housing	(e) Align clutch housing
CLUTCH SLIPPING	(a) Burned, worn, or oil soaked facings.	(a) Replace disc assembly and correct cause of contamination.
	(b) Insufficient pedal free play.	(b) Adjust release fork rod.

21

Condition	Possible Cause	Correction
DIFFICULT GEAR SHIFTING	(c) Weak or broken pressure springs.	(c) Replace cover assembly.
	(a) Excessive pedal free play.	(a) Adjust release fork rod.
	(b) Excessive deflection in linkage or firewall.	(b) Repair or replace linkage.
	(c) Worn or damaged disc assembly.	(c) Replace disc assembly.
	(d) Improperly adjusted cover assembly.	(d) Replace cover assembly.
	(e) Clutch disc splines sticking.	(e) Remove disc assembly and free up splines or replace disc.
	(f) Worn or dry pilot bushing.	(f) Lubricate or replace bushing.
	(g) Clutch housing misaligned.	(g) Align clutch housing.
CLUTCH NOISY	(a) Dry clutch linkage.	(a) Lubricate where necessary.
	(b) Worn release bearing.	(b) Replace release bearing.
	(c) Worn disc assembly.	(c) Replace disc assembly.
	(d) Worn release levers.	(d) Replace cover assembly.
	(e) Worn or dry pilot bushing.	(e) Lubricate or replace bushing.
	(f) Dry contact-pressure plate lugs in cover.	(f) Lubricate very lightly.

MANUAL—TRANSMISSION

Condition	Possible Cause	Correction
HARD SHIFTING	(a) Incorrect clutch adjustment.	(a) Refer to Clutch Group for corrections.
	(b) Improper linkage adjustment.	(b) Perform linkage adjustment
	(c) Synchronizer clutch sleeve damaged.	(c- d-e) Causes noted can only be corrected by disassembling transmission and replacing damaged or worn parts.
	(d) Synchronizer spring improperly installed.	
	(e) Broken or worn synchronizer stop rings.	
TRANSMISSION SLIPS OUT OF GEAR	(a) Linkage interference.	(a) Inspect and remove all linkage interferences.
	(b) Gearshift rods out of adjustment.	(b) Adjust gearshift rods
	(c) Synchronizer clutch teeth worn.	(c) Disassemble transmission and replace parts as necessary.
	(d) Clutch housing bore or face out of alignment.	(d) Refer to Clutch Group for correction procedure.
TRANSMISSION NOISES	(a) Excessive end play in countershaft gear.	(a) Replace thrust washers.
	(b) Loose synchronizer hub spline fit on mainshaft.	(b) Inspect mainshaft and synchronizer hub and replace parts as necessary.
	(c) Damaged, broken or excessively worn gear teeth.	(c) Replace worn gears.
	(d) Rough or pitted bearing races or balls.	(d) Replace worn bearing.

AUTOMATIC TRANSMISSION

Condition	Possible Cause	Correction
HARSH ENGAGEMENT IN D, 1, 2 AND R	(a) Engine idle speed too high.	(a) Adjust engine idle speed Readjust throttle linkage.
	(b) Hydraulic pressures too high or low.	(b) Inspect fluid level, then perform hydraulic pressure tests and adjust to specifications.

22

	(c) Low-reverse band out of adjustment.	(c) Adjust low-reverse band.
	(d) Valve body malfunction or leakage.	(d) Perform pressure tests to determine cause and correct as required.
	(e) Accumulator sticking, broken rings or spring.	(e) Inspect accumulator for sticking, broken rings or spring. Repair as required.
	(f) Low-reverse servo, band or linkage malfunction.	(f) Inspect servo for damaged seals, binding linkage or faulty band lining. Repair as required.
	(g) Worn or faulty front and/or rear clutch.	(g) Disassemble and inspect clutch. Repair or replace as required.
DELAYED ENGAGEMENT IN D, 1, 2 AND R	(a) Low fluid level.	(a) Refill to correct level with Automatic Transmission Fluid,
	(b) Incorrect gearshift control linkage adjustment.	(b) Adjust control linkage.
	(c) Hydraulic pressures too high or low.	(c) Perform hydraulic pressure tests and adjust to specifications.
	(d) Oil filter clogged.	(d) Replace oil filter.
	(e) Valve body malfunction or leakage.	(e) Perform pressure tests to determine cause and correct as required.
	(f) Accumulator sticking, broken rings or spring.	(f) Inspect accumulator for sticking, broken rings or spring. Repair as required.
	(g) Clutches or servos sticking or not operating.	(g) Remove valve body assembly and perform air pressure tests. Repair as required.
	(h) Faulty oil pump.	(h) Perform hydraulic pressure tests. Adjust or repair as required.
	(i) Worn or faulty front and/or rear clutch.	(i) Disassemble and inspect clutch. Repair or replace as required.
	(j) Worn or broken input shaft and/or reaction shaft support seal rings.	(j) Inspect and replace seal rings as required, also inspect respective bores for wear. Replace parts as required.
	(k) Aerated fluid.	(k) Inspect for air leakage into pump suction passages.
RUNAWAY OR HARSH UPSHIFT AND 3-2 KICKDOWN	(a) Low fluid level.	(a) Refill to correct level with Automatic Transmission Fluid,
	(b) Incorrect throttle linkage adjustment.	(b) Adjust throttle linkage.
	(c) Hydraulic pressures too high or low.	(c) Perform hydraulic pressure tests and adjust to specifications.
	(d) Kickdown band out of adjustment.	(d) Adjust kickdown band.
	(e) Valve body malfunction or leakage.	(e) Perform pressure tests to determine cause and correct as required.
	(f) Governor malfunction.	(f) Inspect governor and repair as required.
	(g) Accumulator sticking, broken rings or spring.	(g) Inspect accumulator for sticking, broken rings or spring. Repair as required.
	(h) Clutches or servos sticking or not operating.	(h) Remove valve body assembly and perform air pressure tests. Repair as required.
	(i) Kickdown servo, band or linkage malfunctions.	(i) Inspect servo for sticking, broken seal rings, binding linkage or faulty band lining. Repair as required.
	(j) Worn or faulty front clutch.	(j) Disassemble and inspect clutch. Repair or replace as required.
	(k) Worn or broken input shaft and/or reaction shaft support seal rings.	(k) Inspect and replace seal rings as required, also inspect respective bores for wear. Replace parts as required.
NO UPSHIFT	(a) Low fluid level.	(a) Refill to correct level with Automatic Transmission Fluid

23

	(b) Incorrect throttle linkage adjustment.	(b) Adjust throttle linkage.
	(c) Kickdown band out of adjustment.	(c) Adjust kickdown band.
	(d) Hydraulic pressures too high or low.	(d) Perform hydraulic pressure tests and adjust to specifications.
	(e) Governor sticking or leaking.	(e) Remove and clean governor. Replace parts as necessary.
	(f) Valve body malfunction or leakage.	(f) Perform pressure tests to determine cause and correct as required.
	(g) Clutches or servos sticking or not operating.	(g) Remove valve body assembly and perform air pressure tests. Repair as required.
	(h) Faulty oil pump.	(h) Perform hydraulic pressure tests, adjust or repair as required.
	(i) Kickdown servo, band or linkage malfunction.	(i) Inspect servo for sticking, broken seal rings, binding linkage or faulty band lining. Repair as required.
	(j) Worn or faulty front clutch.	(j) Disassemble and inspect clutch. Repair or replace as required.
	(k) Worn or broken input shaft and/or reaction shaft support seal rings.	(k) Inspect and replace seal rings as required, also inspect respective bores for wear. Replace parts as required.
NO KICKDOWN OR NORMAL DOWNSHIFT	(a) Incorrect throttle linkage adjustment.	(a) Adjust throttle linkage.
	(b) Incorrect gearshift control linkage adjustment.	(b) Adjust control linkage.
	(c) Kickdown band out of adjustment.	(c) Adjust kickdown band.
	(d) Hydraulic pressures too high or low.	(d) Perform hydraulic pressure tests and adjust to specifications.
	(e) Governor sticking or leaking.	(e) Remove and clean governor. Replace parts if necessary.
	(f) Valve body malfunction or leakage.	(f) Perform pressure tests to determine cause and correct as required.
	(g) Clutches or servos sticking or not operating.	(g) Remove valve body assembly and perform air pressure tests. Repair as required.
	(h) Kickdown servo, band or linkage malfunction.	(h) Inspect servo for sticking, broken seal rings, binding linkage or faulty band lining. Repair as required.
	(i) Overrunning clutch not holding.	(i) Disassemble transmission and repair overrunning clutch as required.
SHIFTS ERRATIC	(a) Low fluid level.	(a) Refill to correct level with Automatic Transmission Fluid,
	(b) Aerated fluid.	(b) Inspect tor air leakage into pump suction passages.
	(c) Incorrect throttle linkage adjustment.	(c) Adjust throttle linkage.
	(d) Incorrect gearshift control linkage adjustment.	(d) Adjust control linkage.
	(e) Hydraulic pressures too high or low.	(e) Perform hydraulic pressure tests and adjust to specifications.
	(f) Governor sticking or leaking.	(f) Remove and clean governor. Replace parts if necessary.
	(g) Oil filter clogged.	(g) Replace oil filter.
	(h) Valve body malfunction or leakage.	(h) Perform pressure tests to determine cause and correct as required.
	(i) Clutches or servos sticking or not operating.	(i) Remove valve body assembly and perform air pressure tests. Repair as required.
	(j) Faulty oil pump.	(j) Perform hydraulic pressure tests, adjust or repair as required.
	(k) Worn or broken input shaft and/or reaction shaft support seal rings.	(k) Inspect and replace seal rings as required, also inspect respective bores for wear. Replace parts as required.

24

SLIPS IN FORWARD DRIVE POSITIONS	(a) Low fluid level.	(a) Refill to correct level with Automatic Transmission Fluid,
	(b) Aerated fluid.	(b) Inspect for air leakage into oil pump suction passages.
	(c) Incorrect throttle linkage adjustment.	(c) Adjust throttle linkage.
	(d) Incorrect gearshift control linkage adjustment.	(d) Adjust control linkage.
SLIPS IN ALL POSITIONS	(a) Low fluid level.	(a) Refill to correct level with Automatic Transmission Fluid,
	(b) Hydraulic pressures too low.	(b) Perform hydraulic pressure tests and adjust to specifications.
	(c) Valve body malfunction or leakage.	(c) Perform pressure tests to determine cause and correct as required.
	(d) Faulty oil pump.	(d) Perform hydraulic pressure tests, adjust or repair as required.
	(e) Clutches or servos sticking or not operating.	(e) Remove valve body assembly and perform air pressure tests. Repair as required.
	(f) Worn or broken input shaft and/or reaction shaft support seal rings.	(f) Inspect and replace seal rings as required, also inspect respective bores for wear. Replace parts as required.
NO DRIVE IN ANY POSITION	(a) Low fluid level.	(a) Refill to correct level with Automatic Transmission Fluid,
	(b) Hydraulic pressures too low.	(b) Perform hydraulic pressure tests and adjust to specifications.
	(c) Oil filter clogged.	(c) Replace oil filter.
	(d) Valve body malfunction or leakage.	(d) Perform pressure tests to determine cause and correct as required.
	(e) Faulty oil pump.	(e) Perform hydraulic pressure tests, adjust or repair as required.
	(f) Clutches or servos sticking or not operating.	(f) Remove valve body assembly and perform air pressure tests. Repair as required.
	(g) Torque converter failure.	(g) Replace torque converter.
NO DRIVE IN FORWARD DRIVE POSITIONS	(a) Hydraulic pressures too low.	(a) Perform hydraulic pressure tests and adjust to specifications.
	(b) Valve body malfunction or leakage.	(b) Perform pressure tests to determine cause and correct as required.
	(c) Clutches or servos, sticking or not operating.	(c) Remove valve body assembly and perform air pressure tests. Repair as required.
	(d) Worn or faulty rear clutch.	(d) Disassemble and inspect clutch. Repair or replace as required.
	(e) Overrunning clutch not holding.	(e) Disassemble transmission and repair overrunning clutch as required.
	(f) Worn or broken input shaft and/or reaction shaft support seal rings.	(f) Inspect and replace seal rings as required, also inspect respective bores for wear. Replace parts as required.
NO DRIVE IN REVERSE	(a) Incorrect gearshift control linkage adjustment.	(a) Adjust control linkage.
	(b) Hydraulic pressures too low.	(b) Perform hydraulic pressure tests and adjust to specifications.
	(c) Low-reverse band out of adjustment.	(c) Adjust low-reverse band.
DRIVES IN NEUTRAL	(a) Incorrect gearshift control linkage adjustment.	(a) Adjust control linkage.
	(b) Valve body malfunction or leakage.	(b) Perform pressure tests to determine cause and correct as required.
	(c) Rear clutch inoperative.	(c) Inspect clutch and repair as required.
DRAGS OR LOCKS	(a) Kickdown band out of adjustment.	(a) Adjust kickdown band.

25

	(b) Low-reverse band out of adjustment.	(b) Adjust low-reverse band.
	(c) Kickdown and/or low-reverse servo, band, linkage malfunction.	(c) Inspect servo for sticking, broken seal rings, binding linkage or faulty band lining. Repair as required.
	(d) Front and/or rear clutch faulty.	(d) Disassemble and inspect clutch. Repair or replace as required.
	(e) Planetary gear sets broken or seized.	(e) Inspect condition of planetary gear sets and replaced as required.
	(f) Overrunning clutch worn, broken or seized.	(f) Inspect condition of overrunning clutch and replace parts as required.
GRATING, SCRAPING GROWLING NOISE	(a) Kickdown band out of adjustment.	(a) Adjust kickdown band.
	(b) Low-reverse band out of adjustment.	(b) Adjust low-reverse band.
	(c) Output shaft bearing and/or bushing damaged.	(c) Remove extension housing and replace bearing and/or bushing.
	(d) Governor support binding or broken seal rings.	(d) Inspect condition of governor support and repair as required.
	(e) Oil pump scored or binding.	(e) Inspect condition of pump and repair as required.
	(f) Front and/or rear clutch faulty.	(f) Disassemble and inspect clutch. Repair or replace as required.
	(g) Planetary gear sets broken or seized.	(g) Inspect condition of planetary gear sets and replace as required.
	(h) Overrunning clutch worn, broken or seized.	(h) Inspect condition of overrunning clutch and replace parts as required.
BUZZING NOISE	(a) Low fluid level.	(a) Refill to correct level with Automatic Transmission Fluid,
	(b) Pump sucking air.	(b) Inspect pump for nicks or burrs on mating surfaces, porous casting, and/or excessive rotor clearance. Replace the parts as required.
	(c) Valve body malfunction.	(c) Remove and recondition valve body assembly.
	(d) Overrunning clutch inner race damaged.	(d) Inspect and repair clutch as required.
HARD TO FILL, OIL FLOWS OUT FILLER TUBE	(a) High fluid level.	(a) Drain fluid to correct level.
	(b) Breather clogged.	(b) Inspect and clean breather vent opening in pump housing.
	(c) Oil filter clogged.	(c) Replace oil filter.
	(d) Aerated fluid.	(d) Inspect for air leakage into oil pump suction passages.
TRANSMISSION OVERHEATS	(a) Low fluid level.	(a) Refill to correct level with Automatic Transmission Fluid,
	(b) Kickdown band adjustment too tight.	(b) Adjust kickdown band.
	(c) Low-reverse band adjustment too tight.	(c) Adjust low-reverse band.
	(d) Faulty cooling system.	(d) Inspect the transmission cooling system, clean and repair as required.
	(e) Cracked or restricted oil cooler line or fitting.	(e) Inspect, repair or replace as required.
	(f) Faulty oil pump.	(f) Inspect pump for incorrect clearance, repair as required.
	(g) Insufficient clutch plate clearance in front and/or rear clutches.	(g) Measure clutch plate clearance and correct with proper size snap ring.
STARTER WILL NOT ENERGIZE IN NEUTRAL OR PARK	(a) Incorrect gearshift control linkage adjustment.	(a) Adjust control linkage.
	(b) Faulty or incorrectly adjusted neutral starting switch.	(b) Test operation of switch with a test lamp. Adjust or replace as required.
	(c) Broken lead to neutral switch.	(c) Inspect lead and test with a test lamp. Repair broken lead.

26

REAR AXLE

Condition	Possible Cause	Correction
REAR WHEEL NOISE	(a) Wheel loose.	(a) Tighten loose wheel nuts.
	(b) Spalled wheel bearing cup or cone.	(b) Check rear wheel bearings. If spalled or worn, replace.
	(c) Defective, brinelled wheel bearing.	(c) Defective or brinelled bearings must be replaced. Check rear axle shaft end play.
	(d) Excessive axle shaft end play.	(d) Readjust axle shaft end play.
	(e) Bent or sprung axle shaft flange.	(e) Replace bent or sprung axle shaft.
SCORING OF DIFFERENTIAL GEARS AND PINIONS	(a) Insufficient lubrication.	(a) Replace scored gears. Scoring marks on the pressure face of gear teeth or in the bore are caused by instantaneous fusing of the mating surfaces. Scored gears should be replaced. Fill rear axle to required capacity with proper lubricant. See Specification Section.
	(b) Improper grade of lubricant.	(b) Replace scored gears. Inspect all gears and bearings for possible damage. Clean out and refill axle to required capacity with proper lubricant. See Lubrication section.
	(c) Excessive spinning of one wheel.	(c) Replace scored gears. Inspect all gears, pinion bores and shaft for scoring, or bearings for possible damage. Service as necessary.
TOOTH BREAKAGE (RING GEAR AND PINION)	(a) Overloading.	(a) Replace gears. Examine other gears and bearings for possible damage. Replace parts as needed. Avoid Overloading.
	(b) Erratic clutch operation.	(b) Replace gears, and examine remaining parts for possible damage. Avoid erratic clutch operation.
	(c) Ice-spotted pavements.	(c) Replace gears. Examine remaining parts for possible damage. Replace parts as required.
	(d) Improper adjustment.	(d) Replace gears. Examine other parts for possible damage. Make sure ring gear and pinion backlash is correct.
REAR AXLE NOISE	(a) Insufficient lubricant.	(a) Refill rear axle with correct amount of the proper lubricant. See Specification section. Also check for leaks and correct as necessary.
	(b) Improper ring gear and pinion adjustment.	(b) Check ring gear and pinion tooth contact.
	(c) Unmatched ring gear and pinion.	(c) Remove unmatched ring gear and pinion. Replace with a new matched gear and pinion set.
	(d) Worn teeth on ring gear or pinion.	(d) Check teeth on ring gear and pinion for contact. If necessary, replace with new matched set.

Condition	Possible Cause	Correction
	(e) End play in drive pinion bearings.	(e) Adjust drive pinion bearing preload.
	(f) Side play in differential bearings.	(f) Adjust differential bearing preload.
	(g) Incorrect drive gearlash.	(g) Correct drive gearlash.
LOSS OF LUBRICANT	(a) Lubricant level too high.	(a) Drain excess lubricant by removing filler plug and allow lubricant to level at lower edge of filler plug hole.
	(b) Worn axle shaft oil seals.	(b) Replace worn oil seals with new ones. Prepare new seals before replacement.
	(c) Cracked rear axle housing.	(c) Repair or replace housing as required.
	(d) Worn drive pinion oil seal.	(d) Replace worn drive pinion oil seal with a new one.
	(e) Scored and worn companion flange.	(e) Replace worn or scored companion flange and oil seal.
	(f) Clogged vent.	(f) Remove obstructions.
	(g) Loose carrier housing bolts or housing cover screws.	(g) Tighten bolts or cover screws to specifications and fill to correct level with proper lubricant.
OVERHEATING OF UNIT	(a) Lubricant level too low.	(a) Refill rear axle.
	(b) Incorrect grade of lubricant.	(b) Drain, flush and refill rear axle with correct amount of the proper lubricant. See Specification Section.
	(c) Bearings adjusted too tightly.	(c) Readjust bearings.
	(d) Excessive wear in gears.	(d) Check gears for excessive wear or scoring. Replace as necessary.
	(e) Insufficient ring gear to pinion clearance.	(e) Readjust ring gear and pinion backlash and check gears for possible scoring.
PROPELLER SHAFT VIBRATION	(a) Undercoating or other foreign matter on shaft.	(a) Clean exterior of shaft and wash with solvent.
	(b) Loose universal joint flange bolts.	(b) Tighten bolt nuts to specific torque.
	(c) Loose or bent universal joint flange or high runout.	(c) Install new flange. Tighten to specifications.
	(d) Improper drive line angularity.	(d) Correct angularity. See "Propeller Shaft Angularity."
	(e) Rear spring center bolt not in seat.	(e) Loosen spring U-bolts, reseat center bolt and tighten U-bolts to specified torque.
	(f) Worn universal joint bearings or missing rollers.	(f) Recondition universal joint.
	(g) Propeller shaft damaged (bent tube) or out of balance.	(g) Install new propeller shaft.
	(h) Broken rear spring.	(h) Replace rear spring.
	(i) Excessive runout or unbalance condition.	(i) Reindex propeller shaft 180 degrees, reride and correct as necessary.
UNIVERSAL JOINT NOISE	(a) Propeller shaft flange bolts nuts loose.	(a) Tighten nuts to specified torque.
	(b) Lack of lubrication	(b) Recondition universal joint.

SHOCK ABSORBERS

Condition	Possible Cause	Correction
SHOCK ABSORBER NOISY	(a) Loose bolt or stud. (b) Undercoating on shock absorber reservoir. (c) Bushing excessively worn. (d) Air trapped in system.	(a) Tighten to specifications. (b) Clean undercoating off shock absorber. (c) Replace bushing. (d) Purge shock absorber.
SHOCK ABSORBER DRIPPING OIL	(a) Worn seal. (b) Damaged crimp or reservoir.	(a) Replace shock absorber. (b) Replace shock absorber.

STEERING

Condition	Possible Cause	Correction
HARD STEERING	(a) Low or uneven tire pressure. (b) Insufficient lubricant in the steering gear housing or in steering linkage. (c) Steering gear shaft adjusted too tight. (d) Front wheels out of line.	(a) Inflate tires to recommended pressures. (b) Lubricate as necessary. (c) Adjust according to instructions. (d) Align the wheels. See "Front Suspension."
PULL TO ONE SIDE (Tendency of the Vehicle to veer in one direction only)	(a) Incorrect tire pressure. (b) Wheel bearings improperly adjusted. (c) Dragging brakes. (d) Improper caster and camber. (e) Incorrect toe-in. (f) Grease, dirt, oil or brake fluid in brake linings. (g) Front and rear wheels out of alignment. (h) Broken or sagging rear springs. (i) Bent suspension parts.	(a) Inflate tires to recommended pressures. (b) See "Front Wheel Bearing Adjustment." (c) Inspect for weak, or broken brake shoe spring, binding pedal. (d) See "Front Wheel Alignment Group." (e) See "Front Wheel Alignment Group." (f) Inspect, replace and adjust as necessary. (g) Align the front wheels. See "Front Suspension Group" (h) Replace rear springs. (i) Replace parts necessary.
WHEEL TRAMP (Excessive Vertical Motion of Wheels)	(a) Incorrect tire pressure. (b) Improper balance of wheels, tires and brake drums. (c) Loose tie rod ends or steering connections. (d) Worn or inoperative shock absorbers.	(a) Inflate tires to recommended pressures. (b) Lubricate as necessary. (c) Inspect and repair as necessary. (d) Replace shock absorbers as necessary.
EXCESSIVE PLAY OR LOOSENESS IN THE STEERING WHEEL	(a) Steering gearshaft adjusted too loose or badly worn. (b) Steering linkage loose or worn (c) Front wheel bearings improperly adjusted. (d) Steering arm loose on steering gear shaft.	(a) Replace worn parts and adjust according to instructions. (b) Replace worn parts. See "Front Wheel Alignment." (c) Adjust according to instructions. (d) Inspect for damage to gear shaft and steering arm, replace parts as necessary.

Condition	Possible Cause	Correction
	(e) Steering gear housing attaching bolts loose.	(e) Tighten attaching bolts according to tigntening reference.
	(f) Steering arms loose at steering knuckles.	(f) Tighten according to tightening reference.
	(g) Worn ball joints.	(g) Replace ball joints as necessary. See "Front Suspension."
	(h) Steering gear adjustment too loose.	(h) Adjust

BRAKES

Condition	Possible Cause	Correction
DRAGGING BRAKES (ALL WHEELS)	(a) Brake shoes improperly adjusted.	(a) Adjust brakes.
	(b) Brake pedal linkage binding.	(b) Free up linkage.
	(c) Excessive hydraulic seal friction.	(c) Lubricate seal.
	(d) Compensator port plugged.	(d) Clean out master cylinder.
	(e) Fluid cannot return to master cylinder.	(e) Inspect pedal return.
	(f) Parking brake not returning.	(f) Free up as required.
	(g) Disc brake metering valve malfunction.	(g) Replace metering valve.
	(h) Contaminated brake fluid.	(h) Drain and flush system-replace all rubber parts in hydraulic system.
GRABBING BRAKES	(a) Grease or brake fluid on linings.	(a) Inspect for a leak and replace lining as required.
PEDAL GOES TO FLOOR (OR ALMOST TO FLOOR)	(a) Self-adjusters not operating.	(a) Inspect self-adjuster operations.
	(b) Air in hydraulic system.	(b) Bleed brakes.
	(c) Hydraulic leak.	(c) Locate and correct leak.
	(d) Fluid low in master cylinder.	(d) Add brake fluid.
	(e) Shoe hanging up on rough platform.	(e) Smooth and lubricate platforms.
	(f) Loose disc brake rotor	(f) Check wheel bearing adjustment.
HARD PEDAL (POWER UNIT TROUBLE)	(a) Faulty vacuum check valve.	(a) Replace check valve.
	(b) Collapsed or leaking vacuum hose.	(b) Replace hose.
	(c) Plugged vacuum fittings.	(c) Clean out fittings.
	(d) Leaking vacuum chamber.	(d) Replace unit.
	(e) Diaphragm assembly out of place in housing.	(e) Replace unit.
	(f) Vacuum leak in forward vacuum housing.	(f) Replace unit.
EXCESSIVE PEDAL TRAVEL	(a) Rear brake adjustment required.	(a) Check and adjust rear brakes.
	(b) Air leak, or insufficient fluid in system or caliper.	(b) Check system for leaks.
	(c) Warped or excessively tapered shoe and lining assembly.	(c) Install new shoe and linings.
	(d) Excessive disc runout.	(d) Check disc for runout with dial indicator. Install new disc.
	(e) Loose wheel bearing adjustment.	(e) Readjust wheel bearings to specified torque.
	(f) Improper brake fluid (boil).	(f) Drain and install correct fluid.
	(g) Damaged caliper piston seal.	(g) Install new piston seal.
BRAKE ROUGHNESS OR CHATTER (Pedal Pulsating)	(a) Excessive out-of-parallelism of braking disc.	(a) Check disc for runout with dial indicator. Install new disc.

30

Condition	Possible Cause	Correction
	(b) Rear brake drums out-of-round.	(b) Loosen and retorque wheel nuts to specifications.
	(c) Excessive lateral runout of braking disc.	(c) Check disc for lateral runout with dial indicator. Install new disc.
	(d) Excessive front bearing clearance.	(d) Readjust wheel bearings to specified torque.
	(e) Rear brake drums distorted by improper tightening of nuts.	(e) Check drums for out-of-round and reface if necessary.
EXCESSIVE PEDAL EFFORT	(a) Power brake malfunction.	(a) Replace
	(b) Frozen or seized pistons.	(b) Disassemble caliper and free up pistons. Clean parts.
	(c) Shoe and lining worn below .180 in. (Lining only— .30 in.)	(c) Install new shoe and linings.
	(d) Brake fluid, oil or grease on linings.	(d) Install new shoe linings as required.
	(e) Incorrect lining.	(e) Remove lining and install correct lining.
PULL	(a) Loose calipers.	(a) Tighten caliper mounting bolts from 45 to 60 ft. pounds.
	(b) Frozen or seized pistons.	(b) Disassemble caliper and free up pistons.
	(c) Rear brake pistons sticking.	(c) Free up rear brake pistons.
	(d) Front end out of alignment.	(d) Check and align front end.
	(e) Broken rear spring.	(e) Install new rear spring.
	(f) Out-of-round rear drums.	(f) Check and reface drums if necessary.
	(g) Incorrect tire pressure.	(g) Inflate tires to recommended presures.
	(h) Brake fluid, oil or grease on linings.	(h) Install new shoe and linings.
	(i) Restricted hose or line.	(i) Check hoses and lines and correct as necessary.
	(j) Rear brakes out of adjustment.	(j) Adjust rear brakes.
	(k) Unmatched linings.	(k) Install correct lining.
	(l) Distorted brake shoes.	(l) Install new brake shoes.
NOISE Groan—Brake noise emanating when slowly releasing brakes (creep—groan)	(a) Not detrimental to function of disc brakes—no corrective action required. (Indicate to operator this noise may be eliminated by slightly increasing or decreasing brake pedal efforts).	
Rattle-Brake noise or rattle emanating at low speeds on rough roads, (front wheels only).	(a) Excessive clearance between shoe and caliper.	(a) Install new shoe and lining assemblies.
Scraping—	(a) Loose wheel bearings.	(a) Readjust wheel bearings to correct specifications.
	(b) Braking disc rubbing housing.	(b) Check for rust or mud buildup on caliper mounting and bridge bolt tightness.
	(c) Mounting bolts too long.	(c) Install mounting bolts of correct length.
FRONT BRAKES HEAT UP DURING DRIVING AND FAIL TO RELEASE	(a) Residual pressure valve in master cylinder.	(a) Remove valve from cylinder.
	(b) Frozen or seized piston.	(b) Disassemble caliper, hone cylinder bore, clean seal groove and install new pistons, seals and boots.

31

Condition	Possible Cause	Correction
	(c) Operator riding brake pedal.	(c) Instruct owner how to drive with disc brakes.
	(d) Sticking pedal linkage.	(d) Free up sticking pedal linkage.
	(e) Power brake malfunction.	(e) Replace
LEAKY WHEEL CYLINDER	(a) Corroded bore.	(a) Hone bore and replace boots and cups.
	(b) Damaged or worn caliper piston seal.	(b) Disassemble caliper and install new seal.
	(c) Scores or corrosion on surface of piston.	(c) Disassemble caliper and hone cylinder bore. If neccessary, install new pistons.
GRABBING OR UNEVEN BRAKING ACTION	(a) Causes listed under "Pull."	(a) Corrections listed under "Pull."
	(b) Power brake malfunction.	(b) Replace unit.
BRAKE PEDAL CAN BE DEPRESSED WITHOUT BRAKING EFFECT	(a) Air in hydraulic system or improper bleeding procedure.	(a) Bleed system.
	(b) Leak in system or caliper.	(b) Check for leak and repair as required.
	(c) Pistons pushed back in cylinder bores during servicing of caliper (shoe and lining not properly positioned).	(c) Reposition brake shoe and lining assemblies. Depress pedal a second time and if condition persists, check following causes:
	(d) Leak past piston cups in master cylinder.	(d) Recondition master cylinder.
	(e) Damaged piston seal in one or more of cylinders.	(e) Disassemble caliper and replace piston seals as required.
	(f) Leak in rear brake cylinder.	(f) Hone cylinder bore. Install new piston cylinder cups.
	(g) Rear brakes out of adjustment.	(g) Adjust rear brakes.
	(h) Bleeder screw open.	(h) Close bleeder screw and bleed entire system.

PARKING BRAKES

Condition	Possible Cause	Correction
DRAGGING BRAKE	(a) Improper cable or brake shoe adjustment.	(a) Properly adjust the service brakes, then adjust the parking brake cable.
	(b) Broken brake shoe return spring.	(b) Replace any broken return spring.
	(c) Broken brake shoe retainer spring.	(c) Replace the broken retainer spring.
	(d) Grease or brake fluid soaked lining.	(d) Replace the grease seal or recondition the wheel cylinders and replace both brake shoes.
	(e) Improper stop light switch adjustment	(e) Adjust stop light switch
	(f) Sticking or frozen brake cable.	(f) Replace cable.
	(g) Broken rear spring.	(g) Replace the broken rear spring.
	(h) Bent or rusted cable equalizer.	(h) Straighten, or replace and lubricate the equalizer.
	(h) Heat set parking brake cable spring.	(h) Replace parking brake cable.
BRAKE WILL NOT HOLD	(a) Broken or rusted brake cable.	(a) Replace cable.
	(b) Improperly adjusted brake or cable.	(b) Adjust brakes and cable as necessary.
	(c) Soaked brake lining.	(c) Replace the brake lining.
	(d) Ratchet or pedal mechanism worn.	(d) Replace pedal assembly.

FRONT SUSPENSION AND STEERING LINKAGE

Condition	Possible Cause	Correction
FRONT END NOISE	(a) Ball joint needs lubrication.	(a) Lubricate ball joint.
	(b) Loose shock absorber mounting. Shock absorber inoperative or bushings worn.	(b) Tighten shock absorber mounting nuts. Replace bushings or shock absorber.
	(c) Worn strut bushings.	(c) Replace bushing.
	(d) Loose struts—Lower control arm bolts and nuts,	(d) Tighten all bolts and nuts.
	(e) Loose steering gear on frame.	(e) Tighten the steering gear mounting bolts.
	(f) Worn upper control arm bushings.	(f) Replace worn bushings.
	(g) Worn lower control arm shaft bushings.	(g) Replace worn bushings.
	(h) Worn upper or lower ball joint.	(h) Replace ball joint.
	(i) Worn tie rod ends.	(i) Replace tie rod end.
	(j) Loose or worn front wheel bearings.	(j) Adjust or replace bearings as necessary.
	(k) Steering knuckle arm contacting the lower control arm wheel stop.	(k) Smooth off the contacting area and lubricate with a water resistant grease.
INSTABILITY	(a) Low or uneven tire pressure.	(a) Inflate tires to correct pressure.
	(b) Loose wheel bearings.	(b) Adjust wheel bearing.
	(c) Improper steering cross shaft adjustment.	(c) Adjust steering cross shaft.
	(d) Steering gear not centered.	(d) Adjust steering gear.
	(e) Worn idler arm bushing.	(e) Replace bushing.
	(f) Loose or excessively worn front strut bushings.	(f) Replace bushings.
	(g) Weak or broken rear spring.	(g) Replace spring.
	(h) Incorrect front wheel alignment.	(h) Measure and adjust front wheel alignment.
	(i) Shock absorber inoperative.	(i) Replace shock absorber.
HARD STEERING	(a) Ball joints-require lubrication.	(a) Lubricate ball joints.
	(b) Low or uneven tire pressure.	(b) Inflate tires to recommended pressures.
	(e) Incorrect front wheel alignment (particularly caster) resulting from a bent control arm, steering knuckle or steering knuckle arm.	(e) Replace bent parts and adjust the front wheel alignment.
	(f) Steering gear low on lubricant.	(f) Fill gear to correct level.
	(g) Steering gear not adjusted.	(g) Adjust steering gear.
	(h) Idler arm binding.	(h) Replace idler arm.
CAR PULLS TO ONE SIDE	(a) Low or uneven tire pressure.	(a) Inflate tires to recommended pressure.
	(b) Front brake dragging.	(b) Adjust brakes.
	(c) Grease, lubricant or brake fluid leaking onto brake lining.	(c) Replace brake shoe and lining as necessary and stop all leaks.
	(d) Loose or excessively worn strut bushings.	(d) Tighten or replace strut bushings.
	(e) Power steering control valve out of adjustment.	(e) Adjust steering gear control valve.
	(f) Incorrect front wheel alignment (particularly camber).	(f) Adjust front wheel alignment.
	(g) Broken or weak rear spring.	(g) Replace spring.
EXCESSIVE PLAY IN STEERING	(a) Worn or loose front wheel bearings.	(a) Adjust or replace wheel bearings as necessary.
	(b) Incorrect steering gear adjustment.	(b) Adjust steering gear.
	(c) Loose steering gear to frame mounting bolts.	(c) Tighten steering gear to frame bolts.

33

Condition	Possible Cause	Correction
	(d) Worn ball joints or tie rod.	(d) Replace ball joints or tie rods as necessary.
	(e) Worn steering gear parts.	(e) Replace worn steering gear parts and adjust as necessary.
	(f) Worn upper or lower ball joints.	(f) Replace ball joints.
	(g) Worn idler arm bushing.	(g) Replace bushing.
FRONT WHEEL SHIMMY	(a) Tire, wheel out of balance.	(a) Balance wheel and tire assembly.
	(b) Uneven tire wear, or excessively worn tires.	(b) Rotate or replace tires as necessary.
	(c) Worn or loose wheel bearings.	(c) Replace or adjust wheel bearings as necessary.
	(d) Worn tie rod ends.	(d) Replace tie rod ends.
	(e) Strut mounting bushings loose or worn.	(e) Replace strut mounting bushings.
	(f) Incorrect front wheel alignment (particularly caster).	(f) Adjust front wheel alignment.
	(g) Worn or loose upper control arm ball joints.	(g) Inspect ball joints and replace where required.

STARTER—ELECTRICAL

Condition	Possible Cause	Correction
STARTER FAILS TO OPERATE	(a) Weak battery or dead cell in battery.	(a) Test specific gravity. Recharge or replace battery as required.
	(b) Ignition switch faulty.	(b) Test and replace switch if necessary.
	(c) Loose or corroded battery cable terminals.	(c) Clean terminals and clamps, replace if necessary. Apply a light film of petrolatum to terminals after tightening.
	(d) Open circuit, wire between the ignition — starter switch and ignition terminal on starter relay.	(d) Inspect and test all the wiring.
	(e) Starter relay defective.	(e) Test relay and replace if necessary.
	(f) Faulty starter.	(f) Test and repair as necessary.
	(g) Armature shaft sheared.	(g) Test and repair.
	(h) Open solenoid pull-in wire.	(h) Test and replace solenoid if necessary.
STARTER FAILS AND LIGHTS DIM	(a) Weak battery or dead cell in battery.	(a) Test for specified gravity. Recharge or replace battery as required.
	(b) Loose or corroded battery cable terminals.	(b) Clean terminals and clamps, replace if necessary. Apply a light film of petrolatum to terminals after tightening.
	(c) Internal ground in windings.	(c) Test and repair starter.
	(d) Grounded starter fields.	(d) Test and repair starter.
	(e) Armature rubbing on pole shoes.	(e) Test and repair starter.
STARTER TURNS, BUT ENGINE DOES NOT ENGAGE	(a) Starter clutch slipping.	(a) Replace clutch unit.
	(b) Broken clutch housing.	(b) Test and repair starter.
	(c) Pinion shaft rusted, dirty or dry, due to lack of lubrication.	(c) Clean, test and lubricate.
	(d) Engine basic timing wrong.	(d) check engine basic timing and condition of distributor rotor and cap.
	(e) Broken teeth on engine ring gear.	(e) Replace ring gear. Inspect teeth on starter clutch pinion.

34

Condition	Possible Cause	Correction
STARTER FAILS AND LIGHTS DIM	(a) Weak battery or dead cell in battery.	(a) Test for specified gravity. Recharge or replace battery as required.
	(b) Loose or corroded battery cable terminals.	(b) Clean terminals and clamps, replace if necessary. Apply a light film of petrolatum to terminals after tightening.
	(c) Internal ground in windings.	(c) Test and repair starter.
	(d) Grounded starter fields.	(d) Test and repair starter.
	(e) Armature rubbing on pole shoes.	(e) Test and repair starter.
STARTER TURNS, BUT ENGINE DOES NOT ENGAGE	(a) Starter clutch slipping.	(a) Replace clutch unit.
	(b) Broken clutch housing.	(b) Test and repair starter.
	(c) Pinion shaft rusted, dirty or dry, due to lack of lubrication.	(c) Clean, test and lubricate.
	(d) Engine basic timing wrong.	(d) check engine basic timing and condition of distributor rotor and cap.
	(e) Broken teeth on engine ring gear.	(e) Replace ring gear. Inspect teeth on starter clutch pinion.
STARTER RELAY DOES NOT CLOSE	(a) Battery discharged.	(a) Recharge or replace battery.
	(b) Faulty wiring.	(b) Test for open circuit, wire between starter relay ground terminal post and neutral starter switch (automatic transmission only). Also test for open circuit; wire between ignition-starter switch and ignition terminal and starter relay.
	(c) Clutch start switch or neutral starter switch on automatic transmission faulty.	(c) Test and replace the switch if necessary.
	(d) Starter relay faulty.	(d) Test and replace if necessary.
RELAY OPERATES BUT SOLENOID DOES NOT	(a) Faulty wiring.	(a) Test for open circuit wire between starter-relay solenoid terminal and solenoid terminal post.

ALTERNATOR-REGULATOR—ELECTRICAL

Condition	Possible Cause	Correction
ALTERNATOR FAILS TO CHARGE (No Output or Low Output)	(a) Alternator drive belt loose.	(a) Adjust drive belt to specifications.
	(b) Regulator Base improperly grounded.	(b) Connect regulator to a good ground.
	(c) Worn brushes and/or slip rings.	(c) Install new brushes and/or slip rings.
	(d) Sticking brushes.	(d) Clean slip rings and brush holders. Install new brushes if necessary.
	(e) Open field circuit.	(e) Test all the field circuit connections, and correct as required.
	(f) Open charging circuit.	(f) Inspect all connections in charging circuit, and correct as required.
	(g) Open circuit in stator windings.	(g) Remove alternator and disassemble. Test stator windings. Install new stator if necessary.

35

Condition	Possible Cause	Correction
	(h) Open recitfiers.	(h) Remove alternator and disassemble. Test the recitfiers. Install new recitfiers if necessary.
LOW, UNSTEADY CHARGING RATE	(a) High resistance in body to engine ground lead.	(a) Tighten ground lead connections. Install new ground lead if necessary.
	(b) Alternator drive belt loose.	(b) Adjust alternator drive belt.
	(c) High resistance at battery terminals.	(c) Clean and tighten battery terminals.
	(d) High resistance in charging circuit.	(d) Test charging circuit resistance. Correct as required.
	(e) Open stator winding.	(e) Remove and disassemble alternator. Test stator windings. Install new stator if necessary.
LOW OUTPUT AND A LOW BATTERY	(a) High resistance in charging circuit.	(a) Test charging circuit resistance and correct as required.
	(b) Shorted rectifier. Open recitfier.	(b) Perform current output test. Test the rectifiers and install new rectifiers as required. Remove and disassemble the alternator.
	(c) Grounded stator windings.	(c) Remove and disassemble alternator. Test stator windings. Install new stator if necessary.
	(d) Faulty voltage regulator.	(d) Test voltage regulator.
EXCESSIVE CHARGING RATE TO A FULLY CHARGED BATTERY	(a) Faulty ignition switch.	(a) Install new ignition switch.
	(b) Faulty voltage regulator.	(b) Test voltage regulator. Replace as necessary.
NOISY ALTERNATOR	(a) Alternator mounting loose.	(a) Properly install and tighten alternator mounting.
	(b) Worn or frayed drive belt.	(b) Install a new drive belt and adjust to specifications.
	(c) Worn bearings.	(c) Remove and disassemble alternator. Install new bearings as required.
	(d) Interference between rotor fan and stator leads or rectifiers.	(d) Remove and disassemble alternator. Correct interference as required.
	(e) Rotor or rotor fan damaged.	(e) Remove and disassemble alternator. Install new rotor.
	(f) Open or shorted rectifier.	(f) Remove and disassemble alternator. Test rectifiers. Install new recitfiers as required.
	(g) Open or shorted winding in stator.	(g) Remove and disassemble alternator. Test stator windings. Install new stator if necessary.
EXCESSIVE AMMETER FLUCTUATION	(a) High resistance in the alternator and voltage regulator circuit.	(a) Clean and tighten all connections as necessary.

FUEL SYSTEM .. EMISSION CONTROL 3

INDEX

DESCRIPTION

Four basic carburetors are used for all Corvette models. They are the Rochester 4MV (Quadrajet), the Rochester M4M (Quadrajet), the Holley 2-barrel (Model 2300, 2300C) and the Holley 4-barrel (Model 4150, 4160). The minor differences between the models are covered in the text.

Because many service procedures for the various carburetors are similar, typical illustrations and procedures are used except where specific illustrations or procedures are necessary to clarify the operation. This section covers overhaul, removal, installation, adjustments (on engine) of carburetors. Also covered in this section are maintenance procedures for choke coils, throttle linkage and fuel filters.

Removal

Flooding, stumble on acceleration and other performance complaints are, in many instances, caused by the presence of dirt, water, or other foreign matter in the carburetor. To aid in diagnosing the cause of the complaint, the carburetor should be carefully removed from the engine without draining the fuel from the bowl. The contents of the fuel bowl may then be examined for contamination as the carburetor is disassembled.

Remove air cleaner and gasket. Disconnect fuel and vacuum lines from carburetor. Disconnect choke rod. Disconnect accelerator linkage. If equipped with automatic transmission, disconnect TV linkage. Remove carburetor attaching nuts and/or bolts and remove carburetor.

Installation

Be certain throttle body and intake manifold sealing surfaces are clean. Install new carburetor to manifold flange gasket (if required). Install carburetor over manifold studs. Start vacuum and fuel lines at carburetor. Install attaching nuts and/or bolts and tighten securely. Tighten fuel and vacuum lines. Connect and adjust accelerator and TV linkage. Connect choke rod. Install air cleaner, adjust idle speed and mixture per decal.

Fast idle adjustment

The fast idle adjustment must be set with electrical lead to the Transmission Controlled Spark (TCS) solenoid disconnected (if so equipped), and transmission in "Neutral". Make sure choke is properly adjusted and in wide open position — engine warm. Position fast idle lever on high step (2nd step 1971-72) of fast idle cam. Adjust fast idle screw on Rochester 4MV. On Rochester M4M, turn fast idle screw out until primary throttle valves are closed, then turn screw out to contact lever, then turn screw in specified number of turns.

Rochester M4M Carburetor

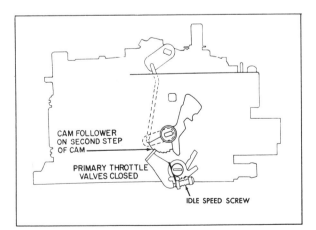

Fast idle adjustment (Rochester 4MV)

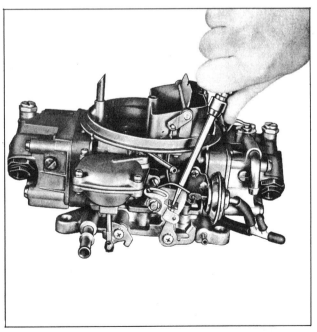

Fast idle adjustment (Holley Models 4150 & 2300)

Float adjustment — Holley

Remove air cleaner then remove the fuel level sight plugs. With parking brake on, and transmission in neutral, start the engine and allow it to idle. With the car on a level surface, the fuel level should be on a level with the threads at the bottom of the sight plug port (plus or minus 1/32 inch).

If necessary to adjust (either or both bowls), loosen inlet needle lock screw and turn the adjusting nut clockwise to lower or counter-clockwise to raise fuel level, then tighten lock screw. 1/6 turn of adjusting nut equals approximately 1/16" fuel level change.

Allow a minute for fuel level to stabilize, and recheck the level at sight plug. Readjust, if necessary, until proper level is obtained, and install sight plug and air cleaner.

To assure proper secondary float level setting it is advisable to accelerate primary throttles slightly and hand operate secondary throttle. This assures a stabilized secondary fuel level.

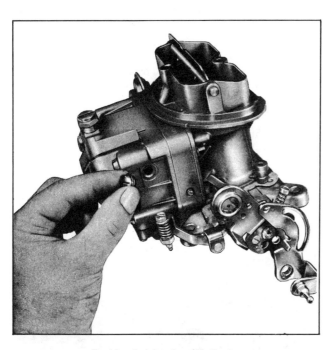

Fuel level sight plug (Holley)

Float adjustment — Rochester M4M

Hold retainer firmly in place. Push float down lightly against needle. Gauge reading should be 15/32" from top of casting to top of float. Gauging point is 3/16" back from end of float at toe. Remove float and bend float arm up or down to adjust. Visually check float alignment after adjusting.

Vacuum break adjustments

Remove air cleaner assembly from vehicle. On vehicles with 'Therm AC' air cleaner, plug the sensor's vacuum take-off port.

Start engine or install a vacuum source to vacuum diaphragm vacuum tube. Remove choke rod from choke lever and install a rubber band to the lever in such a manner as to react on the lever in a direction to close the choke blade. Slowly open accelerator until choke closes (choke will not fully close due to the vacuum break link's reaction on it) and idle is determined by high step of fast idle cam. Release accelerator.

With idle determined by fast idle cam and vacuum break link acting on the choke lever, insert specified gauge (hold vertical) between the carburetor air horn and choke blade. Adjust vacuum break setting as necessary by bending rod or tang.

For the Rochester M4M use the choke valve measuring gage J-26701. Rotate the degree scale until zero is opposite the pointer. Seat choke valve diaphragm using vacuum source. Hold choke valve towards closed position, pushing counterclockwise on inside coil lever. To adjust, turn screw in or out until the bubble is centered. Remove gage.

Carburetor No.	Angle Gage
17058202	.27° below 22,500 miles
17058203	.30° above 22,500 miles
17058204	
17058502	.28° below 30,000 miles
17058504	.31° above 30,000 miles
17058210	.30° below 22,500 miles
17058211	.33° above 22,500 miles
17058228	
17058582	
17058584	

Throttle return check valve adjustment

With carburetor choke closed and fast idle screw on high step of fast idle cam, loosen lock nut and adjust valve to just contact throttle lever. Adjust lock nut to lock valve in position. Tighten nut to 90-130 in. lbs.

With choke wide open, operate throttle lever and observe action of valve plunger — plunger should be depressed when throttle lever returns to closed position.

TYPICAL HOLLEY

Vacuum break adjustment

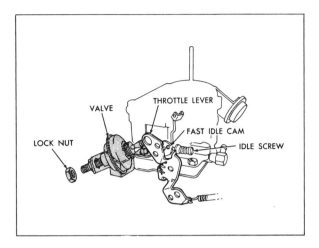

Throttle return check valve

FUEL FILTER
Maintenance
Disconnect fuel line connection at inlet fuel filter nut. Remove inlet fuel filter nut from carburetor with a box wrench or socket. Remove filter element and spring. Check bronze element for restriction by blowing on cone end; element should allow air to pass freely.

Check paper element by blowing on fuel inlet end. If filter does not allow air to pass freely, replace element. No attempt should be made to clean filters. Element should be replaced if plugged or if flooding occurs. A plugged filter will result in a loss of engine power or rough (pulsating) engine feel, especially at high engine speeds.

Install element spring, and install element in carburetor. Bronze filters must have small section of cone facing out. Install new gasket on inlet fitting nut and install nut in carburetor and tighten securely. Install fuel line and tighten connector.

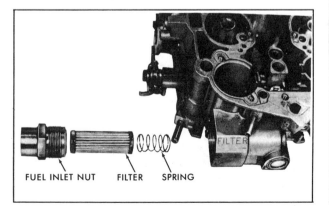

Fuel filter (Paper type)

CHOKE COIL
Replacement
Remove air cleaner and disconnect choke rod upper clip. Remove the choke coil shield by prying with a screw driver in the cut out provided then lift shield carefully over rod. Remove choke rod, bracket screw and choke coil assembly.

For Rochester M4M, remove three attaching screws and retainers from choke cover and coil assembly. Then pull straight outward and remove cover and coil assembly from choke housing. Remove choke cover gasket, if used. Remove choke housing assembly from float bowl by removing retainer screw and washer inside the choke housing. The complete choke assembly can be removed from the float bowl by sliding outward.

Install a new choke coil assembly being sure the locating tab is in the forward hole of the intake manifold and install mounting screw.

Install the choke rod and adjust as necessary (without choke coil shield installed). Disconnect choke rod upper end and lower choke coil shield over choke rod and install over choke coil. Move shield to best fit along manifold and connect upper end of choke rod.

Be sure choke valve moves freely from full open to full closed position. Start and warm up the engine and check operation of the choke. Install the air cleaner.

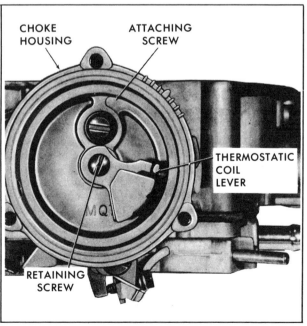

Choke system

IDLE SOLENOID
Removal

Remove engine air cleaner. Disconnect electrical connector at solenoid. Loosen solenoid bracket clamp bolt and remove solenoid.

Installation

Install solenoid in bracket so that it does not contact throttle lever. Set the carburetor in the closed throttle position with the choke off. Start engine and set idle to specifications (refer to tune-up decal) using the carburetor's low idle speed screw.

With solenoid energized and solenoid plunger screw out two turns from being fully bottomed in plunger, slide solenoid into bracket so that the plunger just slightly contacts the carburetor lever. Tighten the bracket strap to secure the solenoid in place. Back out carburetor low idle speed screw approximately one turn (counter-clockwise). The low idle speed screw will be reset in the last step outlined on engine decal.

HOLLEY 2300
SECONDARY CARBURETOR CLOSING ROD
Adjustment

Start engine and allow to warm-up (choke in fully opened position). Disconnect closing rod ends from each secondary carburetor. Set idle to specifications. Turn off engine.

With clevis pin bottomed in primary carburetor throttle slot, adjust rear secondary carburetor closing rod so that it falls short of entering the rear secondary throttle lever hole by 1/2 a rod diameter. Install rod and secure with special clip.

Next, adjust forward secondary carburetor closing rod so that it just enters the forward secondary throttle level hole. Install rod and secure with special clip.

When the adjustment is completed the clevis pin will be at a point just above the bottom of the primary throttle slot, thus insuring that there is freedom in the linkage and a positive secondary throttle closure.

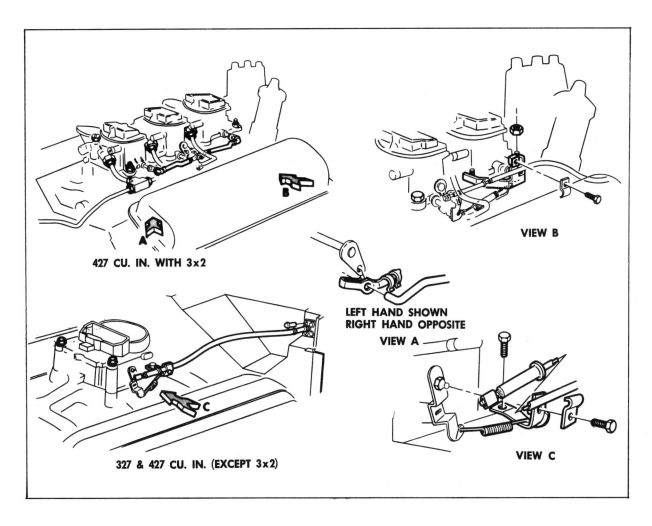

427 CU. IN. WITH 3x2

VIEW B

LEFT HAND SHOWN
RIGHT HAND OPPOSITE
VIEW A

327 & 427 CU. IN. (EXCEPT 3x2)

VIEW C

Throttle linkage

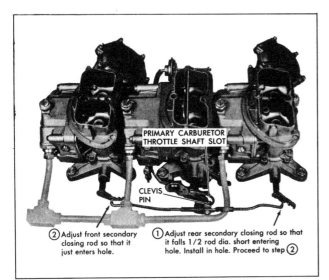

Secondary closing rod adjustment

THROTTLE LINKAGE
Adjustment

Disconnect throttle rod swivel at throttle lever on carburetor or at dash lever. Hold accelerator pedal to floor against stop. Move carburetor throttle lever to wide open position. Tighten cable clamp bolt to 45 in. lbs. of torque.

On automatic transmission equipped vehicles disconnect TV rod at throttle lever. Hold carburetor throttle in wide open position, push throttle rod rearward (to position accelerator pedal at the floor mat) and adjust swivel to just enter hole in throttle lever. Connect swivel to throttle lever and install accelerator return spring.

On vehicles equipped with automatic transmission hold throttle lever in full open position, pull TV rod to full detent position and adjust TV rod to just enter hole on throttle lever, and connect TV rod at throttle lever.

ROCHESTER AIR HORN
Removal

On all except RPO NA9, remove idle vent valve attaching screw then remove idle vent valve assembly. Remove clip from upper end of choke rod, disconnect choke rod from upper choke shaft lever and remove choke rod from bowl. Remove roll pin at pump lever pivot by driving pin inward with proper drift punch.

Remove nine air horn to bowl attaching screws, two attaching screws are located next to the primary venturi. (Two long screws, five short screws, two counter-sunk screws). Remove vacuum break hose at diaphragm and remove diaphragm unit from retaining bracket.

Disconnect choke assist spring at diaphragm plunger and remove air valve-to-dashpot lever at air valve lever. Remove air horn by lifting straight up. Air horn gasket should remain on bowl for removal later.

Care must be taken not to bend two small main well air bleed tubes protruding from air horn. These are permanently pressed into casting. DO NOT REMOVE.

Disassembly

Remove secondary metering rods and hanger by removing hanger attaching screw. Further disassembly of the air horn is not required for cleaning purposes. If part replacement is required, remove choke valve attaching screws then remove choke valve and shaft. Air valves and air valve shaft are calibrated and should not be removed.

Removing air horn screws

Assembly

Install choke shaft, choke valve, and two attaching screws, if removed. Install secondary metering rods and hanger. Rods should be positioned with upper ends through hanger holes and towards each other.

Air horn to bowl installation

Place air horn assembly on bowl, carefully positioning secondary metering rods, vent tubes, and accelerating well tubes through air horn gasket. Do not force air horn assembly on to float bowl as distortion of secondary metering discs will result. A slight sideward movement will center metering rods in metering discs.

42

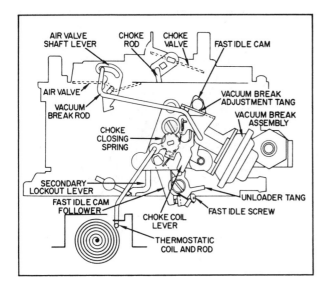

Choke system

Install two long air horn screws, five short screws, and two counter-sunk screws in primary venturi area. All screws must be tightened evenly and securely. Install idle vent actuating rod in pump lever. Connect pump rod in pump lever and retain with spring clip.

Connect choke rod in lower choke lever and retain in upper lever with spring clip. On all except RPO NA9, install idle vent valve, engaging actuating rod and tighten attaching screw. Install air valve-to-dashpot lever in dashpot plunger and air valve shaft lever; then position dashpot in retaining bracket and connect vacuum hose and choke assist spring.

43

Removing air horn

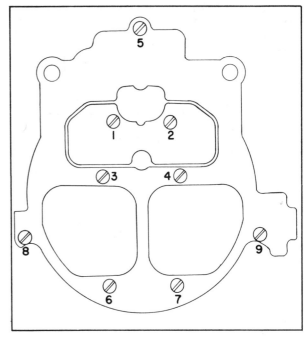

Air horn tightening sequence

FLOAT BOWL
Disassembly — Rochester 4MV, Holley
Remove pump plunger from pump well. Remove air horn gasket from dowels on secondary side of bowl, then remove gasket from around power piston and primary metering rods. Remove pump return spring from pump well. Remove plastic filler over float valve.

Remove power piston and primary metering rods, using needle nosed pliers to pull straight up on metering rod hanger directly over power piston. Remove power piston spring from well. The power piston assembly is held in place for ease in assembly of the air horn, by a plastic retainer at top of power piston bore.

Remove metering rods from power piston by disconnecting tension spring from top of each rod then rotating rod to remove from hanger. Remove float assembly by pulling up slightly on retaining pin until pin can be removed by sliding toward pump well. After pin is removed, slide float assembly toward front of bowl to disengage needle pull clip being careful not to distort pull clip.

Remove pull clip and fuel inlet needle. Remove fuel inlet needle seat with wide blade screw driver. Remove needle seat gasket and discard. Remove primary metering jets. No attempt should be made to remove secondary metering discs. Remove pump discharge check ball retainer and check ball. Remove baffle from secondary side of bowl. Remove retaining screw from choke assembly and remove assembly from float bowl then remove secondary lock out link from bowl.

Remove fast idle cam from choke assembly. Do not place vacuum break assembly in carburetor cleaner. Remove intermediate choke rod and actuating lever from inside of float bowl well. Remove fuel inlet filter nut gasket, filter and spring. Remove throttle body by removing throttle body to bowl attaching screws. Remove throttle body to bowl insulator gaskets.

Assembly

Install new throttle body to bowl insulator gasket being certain the gasket is properly installed over two locating dowels on bowl. Install throttle body making certain throttle body is properly located over dowels on float bowl then install throttle body to bowl screws and tighten evely and securely. Place carburetor on proper holding fixture.

Install fuel inlet filter spring, filter, new gasket and inlet nut and tighten nut securely. Install fast idle cam on choke shaft with cam pick-up lever on underside of cam. Connect choke rod to choke rod actuating lever (plain end) then holding choke rod, with grooved end pointing inward, position choke rod actuating lever in well of float bowl and install choke assembly engaging shaft with hole in actuating lever. Install retaining screw and tighten securely. Remove choke rod from lever for installation later.

Install vacuum hose to tube connection on bowl and vacuum break assembly. Install air deflector in secondary side of bowl with notches towards top. Install pump discharge check ball and retainer in passage next to pump well. Install primary main metering jets. Install fuel inlet needle seat and gasket. Use wide blade screw driver to avoid distortion. Install fuel inlet needle. Using needle nosed pliers to hold needle pull clip, install pull clip on needle. Pull clip is properly positioned with open end towards front of bowl.

Install float by sliding float lever under pull clip from front to back. With float lever in pull clip, hold float assembly at toe and install retaining pin from pump well side. Be careful not to distort pull clip (Do NOT install pull clip thru hole in float arm). Float level adjustment. With an adjustable T-scale, measure from top of float bowl gasket surface (gasket removed) to top of float at toe (located gauging point 3/16" back from toe). Make sure retaining pin is held firmly in place and tang of float needle is seated on float needle. Bend float up or down for proper adjustment.

Install power piston spring in power piston well. If primary main metering rods were removed from hanger re-install making sure that tension spring is connected to top of each metering rod. Install power piston assembly in well with metering rods properly positioned in metering jets. Press down firmly on power piston retainer to insure engagement of retaining clip.

Install plastic filler over float needle, pressing downward until seated properly. Install pump return spring in pump well. Install air horn gasket around primary metering rods and piston. Position gasket over two dowels on secondary side of bowl. Press power piston down firmly to assure correct alignment engagement of pin. Install pump plunger in pump well.

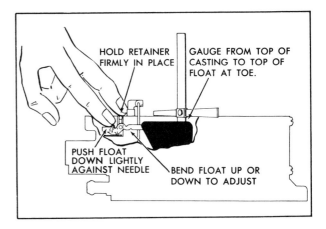

Float level adjustment

Disassembly — Rochester M4M

Remove air horn gasket by lifting out of dowel locating pins and lifting tab of gasket from beneath the power piston hanger, being careful not to distort springs holding the main metering rods. Remove pump plunger from pump well. Remove pump return spring from pump well. Remove power piston and metering rods by depressing piston stem and allowing it to snap free. The power piston can be easily removed by pressing the piston down and releasing it with a snap. This will cause the power piston spring to snap the piston up against the retainer. This procedure may have to be repeated several times. Do NOT remove the power piston by using pliers on the metering rod hanger. Remove the power piston spring from the well.

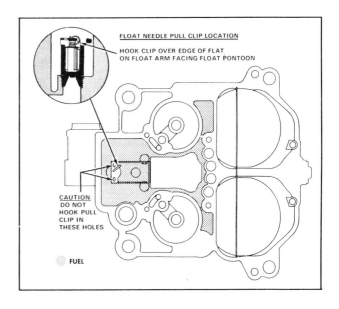

Rochester M4M pull clip location

44

Float Assembly

The A.P.T. metering rod adjustment screw is preset at the factory and no attempt should be made to change this adjustment in the field. if float bowl replacement is required during service, the new bowl assembly will be supplied with an A.P.T. metering rod screw which will be preset as required.

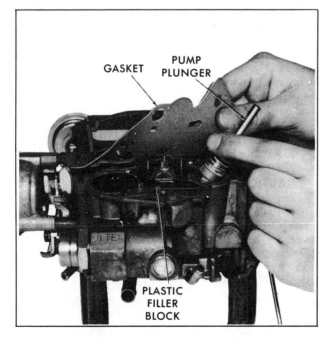

Removing air horn gasket and pump plunger

Remove metering rods from power piston by disconnecting tension spring from top of each rod, then rotate rod to remove from hanger. Use care when disassembling rods to prevent distortion of tension spring and/or metering rods. Note carefully position of tension spring for reassembly.

Remove plastic filler block over float valve. Remove float assembly and float needle by pulling up on retaining pin. Remove float needle seat and gasket. Remove aneroid cavity insert from float bowl. Remove primary main metering jets, only if necessary. No attempt should be made to remove the secondary metering jet. These jets are fixed and, if damaged, bowl replacement is required. Remove pump discharge check ball retainer and check ball. Remove secondary air baffle, if replacement is required. Remove pump well fill slot baffle.

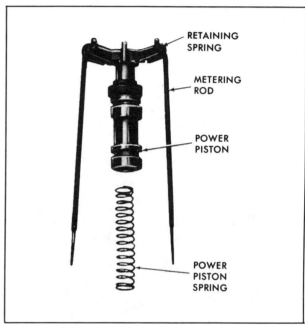

Power piston and metering rods

THROTTLE BODY
Disassembly — Rochester 4MV, Holley
Remove pump rod from throttle lever by rotating rod out of primary throttle lever. Remove idle mixture screws and springs. Extreme care must be taken to avoid damaging secondary throttle valves. No further disassembly of the throttle body is required.

Cleaning and inspection
The carburetor should be cleaned in a cold immersion type cleaner. Thoroughly clean carburetor castings and metal parts in an approved carburetor cleaner. Any rubber parts, plastic parts, diaphragms and pump plungers, should not be immersed in carburetor cleaner. However, the Delrin cam on the air valve shaft will withstand normal cleaning in carburetor cleaner. Blow out all passages in castings with compressed air. Do not pass drills through jets or passages.

Inspect idle mixture needles for damage. Examine float needle and seat for wear. Replace if necessary with new float needle assembly. Inspect upper and lower surfaces of carburetor castings for damage. Inspect holes in levers for excessive wear or out of round conditions. If worn, levers should be replaced. Examine fast idle cam for wear or damage. Check air valve for binding conditions. If air valve is damaged, air horn assembly must be replaced. Check all throttle levers and valves for binds or other damage.

Assembly
Install idle mixture needles and springs until lightly seated. Back out needles two turns as a preliminary idle adjustment. Install pump rod in hole of throttle lever by rotating rod.

Disassembly — Rochester M4M
Remove pump rod from throttle lever. Do NOT remove idle mixture limiter caps, unless it is necessary to replace the mixture needles or normal soaking and air pressure fails to clean the idle passages. If the idle mixture needles are removed, destroy the idle mixture plastic limiter caps. Do not replace the caps as a bare mixture screw is sufficient to indicate that the mixture has been readjusted.

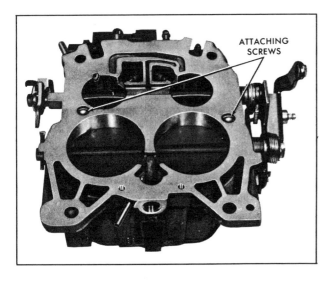

Removing throttle body

CHOKE ROD
Adjustment — Rochester 4MV, Holley
With the cam follower on second step of fast idle cam and against the high step, rotate the choke valve toward the closed position by turning the external choke lever counter-clockwise. Dimension between the lower edge of choke valve, at choke lever end, should be as specified. Bend choke rod to adjust.

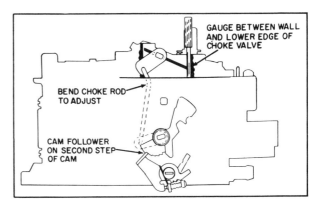

Choke rod adjustment

Adjustment — Rochester M4M
Make fast idle adjustment. Using gage J-26701, rotate degree scale until zero is opposite pointer. With choke valve completely closed, place magnet squarely on top of choke valve. Rotate bubble until it is centered. Rotate scale so 46° is opposite pointer. Place cam follower on second step of cam next to high step. Close choke by pushing upward on choke coil lever. To adjust, bend tang on fast idle cam until bubble is centered. Remove gage.

VACUUM BREAK
Adjustment
Hold the choke valve in the closed position using a rubber band on the external choke lever. Hold vacuum break diaphragm stem against its seat. The dimension between the lower edge of choke valve and air horn, at choke lever end, should be as specified. Bend tang to adjust. Recheck adjustment with carburetor on engine and engine running.

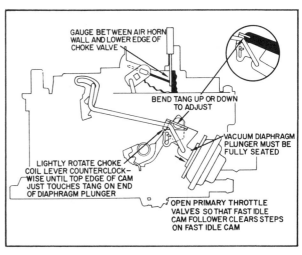

Vacuum break adjustment

AIR VALVE DASHPOT
Adjustment

With the vacuum break diaphragm seated, there must be clearance as specified between the dashpot rod and end of slot in air valve lever. Bend rod, at air valve end, to adjust.

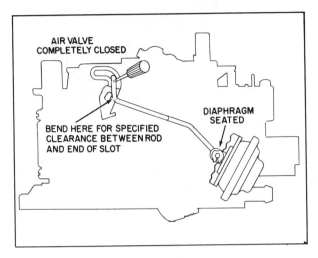

Air valve dashpot adjustment

HOLLEY 2300 AND 2300C
Disassembly

Before disassembly, loosen the fuel inlet fitting, fuel bowl sight plugs and needle and seat assembly lock screws.

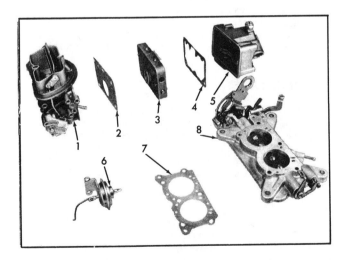

Subassemblies primary — Holley

1. Carburetor body
2. Metering body gasket
3. Metering body
4. Fuel bowl gasket
5. Fuel bowl assembly
6. Vacuum break
7. Throttle body gasket
8. Throttle body

Remove primary fuel bowl screws and remove fuel bowl, metering body (on primary carburetor), splash shield and gasket. On secondary carburetors remove metering block screws and remove metering body and gasket. On primary carburetor disconnect vacuum break hose at vacuum break. On secondary carburetors disconnect secondary diaphragm housing assembly at throttle lever. Remove secondary diaphragm housing assembly and gasket from secondary carburetor body. Remove throttle body to main body screws and then remove throttle body and gasket.

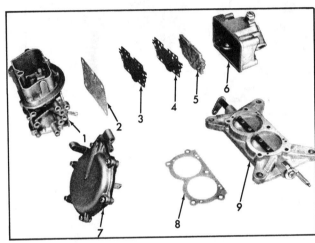

Subassemblies secondary

1. Carburetor body
2. Fuel bowl gasket
3. Metering body plate
4. Metering body gasket
5. Metering body
6. Fuel bowl assembly
7. Diaphragm housing assembly
8. Throttle body gasket
9. Throttle body

Assembly

Invert the main body, align new throttle body to main body gasket, then install screws and tighten securely. On secondary carburetors install throttle operating assembly with new gasket and connect rod to throttle lever. On primary carburetor connect vacuum break hose to vacuum break. On secondary carburetor install new fuel bowl gaskets on main body, align metering body to main body and install six retaining screws tightening securely. On secondary carburetor install fuel bowl with new retaining screw gaskets. On primary carburetor place fuel bowl gasket splash shield, metering body and metering body gasket in alignment. Insert screws and new gaskets through fuel bowl and metering body. Install fuel and metering body to main body keeping pump lever in alignment under operating lever duration spring then finish installing fuel bowl retaining screws and tighten securely.

FUEL BOWLS
Disassembly

Remove fuel inlet baffle. Remove float hinge screws and remove float from bowl. Primary carburetor uses a brass float and secondary carburetors use a plastic float. Loosen inlet needle and seat lock screw, then turn adjusting nut counter-clockwise to remove the needle and seat assembly.

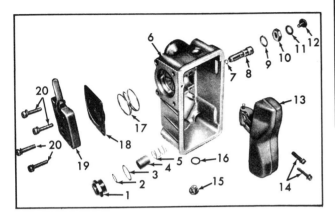

Fuel bowl assembly (primary) – Holley

1. Nut-fuel inlet
2. Gasket-fuel filter
3. Gasket-inlet nut
4. Fuel filter
5. Spring fuel filter
6. Fuel bowl
7. Seal-inlet needle and seat assembly
8. Inlet needle and seat assembly
9. Gasket-inlet adjusting nut
10. Nut-inlet adjusting
11. Gasket-inlet lock screw
12. Screw-inlet lock
13. Float assembly
14. Screw-float hinge
15. Fuel level sight plug
16. Gasket-sight plug
17. Spring-pump diaphragm
18. Pump diaphragm
19. Cover assembly-pump diaphragm
20. Screw-pump diaphragm cover
25. Fuel line
26. Purge valve

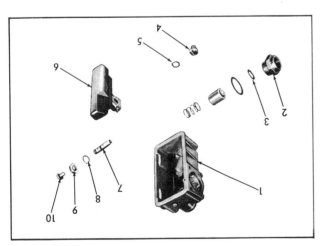

Fuel bowl assembly (secondary) – Holley

1. Float bowl
2. Inlet nut
3. Inlet nut gasket
4. Sight plug
5. Sight plug gasket
6. Float
7. Needle and seat assembly
8. Gasket
9. Adjusting nut
10. Lockscrew

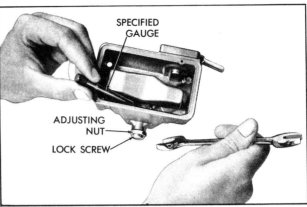

Preliminary float adjustment

Remove sight plug and gasket. Remove fuel inlet fitting, fuel filter, spring and gasket. On primary carburetor remove pump diaphragm cover screws and then pump cover, diaphragm, and spring. On primary carburetor check pump inlet ball for free movement and damage. Damage to ball, passage or retainer requires a new bowl assembly.

Assembly

Install fuel level sight plug with new gasket. Install fuel inlet needle and seat with a new gasket. Leave lock screw loose. Install fuel inlet fitting, fuel filter, spring and new gasket. Place fuel float in bowl and install float hinge screws. Install fuel inlet baffle. On primary carburetor install accelerator pump spring, diaphragm and cover.

Adjust float

Make a preliminary float adjustment by inverting fuel bowl and turn adjustable needle and seat until top of float is specified distance from top of fuel bowl. Do not fully tighten lock screw. Snug screw to temporarily retain adjustment. Final adjustment of the float is made on the vehicle.

METERING BODIES
Disassembly

On primary carburetor remove main metering jets (with a wide blade screw driver to avoid damage to the jets), and the power valve (using 1" 12 point socket). On primary carburetors remove the vacuum fitting and the idle mixture needles and seals. On secondary carburetors remove plate and gasket from secondary metering body dowel pins.

Assembly

On primary carburetor install vacuum fitting idle mixture needles and new needle seals. Make a preliminary adjustment of idle mixture screws by turning needles lightly to their seats and back out one turn. Do not turn screw tightly against seat or damage may result. On secondary carburetor install a new gasket and then metering plate over metering body dowel pins.

SECONDARY THROTTLE OPERATING ASSEMBLY   48
Disassembly

Remove diaphragm cover. Remove spring and diaphragm.

Assembly secondary diaphragm housing
Install diaphragm assembly and spring in housing. Install diaphragm cover and tighten securely. Diaphragm may be checked for leaks by pushing in on rod then holding finger over vacuum hole. Rod should remain in.

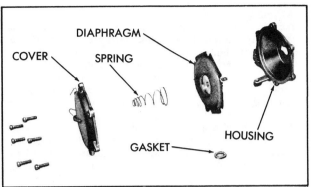

Secondary diaphragm housing assembly

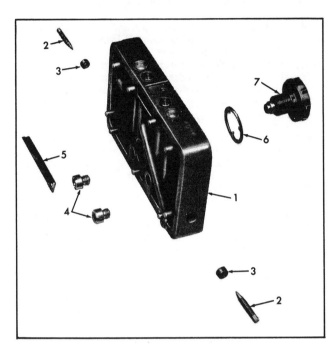

Metering body assembly (primary)

1. Metering body
2. Needle-idle mixture
3. Seal-idle mixture needle
4. Metering jet
5. Splash shield
6. Gasket-power valve
7. Power valve

MAIN BODY (PRIMARY CARBURETOR)
Disassembly
Secondary carburetor main bodies cannot be disassembled.

Remove choke vacuum break retaining screws and remove choke vacuum break by disconnecting link at choke lever. Remove choke lever retaining clip then remove choke lever and fast idle cam. Remove pump discharge nozzle screw, nozzle and gasket then up end the body assembly to remove pump discharge check valve.

The choke rod seal will withstand normal cleaning in carburetor cleaner, therefore, further disassembly of the main body is not required for cleaning purposes. If parts replacement is required, file off the staked ends of choke shaft screws then remove screws.

Remove choke rod (upward through seal) and remove seal from main body. Remove valve from shaft slot and slide shaft from main body.

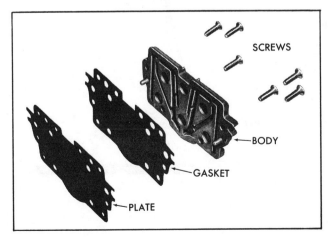

Metering body assembly (Secondary) – Holley

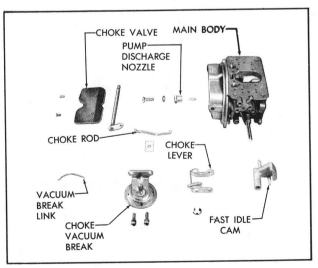

Main body assembly

Assembly

Install pump discharge valve. Install pump discharge nozzle. Use new gaskets. Install choke rod seal and choke rod in main body. Install choke shaft in main body and connect upper end of choke rod. Install choke rod on choke shaft (do not tighten retaining screws). Center choke valve on shaft by holding valve closed while tigtening screws. Stake screw ends with pliers.

The choke valve is offset and should fall freely to wide open position from its own weight. Install choke lever and fast idle cam then retainer. Connect vacuum break link to choke lever then install vacuum break with two retaining screws.

THROTTLE BODY
Disassembly

Ordinarily the throttle body need not be disassembled for cleaning and inspection purposes. If necessary, remove pump operating lever assembly. Disassemble spring, bolt, and nut if needed. Remove idle speed screw and spring. Remove fast idle stop and choke unloader lever from throttle shaft. File off the staked ends of the throttle plate attaching screws, then remove the screws and throttle plates. Remove burrs from throttle shaft and slide the shafts out of flange. Remove accelerator pump cam from throttle lever. Remove vacuum break hose from fitting.

Cleaning and inspection

The most frequent causes of carburetor malfunction are gum, carbon, and water. Carefully clean and inspect all parts and castings as the carburetor is being serviced.

Clean throttle body (if not disassembled), vacuum break and all non-metallic parts in alcohol or gasoline. Wash all other parts in cleaning solvent. Inspect holes in all operating levers and castings for excessive wear. Inspect bearing surfaces of all shafts for excessive clearance. It is not necessary to remove shafts and plates to inspect. If wear is excessive to the extent of improper operation of the carburetor, the worn parts should be replaced.

Inspect primary float for bad dents and/or possible leaks. Inspect pump diaphragm for damage. Inspect float needle and seat for burrs and ridges; if present, replace both needle and seat. Never replace either alone as these are an assembly. Inspect the edges of the throttle valves for gouges and any other deformation. If these or any other condition exists which would prevent full seating, replace the faulty valve.

Inspect all mating surfaces of fuel bowl, main body, and throttle body for burrs, gouges, or other surface irregularities. All surfaces must be smooth and square to prevent leaks. Check secondary carburetor throttle operating diaphragm for free operation and leakage by moving diaphragm rod to the up position then covering vacuum passage opening in housing with thumb. The diaphragm should hold upward. Remove thumb from vacuum passage and diaphragm rod should move down readily.

After washing in solvent, clear all passages in the metering body and main body with compressed air. Check fuel filter element for restriction by blowing on cone end, element should allow air to pass freely. Clean element by washing in solvent and blowing out. Blow in opposite direction of fuel flow.

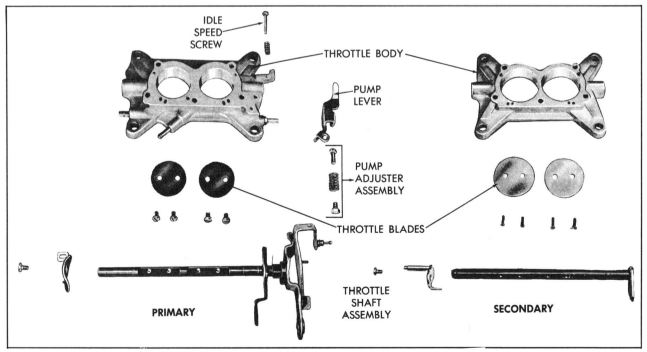

Throttle body assemblies — Holley

50

Assembly

On primary carburetor install accelerator cam on throttle lever shaft. Install throttle shaft in throttle body. Install throttle valves on throttle shaft. Do not tighten retaining screws. Center the throttle valves on the shafts by holding the valves closed while tightening the screws. The throttle valves are installed with identification numbers down (to manifold side).

Support the throttle shafts and stake the throttle valve screws. Install secondary throttle diaphragm lever on secondary carburetor. On primary carburetor install fast idle cam lever on vacuum break hose on fitting. Install idle speed screw and spring on primary carburetor. Turn idle speed screw clockwise until contact with throttle lever is made and turn 1-1/2 additional turns for preliminary adjustment. Install accelerator pump actuating lever assembly and assemble screw, spring, and nut if disassembled.

Idle speed adjustment — preliminary

Turn idle screw in until contact is made with throttle lever, then turn screw in 1-1/2 additional turns for preliminary idle speed adjustment.

Adjust fast idle cam (primary carburetor)

Open throttle slightly, close choke plate positioning fast idle lever against top step of fast idle cam. Adjust fast idle to give specified opening of throttle plate on idle transfer slot side of carburetor (see specifications). Bend fast idle lever to adjust.

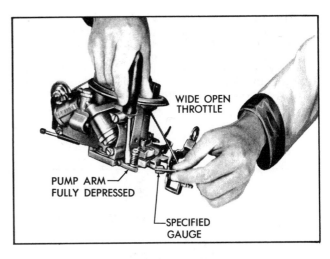

Accelerator pump adjustment

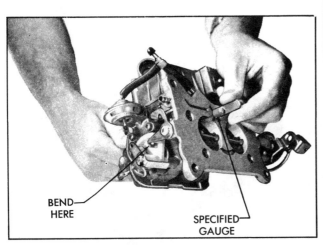

Fast idle cam adjustment

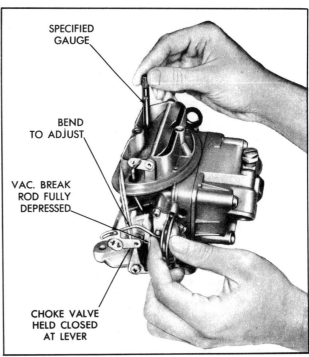

Vacuum break adjustment

Adjust accelerator pump (primary carburetor)

Hold throttle lever in wide open position with a rubber band and hold pump lever fully compressed (down). Measure the clearance between spring adjusting nut and arm of the pump lever. Adjust to specified clearance by turning nut or screw as required while holding opposite end. (The pump operating lever is not threaded). After adjustment is made, rotate the throttle lever to fully closed and partly open again. Any movement of the throttle lever should be noticed at operating lever spring end, indicating correct pump tip-in.

Adjust vacuum break (primary carburetor)

Hold choke valve closed with rubber band. Hold vacuum break in against stop. Measure distance between choke valve lower edge and main body. Bend vacuum break rod to adjust. Recheck adjustment with carburetor on engine and engine running.

Adjust choke unloader (primary carburetor)

Hold throttle lever in wide open throttle position with rubber band. Hold choke valve toward closed position against unloader tang of throttle shaft, then measure opening between choke valve lower edge and main body (see specifications). If necessary to adjust, bend choke rod (at off-set bend). Recheck after adjusting.

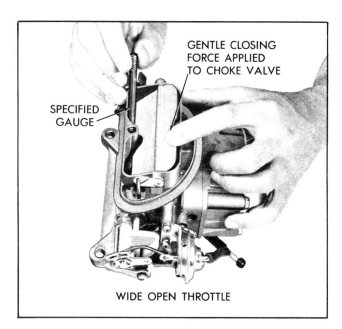

Choke unloader adjustment

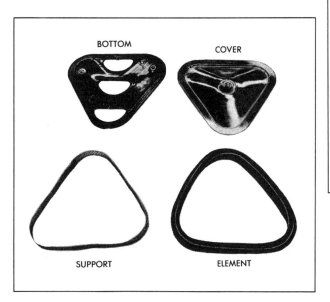

Polyurethane element air cleaner

AIR CLEANERS
Description

Air cleaners on all models operate primarily to remove dust and dirt from the air that is taken into the carburetor and engine. The air cleaner is also effective in reducing engine air inlet noise.

Three basic types of air cleaners are used. They are the oiled paper, oiled paper with polyurethane band, and the wire mesh with polyurethane band.

On most vehicles an automatic air inlet temperature control device is used. Air temperature is automatically controlled by a thermostatic valve which selects warmed air from the heat stove and/or cooler air from the engine compartment. The oiled paper filters used in most air cleaner assemblies have both ends of the paper element bonded with plastisol sealing material. Oil on the paper causes the element to become discolored by a small amount of dirt that does not necessarily indicate that the element is plugged or reduced in efficiency. It is advisable to rotate the air cleaner element 180° at 12,000 miles and to replace the air cleaner element every 24,000 miles. The preceding maintenance operations should be performed more often when subject to dusty or otherwise adverse driving conditions.

POLYURETHANE ELEMENT
Maintenance

Remove cover wing nut, cover and filter element. Visibly check the element for tears or rips and replace if necessary. Clean all accumulated dirt and grime from air cleaner bottom and cover. Discard air horn to air cleaner gaskets.

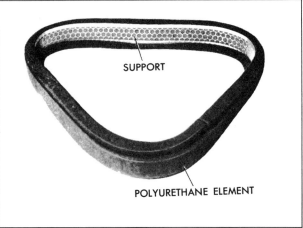

Polyurethane support

Remove support screen from element and wash element in kerosene or mineral spirits; then squeeze out excess solvent. Never use a hot degreaser or any solvent containing acetone or similar solvent.

Dip element into light engine oil and squeeze out excess oil. Never shake, swing or wring the element to remove excess oil or solvent as this may tear the polyurethane material. Instead, "squeeze" the excess from the element.

Install element on screen support. Using new gaskets, replace air cleaner body over carburetor air horns.

Replace the element in the air cleaner. Care must be taken that the lower lip of the element is properly placed in the assembly and that the filter material is not folded or creased in any manner that would cause an imperfect seal. Take the same precautions when replacing the cover that the upper lip of the element is in proper position. Replace cover and wing nut.

OIL WETTED PAPER ELEMENT
Replacement
Remove wing nut and cover. Remove paper element and discard. Remove bottom section of air cleaner and gasket on air horn of carburetor. Discard air horn gasket. Clean bottom section of air cleaner and cover pieces thoroughly, to remove dust and grime. Do not submerge bottom section of thermostatic air cleaners in solvent. Check air cleaner cover seal for tears or cracks. Install a new gasket on carburetor air horn and set bottom section of air cleaner on carburetor. Install new paper element on bottom section of air cleaner with either end up. Install cover and wing nut.

Inspection
Remove air cleaner element as previously outlined. Inspect top and bottom seals for deformation or cracking. These surfaces must be smooth and uniform. Inspect element for punctures or splits by looking through the element towards a light source. Internal portions of air cleaner cover and bottom should be clean. If washed with solvent surfaces should be dried thoroughly. Do not submerge bottom section of thermostatic air cleaners in solvent. Install air cleaner element as previously outlined.

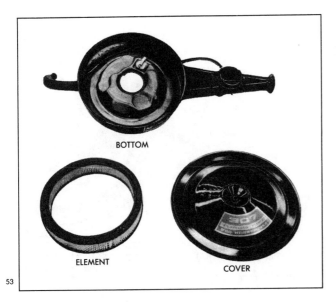

53

Paper element air cleaner

THERMOSTATICALLY CONTROLLED AIR CLEANER
Description
The thermostatic air cleaner system includes a temperature sensor, a vacuum motor and control damper assembly mounted in the air cleaner, vacuum control hoses, manifold heat stove and connecting pipes. The vacuum motor is controlled by the temperature sensor. The vacuum motor operates the air control damper assembly to regulate the flow of hot air and under hood air to carburetor. The hot air is obtained from the heat stove on the exhaust manifold.

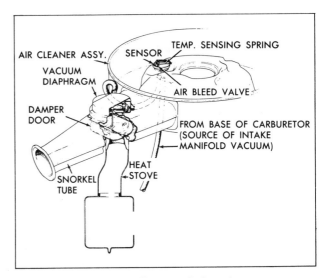

Thermostatically controlled air cleaner

Inspection
Check for proper, secure connections of heat pipe and hoses. Check for kinked or deteriorated hoses. Repair or replace as required.

Remove air cleaner cover and install a temperature gauge as close as possible to sensor. Reinstall cover without wing nut. (Temperature must be below 85°F before proceeding.) With the engine "Off", observe damper door position through snorkel opening. Snorkel passage should be open. If not, check for binds in linkage.

Start and idle engine. With air temperature below 85°F, snorkel passage should be closed. When damper door begins to open snorkel passage, remove air cleaner cover and observe thermometer reading. It should be between 85°F and 115°F.

If damper door does not close completely or does not open at correct temperature, turn off engine. Disconnect diaphragm assembly vacuum hose at sensor unit. Apply at least 9 in. Hg. of vacuum to diaphragm assembly through the hose. This can be done by mouth. Damper door should completely close snorkel passage when vacuum is applied. If not check to see if linkage is hooked up

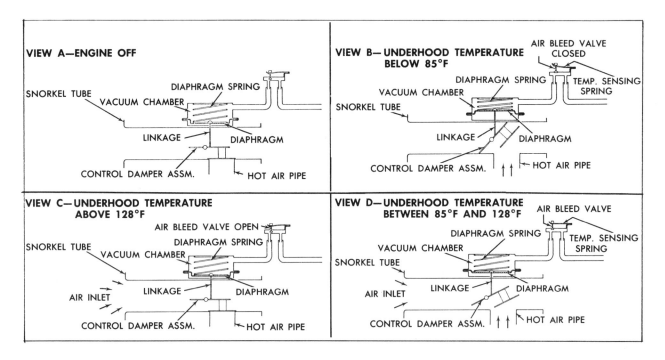

Operational diagrams

correctly and for a vacuum leak. With vacuum applied, bend or clamp hose to trap vacuum in diaphragm assembly. Damper door should remain in position (closed snorkel passage). If it does not, there is a vacuum leak in diaphragm assembly. Replace diaphragm assembly. If vacuum motor check is found satisfactory, replace sensor unit.

VACUUM MOTOR
Removal
Remove air cleaner from engine. Drill out spot welds fastening vacuum motor retaining strap to snorkel tube. Remove vacuum motor by lifting and unhooking linkrod from damper door.

Replacement
Drill 7/64" hole in snorkel tube at center of vacuum

Checking vacuum diaphragm

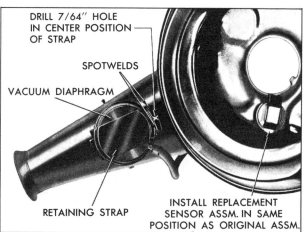

Vacuum diaphragm replacement

54

motor retaining strap. Connect vacuum motor linkage to damper door. Fasten retaining strap to air cleaner with sheet metal screw. Replace air cleaner on engine and check operation of vacuum motor and control damper assembly.

TEMPERATURE SENSOR
Removal
Remove air cleaner from engine and disconnect vacuum hoses at sensor. Pry up tabs of sensor retaining clip. Observe position of sensor, new sensor must be installed in this same position. Remove clip and sensor from air cleaner.

Replacement
Install sensor and gasket assembly in air cleaner in position as noted above. Press retaining clip on sensor. Support the sensor on its sides to prevent damage to the control mechanism at the center. Install air cleaner on engine and connect vacuum hoses.

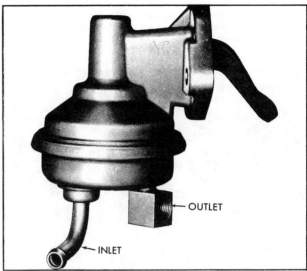

Fuel pump (nonserviceable)

Tighten any loose line connections and look for bends or kinks in lines. Disconnect fuel pipe at carburetor. Disconnect distributor to coil primary wire so that engine can be cranked without firing. Place suitable container at end of pipe and crank engine a few revolutions. If little or no gasoline flows from open end of pipe then fuel pipe is clogged or pump is inoperative. Before removing pump disconnect fuel pipe at gas tank and outlet pipe and blow through them with an air hose to make sure they are clear. Reconnect pipes and retest while cranking engine. Whenever the engine is cranked remotely at the starter, with a special jumper cable or other means the distributor primary lead must be disconnected from the negative post on the coil and the ignition switch must be in the "ON" position. Failure to do this will result in a damaged grounding circuit in the ignition switch.

If fuel flows from pump in good volume from pipe at carburetor, check fuel delivery pressure to be certain that pump is operating within specified limits. Attach a fuel pump pressure test gauge to disconnect end of pipe. Run engine at approximately 450-1,000 rpm (on gasoline in carburetor bowl) and note reading on pressure gauge.

If pump is operating properly the pressure will be within specifications and will remain constant at speeds between 450-1,000 rpm. If pressure is too low, too high, or varies significantly at different speeds, the pump should be replaced.

Removal
Disconnect fuel inlet and outlet pipes at fuel pump. Remove fuel pump mounting bolts and remove pump and gasket. On V8 engines, if push rod is to be removed, remove pipe plug and push rod (396, 427 and 454 cu. in. engines), and fuel pump adapter and gasket and push rod 350 cu. in. engines.

Installation
On V8 engines, if fuel pump push rod has been removed, install push rod and pipe fitting or fuel pump adapter using gasket sealer on gasket or pipe fitting. Install fuel pump using a new gasket and tighten securely. Use sealer on fuel pump mounting bolt threads.

Removing sensor unit

FUEL PUMP
Description
The fuel pump used on all vehicles is the diaphragm type. The pump is actuated by an eccentric located on the engine camshaft. On in-line engine, the eccentric actuates the rocker arm. On V-8 engines, a push rod (located between the camshaft eccentric and fuel pump) actuates the pump rocker arm. Because of design, this pump is serviced as an assembly only.

Inspection
The fuel pump should be checked to make sure the mounting bolts and inlet and outlet connections are tight.

Test
Always test pump while it is mounted on the engine and be sure there is gasoline in the tank. The line from the tank to the pump is the suction side of the system and the line from the pump to the carburetor is the pressure side of the system. A leak on the pressure side, therefore, would be made apparent by dripping fuel, but a leak on the suction would not be apparent except for its effect of reducing volume of fuel on the pressure side.

On V8 engines, a pair of mechanical fingers or heavy grease may be used to hold fuel pump push rod up while installing fuel pump. Connect fuel pipes to pump. Start engine and check for leaks.

Installing fuel pump

FUEL TANKS
Description

All fuel tanks are of steel construction, and are mounted behind the differential between the frame rails. On models to 1972 the fuel tanks are vented to the atmosphere with a vented, anti-surge type filler cap. Later models with evaporative emission control systems use a non-vented filler cap.

The tank gauge is located on the lower side of the tank. The metering unit can be serviced without first removing the fuel tank assembly. The tank straps are hinged at the tank support and attached at the rear end with bolts.

FUEL TANK 1975 AND LATER
General

The fuel tank consists of a conventional metal tank with a nylon reinforced rubber bladder. It is similar to bladders used for aircraft and racing applications.

The bladder may be removed from the tank by first removing the tank from the car; then removing the top plate, and the bladder carefully pulled out. The bladder, inside or outside the tank, should not be stored dry for an extended period of time. If the bladder is to be stored out of the car, the inside of it should be coated with a light coat of non-detergent 10W oil.

Metering unit

The metering unit is part of the plate assembly. To replace it, remove the tank from the car.

Draining

The absence of a gas tank drain plug makes it necessary to siphon fuel from the tank when draining is needed. Obtain approximately 10 feet of 3/8" I.D. hose and cut a flap-type slit 18" from one end. Make this cut in the direction of the shorter end of hose.

Before draining, be sure that the fuel tank gauge unit wire or battery negative is disconnected. Always drain gasoline from complete duel system including carburetor, fuel pump, all fuel lines and fuel tank if the vehicle is to be stored for any appreciable length of time. This precaution will prevent accumulation of gum formation and resultant poor engine performance.

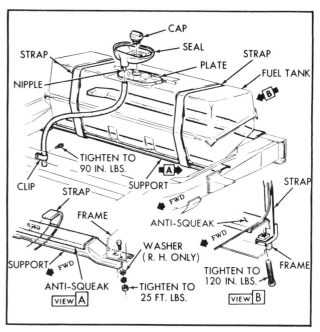

1976 fuel tank support (earlier model similar)

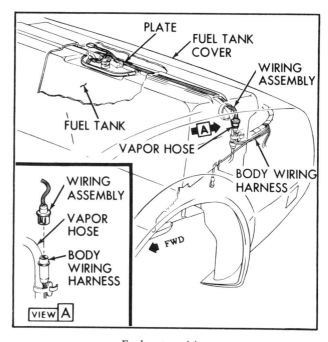

Fuel meter wiring

Removal and installation

Remove battery ground cable. Remove gas cap, filler neck boot from top of tank and disconnect drain line. Raise vehicle on hoist. Remove the spare tire from tire carrier. Remove spare tire carrier bolted attachments and remove carrier. Loosen "U" clamps and separate exhaust systems at transmission support.

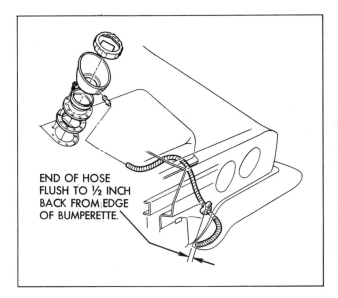

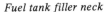

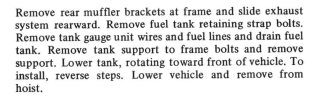

Fuel tank filler neck

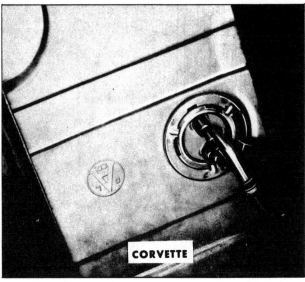

Metering units — typical

Remove rear muffler brackets at frame and slide exhaust system rearward. Remove fuel tank retaining strap bolts. Remove tank gauge unit wires and fuel lines and drain fuel tank. Remove tank support to frame bolts and remove support. Lower tank, rotating toward front of vehicle. To install, reverse steps. Lower vehicle and remove from hoist.

FUEL GAUGE UNIT
Removal and installation
Remove spare tire from carrier. Remove spare tire carrier bolted attachments and let carrier hand; then remove plastic cover. Disconnect tank gauge wires. Drain fuel tank. Disconnect fuel lines at gauge unit. Use Tool J-23346 to remove cam lock. Remove unit and rubber gasket. Remove unit carefully to avoid damage to screen at end of pickup pipe. Clean strainer by blowing out with compressed air. Reverse removal procedure to install. Always use a new gasket when replacing or installing a new gauge unit.

FUEL LINES
Description
The fuel lines, extending from the fuel tank to fuel pump, are routed along the right frame rail. Fuel lines are secured to the frame or underbody with clamp and screw assemblies. When replacing a fuel line, only seamless steel tubing is to be used. Also, the ends of the tubing must be double-flared using commercially available double flaring tools. All fuel lines must be properly routed and retained. Flexible hoses are located at the fuel tank pickup and at the fuel pump. The fuel lines should be inspected occasionally for leaks, kinks or dents. If evidence of dirt is found in the carburetor, fuel pump or on a disassembly, the lines should be disconnected and blown out. Check the fuel strainer in the tank for damage or omission.

EARLY FUEL EVAPORATION SYSTEM (EFE)
Description and operation
The system improves cold engine warm-up by routing hot

exhaust gases under the base of the carburetor which results in better fuel atomization, and helps reduce exhaust emissions during cold drive away. The system consists of a vacuum controlled actuator linked to a stainless steel exhaust heat valve. The system is checked by starting the engine when it is cold, and observing the movement of the actuator valve. The engine vacuum should pull the valve closed. As the engine heats up a switch shuts off vacuum to the actuator, so the valve can open.

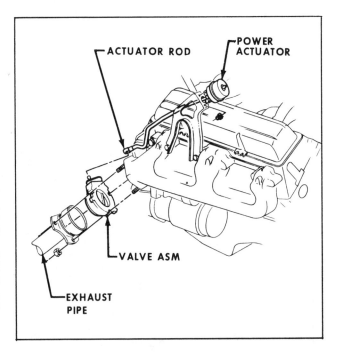

Typical EFE valve 1975-76

EVAPORATIVE EMISSION CONTROL SYSTEM
Description
All models use a pressure-vacuum gasoline tank filler neck cap. No other type of gasoline tank filler neck cap is to be used on vehicles incorporating this system. The gasoline tanks incorporate special internal fill limiters and vents and also external hose connections. The separator mounts in front of the gasoline tank. If service is necessary, the entire assembly must be replaced.

The canister is mounted on the left side of the engine compartment. A filter is mounted in the bottom of the canister. It is to be replaced according to the recommended maintenance schedule. When replacing any evaporative emission hose, use only replacement hose marked "EVAP". No other type of hose is to be used.

CANISTER AND/OR CANISTER FILTER
Removal
Raise vehicle on hoist. Note installed position of hoses on canister. Disconnect hoses from top of canister. Loosen clamps and remove canister. If replacing filter, remove bottom of canister and pull out filter.

Inspection
Check hose connection openings. Assure that they are open. On four barrel carburetor models, check operation of purge valve by applying vacuum to the valve. A good valve will hold vacuum.

Installation
Install new filter. Assemble bottom of canister to canister body. Install canister and tighten clamp bolts. Connect hoses to top of canister in same position as above.

CANISTER PURGE VALVE
Disassembly
Disconnect lines at valve. Snap off valve cap (Slowly remove cap as diaphragm is under spring tension). Remove diaphragm, spring retainer and spring. Replace parts as necessary. Check orifice openings.

Assembly
Install spring, spring retainer, diaphragm and cap. Reinstall canister as previously outlined.

SEPARATOR
Removal
Raise vehicle on hoist. Disconnect lines from separator. Remove retaining screw and remove separator.

Installation
Install separator and its retaining screw. Connect lines to separator. Lower vehicle and remove from hoist.

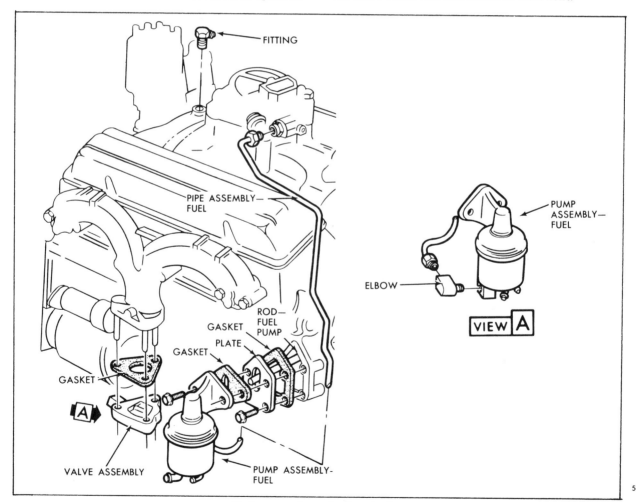

Fuel pump, pipe and exhaust valve

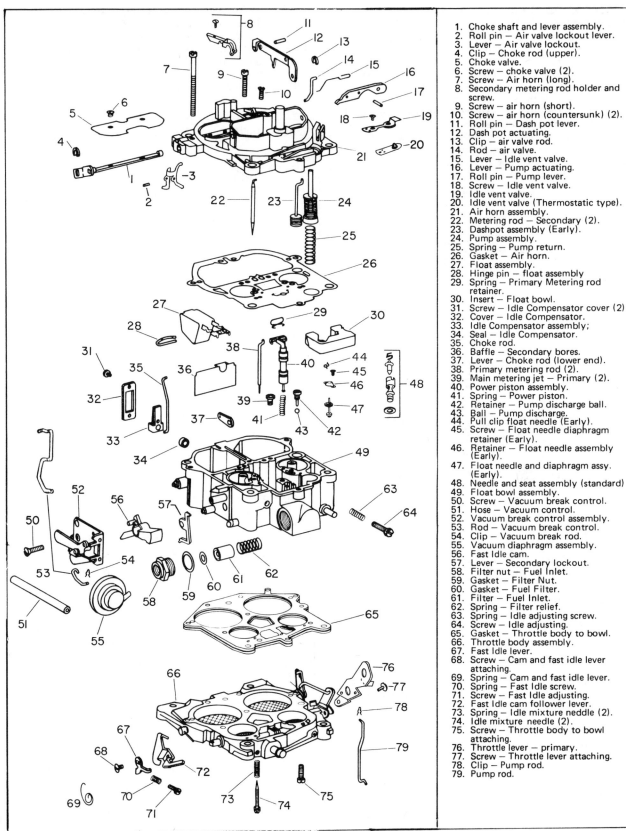

1. Choke shaft and lever assembly.
2. Roll pin — Air valve lockout lever.
3. Lever — Air valve lockout.
4. Clip — Choke rod (upper).
5. Choke valve.
6. Screw — choke valve (2).
7. Screw — Air horn (long).
8. Secondary metering rod holder and screw.
9. Screw — air horn (short).
10. Screw — air horn (countersunk) (2).
11. Roll pin — Dash pot lever.
12. Dash pot actuating.
13. Clip — air valve rod.
14. Rod — air valve.
15. Lever — Idle vent valve.
16. Lever — Pump actuating.
17. Roll pin — Pump lever.
18. Screw — Idle vent valve.
19. Idle vent valve.
20. Idle vent valve (Thermostatic type).
21. Air horn assembly.
22. Metering rod — Secondary (2).
23. Dashpot assembly (Early).
24. Pump assembly.
25. Spring — Pump return.
26. Gasket — Air horn.
27. Float assembly.
28. Hinge pin — float assembly
29. Spring — Primary Metering rod retainer.
30. Insert — Float bowl.
31. Screw — Idle Compensator cover (2).
32. Cover — Idle Compensator.
33. Idle Compensator assembly;
34. Seal — Idle Compensator.
35. Choke rod.
36. Baffle — Secondary bores.
37. Lever — Choke rod (lower end).
38. Primary metering rod (2).
39. Main metering jet — Primary (2).
40. Power piston assembly.
41. Spring — Power piston.
42. Retainer — Pump discharge ball.
43. Ball — Pump discharge.
44. Pull clip float needle (Early).
45. Screw — Float needle diaphragm retainer (Early).
46. Retainer — Float needle assembly (Early).
47. Float needle and diaphragm assy. (Early).
48. Needle and seat assembly (standard)
49. Float bowl assembly.
50. Screw — Vacuum break control.
51. Hose — Vacuum control.
52. Vacuum break control assembly.
53. Rod — Vacuum break control.
54. Clip — Vacuum break rod.
55. Vacuum diaphragm assembly.
56. Fast Idle cam.
57. Lever — Secondary lockout.
58. Filter nut — Fuel Inlet.
59. Gasket — Filter Nut.
60. Gasket — Fuel Filter.
61. Filter — Fuel Inlet.
62. Spring — Filter relief.
63. Spring — Idle adjusting screw.
64. Screw — Idle adjusting.
65. Gasket — Throttle body to bowl.
66. Throttle body assembly.
67. Fast Idle lever.
68. Screw — Cam and fast idle lever attaching.
69. Spring — Cam and fast idle lever.
70. Spring — Fast Idle screw.
71. Screw — Fast Idle adjusting.
72. Fast Idle cam follower lever.
73. Spring — Idle mixture neddle (2).
74. Idle mixture needle (2).
75. Screw — Throttle body to bowl attaching.
76. Throttle lever — primary.
77. Screw — Throttle lever attaching.
78. Clip — Pump rod.
79. Pump rod.

DUCTED HOOD AIR DOOR
Description
The domed air induct on hood improves high-output engine operation. The air door opens when the accelerator is positioned to the flow. A kickdown switch, normally open, is tripped by a pedal linkage. Current then passes through the radio fuse, bulkhead connector and to the forward lamp harness and through wires along the underside of the load to a solenoid (normally retracted plunger) mounted to the air door linkage. The mirror mounted map lamp and radio share a fuse with the air door circuit. If the vehicle is equipped with either of these accessories and they operate, then the electrical fuel circuit, including the fuse, is good.

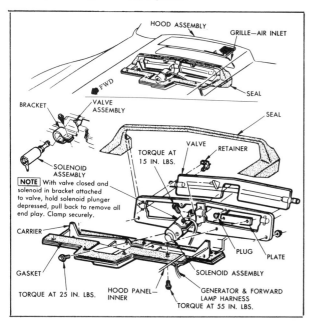

Ducted hood air door (Carburetor air control)

Checking
Check for good radio fuse. Be sure all electrical connectors are plugged in at the solenoid (front of hood) and at kickdown switch under the instrument cluster. Rig a jumper wire from the horn relay to the solenoid. If the solenoid does not click, replace.

Check the binding linkage on the air door. Lubricate if necessary.

Check continuity of kickdown switch.

EXHAUST SYSTEM
The exhaust system through 1974 models is of conventional design. Some models have the muffler outlet flange notched and mated to a welded tab located on the outside diameter of the tailpipe. The exhaust pipes and muffler use locater tabs for alignment.

Starting with 1975 models the exhaust system was modified to include a catalyctic converter between the front exhaust pipe and the tailpipe. Cars equipped with a catalytic converter must use unleaded gasoline. The converter requires no specific service and none is recommended. If necessary, the catalyst in the converter can be replaced on the car. Special equipment is required.

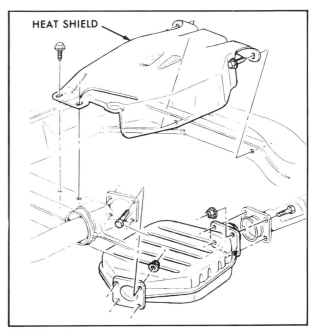

Catalytic converter

CARBURETOR AIR INTAKE (1976)
The air induction system uses a duct over the radiator instead of the inlet at the rear of the hood as used on earlier models. This system requires no service.

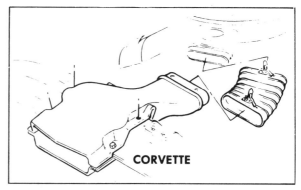

Carburetor air intake

IGNITION 4

INDEX

DESCRIPTION

Two ignition systems are used. The breaker point type and the optional transistor controlled breakerless ignition system (magnetic pulse type). The transistor ignition system features a specially designed distributor, ignition pulse amplifier, and a special coil. Two resistance wires are also used in the circuit; one as a ballast between the coil negative terminal and ground, while the other reistance wire provides a voltage drop for the engine run circuit and is by-passed at cranking. The other units in the system (ignition switch, spark plugs, and battery) are of standard design.

Although the outside appearance of the distributor resembles a standard distributor, the inside is quite different. An iron timer core replaces the conventional breaker cam. The timer core has the same number of equally-spaced projections, or vanes as engine cylinders.

The timer core rotates inside a magnetic pickup assembly, which replaces the conventional breaker plate, contact point set, and condenser assembly. The magnetic pickup assembly consists of a ceramic permanent magnet, a pole piece, and a pickup coil. The pole piece is a steel plate having equally spaced internal teeth, one tooth for each cylinder of the engine.

The magnetic pickup assembly is mounted over the main bearing of the distributor housing, and is made to rotate by the vacuum control unit, thus providing vacuum advance. The timer core is made to rotate about the shaft by conventional advance weights, thus providing centrifugal advance.

BREAKER POINT SYSTEM
Service operations

The distributor breaker points and spark plugs are the only ignition system components that require periodic service. The remainder of the ignition system requires only periodic inspection to check operation of the units, tightness of the electrical connections, and condition of the wiring. When checking the coil, test with a reputable tester.

Breaker type distributors are equipped with cam lubricator and should have the wick replaced at the same time contact point set is replaced. It is not necessary to lubricate the breaker cam when using a cam lubricator. *Do not attempt to lubricate the wick — Replace when necessary.* When installing a new wick, adjust its position so the end of the wick just touches the lobe of the breaker cam.

Distributor shaft lubrication is accomplished by a reservoir of lube around the mainshaft in the distributor body.

Dirty contact points should be dressed with a few strokes of a clean, fine-cut contact file. The file should not be used for other metals and should not be allowed to become greasy or dirty. *Never use emery cloth to clean contact points.* Contact surfaces, after considerable use, may not appear bright and smooth, but this is not necessarily an indication that they are not functioning satisfactorily. Do not attempt to remove all roughness nor dress the point surfaces down smooth; merely remove scale or dirt.

Badly burned or pitted contact points should be replaced and the cause of trouble determined so it can be eliminated. High resistance or loose connections in the condenser circuit, oil or foreign materials on the contact surfaces, improper point adjustment or high voltages may cause oxidized contact points. Check for these conditions where burned contacts are experienced. An out-of-balance condition in the ignition system, often the result of too much or too little condenser capacity, is indicated where point pitting is encountered.

Replacement

Contact points utilizing the push-in type terminal for the condenser and primary leads are recommended when contact point replacement is required on distributors containing the radio frequency inference shield. Point sets utilizing a lock screw at the condenser and primary lead terminal, if not carefully installed, may short out due to the head of the lock screw or lead clips contacting the shield for insulating tape.

The contact point set is replaced as one complete assembly and only dwell angle requires adjustment after replacement. *Breaker lever spring tension and point alignment* are factory set.

Remove the distributor cap by placing a screwdriver in the slot head of the latch, press down and turn 1/4 turn in either direction. Remove R.F.I. shield attaching screws and 2-piece shield (where applicable). Remove the two attaching screws which hold the base of the contact set assembly in place. Remove the primary and condenser leads from their nylon insulated connection, in contact set. Reverse the procedure to install new contact set. *Install the primary and consenser leads. Improper installation will cause lead interference between the cap, weight base and breaker advance plate.*

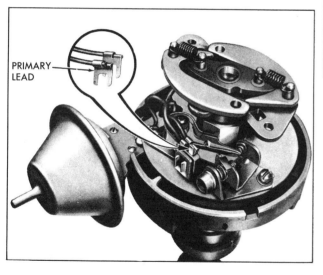

PRIMARY LEAD

Distributor lead arrangements

If car has 20,000 to 25,000 miles (or sooner if desired) the cam lubricator wick should be changed. Using long nosed pliers squeeze assembly together at base and lift out. Remove all old lubricant from cam surface. Replace in same manner. End of cam lubricant wick should be adusted to just touch cam lobes. Over lubrication of cam resulting in grease on contact points can be caused by cam lubrication wick bearing too hard against cam surface. A correctly adjusted cam lubricator wick will provide adequate lubrication for cam. Do not apply additional grease to cam surface. Start engine and check point dwell and ignition timing.

Condenser replacement

Remove distributor cap. Loosen condenser lead attaching screw and lift out condenser lead clip. Remove screw holding condenser bracket to breaker plate and slide condenser from bracket. To replace condenser, reverse the above procedure. Make sure that new condenser lead is installed in proper position.

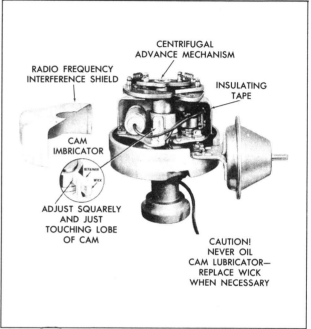

CENTRIFUGAL ADVANCE MECHANISM

RADIO FREQUENCY INTERFERENCE SHIELD

INSULATING TAPE

CAM IMBRICATOR

RETAINER
WICK

ADJUST SQUARELY AND JUST TOUCHING LOBE OF CAM

CAUTION! NEVER OIL CAM LUBRICATOR— REPLACE WICK WHEN NECESSARY

Top view of distributor

Setting dwell angle, on car

With engine running at idle and operating temperatures normalized, the dwell is adjusted by first raising the window provided in the cap and inserting a "Hex" type wrench into the adjusting screw head. Turn the adjusting screw until the specified dwell angle is obtained as measured in degrees (28° to 32°, 30° preferred) by a dwell angle meter. To adjust without dwell meter, turn adjusting screw in slowly (clockwise) until the engine stalls, then turn screw 1/2 turn in the opposite direction (counter-clockwise). This will give the approximate dwell angle required. (Use only when meter is not available.)

Setting dwell angle, off car

Distributor Test Method — With the distributor mounted on a distributor testing machine, connect the dwell meter to the distributor primary lead. Turn the adjusting screw to set the dwell angle to 30 degrees.

Test Light Method — With the distributor mounted in a vise connect a testing lamp to the primary lead. Rotate the shaft until one of the circuit breaker cam lobes is under the center of the rubbing block of the breaker lever. Turn the adjusting screw clockwise until the lamp lights, then give the wrench 1/2 turn in the opposite direction (counter-clockwise) to obtain the proper dwell angle.

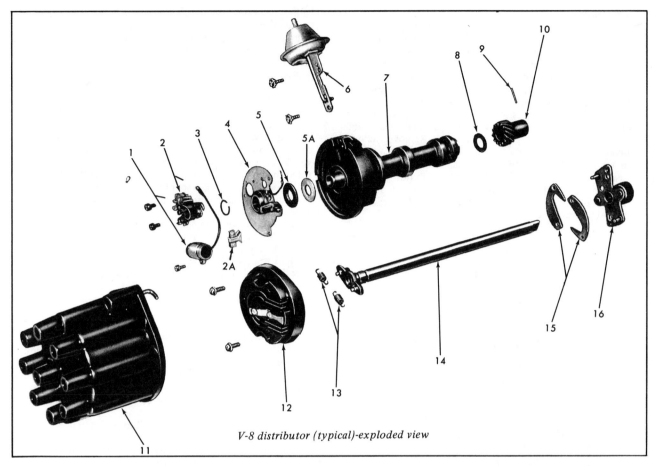

V-8 distributor (typical)-exploded view

1. Condenser
2. Contact point assembly
2a. Cam lubricator
3. Retaining ring
4. Breaker plate
5. Felt washer

5a. Plastic seal
6. Vacuum advanceunit
7. Housing
8. Shim washer
9. Drive gear pin
10. Drive gear

11. Cap
12. Rotor
13. Weight springs
14. Mainshaft
15. Advance weights
16. Cam weight base assembly

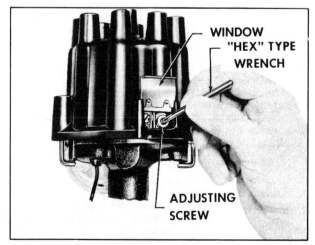

Adjusting dwell angle

Removal

On radio equipped vehicles, remove ignition shield from over distributor and coil. One bolt is accessible from top of shield, the other two are at rear of shield, facing firewall. Release the distributor cap hold-down screws, remove the cap and place it clear of the work area. If necessary, remove secondary leads from the distributor cap after first marking the cap tower for the lead to No. 1 cylinder. This will aid in the reinstallation of leads in the cap.

Disconnect the distributor primary lead from the coil terminal. Scribe a realignment mark on the distributor bowl and engine in line with the rotor segment. Disconnect vacuum line to distributor and tachometer drive cable. Remove the distributor hold-down bolt and clamp and remove the distributor from the engine. Note position of vacuum advance mechanism relative to the engine. *Avoid rotating the engine with the distributor removed as the ignition timing will be upset.*

Disassembly

Remove the rotor. Remove attaching screws and two-piece R.F.I. shield (where applicable). Remove both weight springs and advance weights. Remove roll pin retaining driven gear to distributor shaft, slide the gear and spacers from the shaft. Remove tachometer drive gear. Before sliding the distributor shaft from the housing, check for and remove any burrs on the shaft. This will prevent damage to the seals and bushing still positioned in the housing. Slide the distributor mainshaft and cam-weight base assembly from the housing. Remove vacuum advance mechanism retaining screws, remove the vacuum advance assembly. Remove the spring retainer, remove the breaker plate assembly from the distributor housing. Remove the contact point and condenser from the breaker plate. Remove the felt washer and plastic seal located beneath the breaker plate.

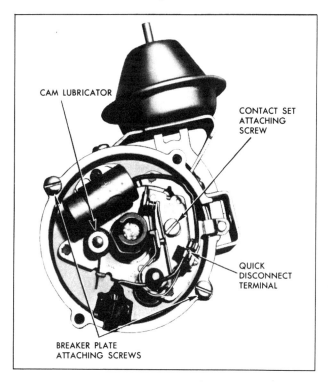

Breaker plate and attaching parts

Cleaning and inspection

Wash all parts in cleaning solvent except cap, rotor, condenser, breaker plate assembly and vacuum control unit. Degreasing compounds may damage insulation of these parts or saturate the lubricating felt in the case of the breaker plate assembly. Inspect the breaker plate assembly for damage or wear and replace if necessary. Inspect the shaft for wear and check its fit in the bushings in the distributor body. If the shaft or bushings are worn, the parts should be replaced. Mount the shaft in "V" blocks and check the shaft alignment with a dial gauge. The run-out should not exceed .002". Inspect the advance weights for wear or burrs and free fit on their pivot pins. Inspect the cam for wear or roughness. Then check its fit on the end of the shaft. It should be absolutely free without any roughness. Inspect the condition of the distributor points. Dirty points should be cleaned and

badly pitted points should be replaced. Test the condenser for series resistance, microfarad capacity (.18 to .23) and leakage or breakdown, following the instructions given by the manufacturer of the test equipment used. Inspect the distributor cap and spark plug wires for damage and replace if necessary.

Assembly

Fill housing lubricating cavity with SAE 20 oil, press in new plastic seal and install felt washer. Replace the vacuum advance unit, the breaker plate in housing and the spring retainer on the upper bushing. Lubricate and slide weight cam over mainshaft and install weights and spring. Insert mainshaft into housing, indexing it with drive gear and washers. Install tachometer drive gear. Slide distributor drive gear shims and gear over shaft and install new pin. Tap new pin through gear and mainshaft. Check shaft for free rotation. Mainshaft end clearance should be .002"-.007". Add or remove shims as necessary. Install contact point set and condenser to breaker plate. Connect leads. Contact point spring tension is factory-set above specifications to assure ease of final adjustment. Correct tension is 19-23 oz. Position R.F.I. shield and install attaching screws. Install rotor to cam assembly, indexing round and square pilot holes.

Installation — engine not disturbed

Turn the rotor about 1/8 turn in a clockwise direction past the mark previously placed on the distributor housing to locate rotor. Push the distributor down into position in the block with the housing in a normal "installed" position. It may be necessary to move rotor slightly to start gear into mesh with camshaft gear, but rotor should line up with the mark when distributor is down in place.

Tighten the distributor clamp bolt snugly and connect vacuum line. Connect primary wire to coil terminal and install cap. Also install spark plug and high tension wires if removed. It is important that the spark plug wires be installed in their proper location in the supports. Time ignition.

Installation — engine disturbed

Remove No. 1 spark plug and, with finger on plug hole, crank engine until compression is felt in the No. 1 cylinder. Continue cranking until timing mark on crankshaft pulley lines up with timing tab attached to engine front cover, or remove rocker cover (left bank) and crank engine until No. 1 intake valve closes and continue to crank slowly about 1/3 turn until timing mark on pulley lines up with timing tab. Position distributor to opening in block in normal installed attitude, noting position of vacuum control unit. Position rotor to point toward front of engine (with distributor housing held in installed attitude), then turn rotor counter-clockwise approximately 1/8 turn more toward left cylinder bank and push distributor down to engine camshaft. It may be necessary to rotate rotor slightly until camshaft engagement is felt.

While pressing firmly down on distributor housing, kick starter over a few times to make sure oil pump shaft is engaged. Install hold-down clamp and bolt and snug up bolt. Turn distributor body slightly until points just open and tighten distributor clamp bolt. Place distributor cap in position and check to see that rotor lines up with terminal for No. 1 plug.

Install cap, check all high tension wire connections and connect spark plug wires if they have been removed. It is important that the wires be installed in their location in the supports. The brackets are numbered to show the correct installation. Wires must be installed as indicated to prevent cross firing. Connect vacuum line to distributor and distributor primary wire to coil terminal. Start engine and set timing.

BREAKERLESS SYSTEM (Early model)

Description and service

Since there are no moving parts in the ignition pulse amplifier unit mounted forward of the radiator bulkhead, and the distributor shaft and bushings have permanent type lubrication, no periodic maintenance is therefore required for the breakerless ignition system. The distributor lower bushing is lubricated by engine oil through

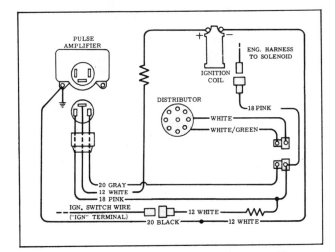

Breakerless ignition system

a splash hole in the distributor housing, and a housing cavity next to the upper bushing contains a supply of lubricant which will last between overhaul periods. At time of overhaul, the upper bushing may be lubricated by removing the plastic seal and then adding SAE 20 oil to the packing in the cavity. A new plastic seal will be required since the old one will be damaged during removal.

Tachometer readings for test purposes can be made on the primary circuit of the breakerless ignition system in the same manner as on the conventional ignition system, however, before attempting to connect a test tachometer into the primary circuit, check with your instrument supplier to insure that safisfactory readings can be obtained and the breakerless system will not be damaged by the tachometer that is to be used.

Removal

If vehicle is equipped with radio, remove three bolts securing ignition shield over distributor and coil. One bolt is accessible from the top of shield, the other two are at rear of shield, facing firewall. Disconnect tachometer drive cables from distributor housing. Disconnect pickup coil leads at connector.

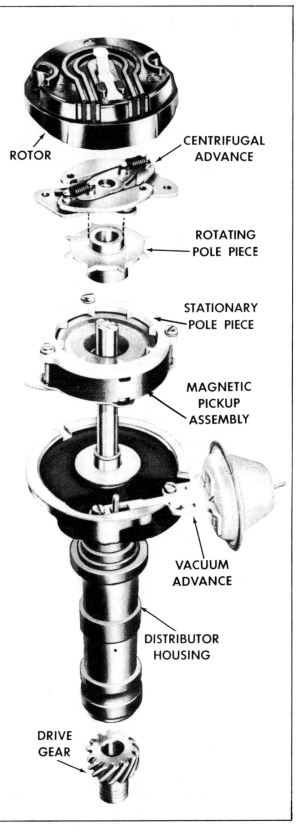

Magnetic pulse distributor components

Remove distributor cap. Crank engine so rotor is in position to fire No. 1 cylinder and timing mark on harmonic balancer is indexed with pointer. Remove vacuum line from distributor. Remove distributor clamping screw and hold-down clamp. Remove distributor and distributor-to-block gasket. It will be noted that the rotor will rotate as the distributor is pulled out of the block. Mark the relationship of the rotor and the distributor housing after removal so that the rotor can be set in the same position when the distributor is being installed.

Disassembly and assembly

Remove screws securing rotor and remove rotor. Remove centrifugal weight springs and weights. Remove the tachometer drive gear from the distributor. Remove roll pin, then remove distributor drive gear and washer. *To prevent damage to the permanent magnet, support drive gear when driving out roll pin.* Remove drive shaft assembly. Remove centrifugal weight support and timer core from drive shaft. Remove connector from pickup coil leads. Remove retaining ring which secures magnetic core support plate to distributor shaft bushing in housing. As a unit, remove the entire magnetic pickup assembly from the distributor housing. Remove brass washer and felt pad. Remove vacuum advance unit. To reassemble distributor reverse the procedure.

Installation

Check to see that the engine is at firing position for No. 1 cylinder (timing mark on harmonic balancer indexed with pointer). Position a new distributor-to-block gasket on the block. Before installing distributor, index rotor with housing as noted when distributor was removed. Install distributor in block so that vacuum diaphragm faces approximately 45° forward on the right side of the engine and the rotor points toward contact in cap for No. 1 cylinder. Replace distributor clamp leaving screw loose enough to allow distributor to be turned for timing adjustment. Install spark plug wires in distributor cap. Place wire for No. 1 cylinder in tower (marked on old cap during assembly) then install remaining wires clockwise around the cap according to the firing order (1-8-4-3-6-5-7-2). Attach distributor to coil primary wires. Replace distributor cap. Adjust timing and then fully tighten distributor clamp screw. Attach vacuum line to distributor. Connect tachometer drive cables to distributor body. Replace ignition shields.

IGNITION PULSE AMPLIFIER
Disassembly

Remove the bottom plate from the amplifier. To aid in reassembly, note the locations of the lead connections to the panel board. Remove the three panel board attaching screws, and lift the assembly from the housing. To aid in reassembly, note any identifying markings on the two transistors and their respective locations on the panel board and heat sink assembly. Note the insulators between the transistors and the heat sink, and the insulators separating the heat sink from the panel board. Remove the transistor attaching screws, and separate the two transistors and heat sink from the panel board. Carefully examine the panel board for evidence of damage.

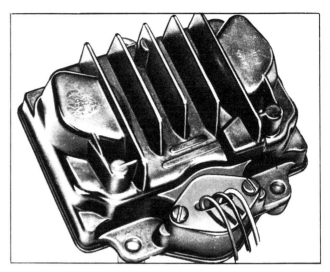

Ignition pulse amplifier unit

Component Checks

With the two transistors separated from the assembly, an ohmmeter may be used to check the transistors and components on the panel board for defects. An ohmmeter having a 1-1/2 volt cell, which is the type usually found in service stations, is recommended. The low range scale on the ohmmeter should be used except where specified otherwise. A 25 watt soldering gun is recommended, and a 60% tin 40% lead solder should be used when re-soldering. Avoid excessive heat which may damage the panel board. Clip away any epoxy involved, and apply new epoxy which is commercially available.

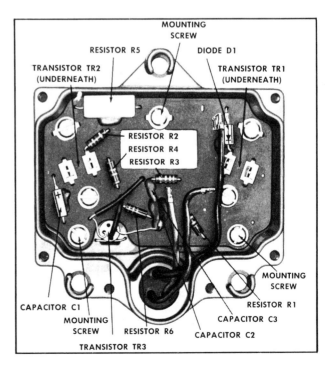

Pulse amplifier panel board

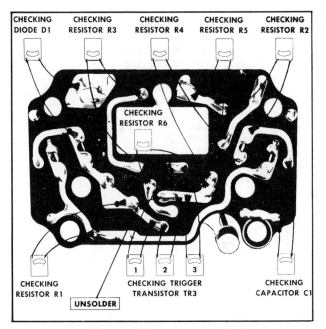

Pulse amplifier component checks

IGNITION COIL 66-74
Check

The ignition coil primary can be checked for an open condition by connecting an ohmmeter across the two primary terminals with the battery disconnected. Primary resistance at *75°F. should be between .35 and .55 ohm.* An infinite reading indicates the primary is open. For the engine to run but miss at times, the primary open may be of the intermittent type.

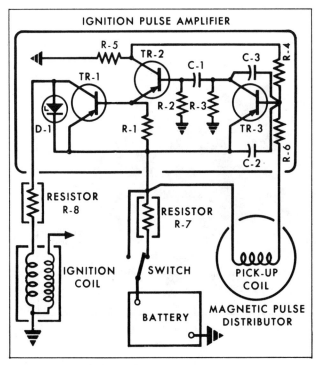

Pulse amplifier internal circuitry

The coil secondary can be checked for an open by connecting an ohmmeter from the high tension center tower to either primary terminal. To obtain a reliable reading, a scale on the ohmmeter having the 20,000 ohm value within, or nearly within, the middle third of the scale should be used. *Secondary resistance at 75°F. should be between 8,000 and 12,500 ohms.* If the reading is infinite, the coil secondary winding is open.

A number of different types of coil testers are available from various test equipment manufacturers. When using these testers, follow the procedure recommended by the tester manufacturer. Make sure the tester will properly check this special coil.

Replacement

Disconnect ignition switch and distributor leads from terminals on coil. On Corvettes equipped with radio, remove bolts securing ignition shield over distributor and coil. Pull high tension wire from center terminal of coil. Remove the two coil support mounting bolts or loosen friction clamp screw and remove coil. Place new coil in position and install attaching bolts or tighten clamp screw. Place high tension lead securely in center terminal of coil and connect ignition switch and distributor primary leads to terminals on coil. Replace ignition shield. Start engine and test coil operation.

Assembly

During assembly, coat with silicone grease both sides of the flat insulators used between the transistors and heat sink, and also the heat sink on the side on which the transistors are mounted. The silicone grease, which is available commercially, conducts heat and thereby provides better cooling.

BREAKERLESS IGNITION SYSTEM (1975-1980)
Description and operation

The High Energy Ignition (HEI) system used as standard equipment on 1975-1979 models is a transistor controlled, pulse triggered, inductive discharge ignition system. It is similar in operation to the earlier system offered as an option. The HEI system represents refinements and advancements in the state of electronics by reducing several of the components in physical size. The large pulse amplifier has been replaced by an internal electronic module. The coil has been designed to be a part of the distributor cap; the iron core is on the outside, instead of the more conventional center of the coil.

The HEI system is built around a rotating timer core with external teeth turned by the distributor shaft, a stationary pole piece with internal teeth, and a pick-up coil and bottom plate. When the teeth of the timer core, rotating inside the pole piece line up with the teeth of the pole piece an induced voltage in the pick-up coil signals the electronic module to open the coil primary circuit. The primary current decreases and a high voltage is induced in the ignition coil secondary winding. This is directed through the rotor and the high voltage leads to fire the correct spark plug. The capacitor in the distributor is for radio noise suppression only.

The module automatically controls the dwell period, stretching it with increased engine speed. The HEI system provides a longer spark duration, which is desirable for firing lean and EGR diluted mixtures.

Because of the higher voltages a larger diameter (8mm) spark plug wire with silicone insulation is used. It is more heat resistant, less vulnerable to deterioration, and is grey in color. The insulation is soft and must be handled with reasonable care.

The system is designed to eliminate periodic maintenance. All components which prove defective are to be replaced.

Vacuum and centrifugal advance operations are the same as for conventional distributors with points.

DISTRIBUTOR
Removal
Disconnect wiring harness connectors at side of distributor cap. Remove distributor cap and position out of way. Disconnect vacuum advance hose from vacuum advance mechanism. Scribe a mark on the engine in line with rotor. Note approximate position of distributor housing in relation to engine. Remove distributor hold-down nut and clamp. Lift distributor from engine.

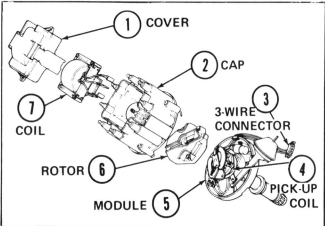

Exploded view H.E.I. distributor

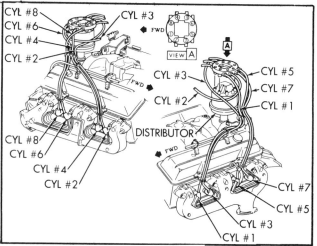

8 cylinder H.E.I. ignition wiring

Tachometer connections
In the distributor cap connector is a "tach" terminal. Connect the tachometer to this terminal. Follow tachometer manufacturer's instructions, as some require a ground lead while others require a hot lead. CAUTION: Grounding the "tach" terminal could damage the electronic module.

Timing light connections
Timing light connections should be made in parallel using an adapter at the distributor number one terminal.

Installation
Install distributor using same procedure as for standard distributor. Install distributor hold-down clamp and snugly install nut. Move distributor housing to approximate position relative to engine noted during removal. Position distributor cap to housing with tab in base of cap aligned with notch in housing and secure with four latches. Connect wiring harness connector to terminals on side of distributor cap. Connector will fit only one way. Adjust ignition timing as described in Specification Section of this manual.

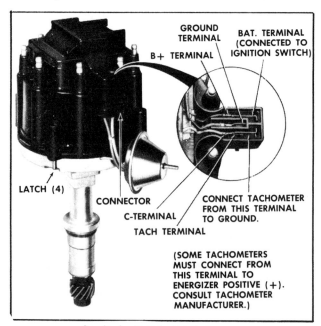

8 cylinder H.E.I. distributor

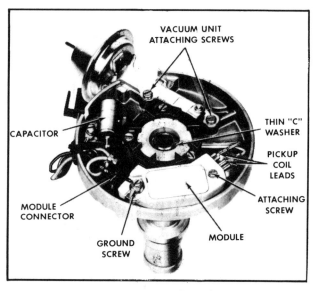

Top view of distributor housing

Disassembly

Remove distributor as described above. Remove rotor from distributor shaft by removing two screws. Remove two advance springs, weight retainer, and advance weights. Remove two screws holding module to housing and move module to a position where connector may be removed from 'B' and 'C' terminals. Remove wires from "W" and "G" terminals of module.

Remove roll pin from drive gear. CAUTION: Distributor gear should be supported in such a way that no damage will occur to distributor shaft while removing pin. Remove gear, shim and tanged washer from distributor shaft. Remove any burrs that may have been caused by removal of pin. Remove distributor shaft from housing.

Remove washer from upper end of distributor housing. Remove three screws securing pole piece to housing and remove pole piece, magnet and pick-up coil. Remove lock ring at top of housing and remove pick-up coil retainer, shim and felt washer. NOTE: No attempt should be made to service the shaft bushings in the housing. Remove vacuum advance mechanism by removing two screws. Disconnect capacitor lead and remove capacitor by removing one screw. Remove wiring harness from distributor housing.

Assembly

Position vacuum advance unit to housing and secure with two screws. Position felt washer over lubricant reservoir at top of housing. Position shim on top of felt washer.

Position pick-up coil retainer to housing with vacuum advance arm over actuating pin of vacuum advance mechanism. Secure with lock ring. Install pick-up coil magnet and pole piece. Loosely install three screws holding pole piece. Install washer to top of housing. Install distributor shaft and rotate to check for even clearance all around between pole piece and shaft projections. Move pole piece to provide even clearance and secure with three screws. Install tanged washer, shim and drive gear (teeth up) to bottom of shaft. Align drive gear and install new roll pin.

Position capacitor to housing and loosely install one mounting screw. Install connector to "B" and "C" terminals on module with tab on top. Apply special silicone lubricant liberally to bottom of module and secure with two screws. Position wiring harness with grommet in housing notch. Connect pink wire to capacitor stud, and black wire to capacitor mounting screw. Tighten screw. Connect white wire from pick-up coil to terminal "W" module. Connect green wire from pick-up coil to terminal "G" of module. Install centrifugal advance weights, weight retainer (dimple facing down), and springs. Install rotor and secure with two screws. CAUTION: Notch on side of rotor must engage tab on cam weight base. Install distributor.

TROUBLESHOOTING THE H.E.I. SYSTEM
Engine cranks but will not start

Check for moisture and dust inside and outside the distributor. Dry or clean if necessary. With the ignition turned on, check if current is going into distributor, by inserting a test light probe first in battery lead; then in red B+ wire at distributor connector. If no current, follow wire back to ignition switch; check switch. Replace if defective.

Check for high voltage coming out of distributor by removing one spark plug wire and hold it near engine while turn-

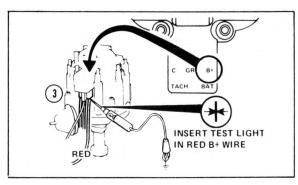

Checking current to distributor

ing engine over. Because of the high voltage generated by the HEI system it would be better to lay the wire on the engine or valve cover in such a way that a spark could jump without someone actually holding the wire. If a spark does jump, the trouble is probably in the spark plugs, fuel system, or the engine is flooded.

If current is going into the distributor but no spark jumps, then the pick-up coil and the module should be checked. To check the pick-up coil, remove green and white leads from module with a pair of needle nose pliers. Do not pull the wires. Connect an ohmmeter across ground and either green or white lead. If reading is less than infinity on 1000 scale, replace pick-up coil. If reading is infinite then connect ohmmeter across the two leads. If reading is less than 500 or more than 1500 ohms, replace the pick-up coil. To test the module, an appropriate tester is required.

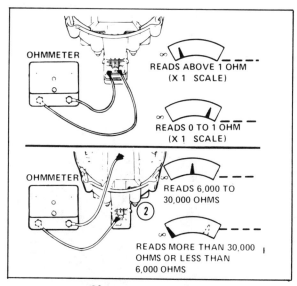

Ohmmeter connections

Engine runs rough or cuts out.

Follow same procedures as for checking HEI system when engine will not start. In addition check coil. Remove the distributor cap and connect ohmmeter leads across "B+" and "C" terminals in cap. A reading above 1 ohm, on 1 ohm scale, indicates a bad primary winding. If reading is less than 1 ohm, check secondary winding by connecting ohmmeter leads across "B+" terminal and high voltage terminal in center of cap. A reading of less than 6000 or more than 30,000 ohms indicates a bad secondary winding. Coil should be replaced.

SPARK PLUG AND WIRE SERVICE

Removal and inspection

To disconnect wires, pull only on the boot. Pulling on the wire might cause separation of the core of the wire. Remove spark plugs and gaskets using a 13/16" deep socket on 13/16" hex plugs, or a 5/8" deep socket on the 5/8" hex tapered plugs. Use care in this operation to avoid cracking spark plug insulators.

Carefully inspect the insulator and electrodes of all spark plugs. Replace any spark plug which has a cracked or broken insulator. If the insulator is worn away around the center electrode, or the electrodes are burned or worn, the spark plug is worn out and should be discarded. Spark plugs which are in good condition except for carbon or oxide deposits should be thoroughly cleaned and adjusted.

The spark plug wires are of a special resistance type. The core is carbon-impregnated linen. This wire is designed to eliminate radio and television interference radiation, but is also superior in resistance to cross fire. The resistance type wire, however, is more easily damaged than copper core wire. For this reason care must be taken that the spark plug wires are removed by pulling on the spark plug boots rather than on the wire insulation. Also, when it is necessary to replace a spark plug boot, the old boot should be carefully cut from the wire and a small amount of silicone lubricant used to aid in installing the new boot. If the wire is stretched, the core may be broken with no evidence of damage on the outer insulation. The terminal may also pull off the wire. If the core is broken, it will cause missing. In the case of wire damage, it is necessary to replace the complete wire assembly as a satisfactory repair cannot be made. Wipe ignition wires with cloth moistened with kerosene, and wipe dry. Carefully bend wires to check for brittle, cracked, or loose insulation. Defective insulation will permit missing or cross-firing of engine, therefore defective wires should be replaced. If the wires are in good condition, clean any terminals that are corroded and replace any terminals that are broken or distorted. Replace any broken or deteriorated cable nipples or spark plug boots.

Adjusting spark plug gap

Use round wire feeler gages to check the gap between spark plug electrodes of used plugs. Flat feeler gages will not give a correct measurement if the electrodes are worn; file center electrode tip flat before adjusting. Adjust gap by bending the side electrodes only; bending the center electrode will crack the insulator. Adjust gaps to .035 in. Setting spark plug gap to any other dimension to effect changes in engine performance is not recommended.

Installation of spark plugs

When installing spark plugs, make sure that all surfaces on plugs and in cylinder heads are clean. When installing 13/16" hex spark plugs, tighten to 25 lb. ft. torque, using a 13/16" deep socket, an extension and a torque wrench. When installing the 5/8" hex tapered seat spark plugs, tighten to 15 lb. ft., using a 5/8" deep socket, an extension and a torque wrench. *If tapered seat spark plugs are over-tightened, they will be more difficult to remove at the next tune-up.*

Installation of spark plug wires

No. 1 spark plug wire is installed in the first distributor cap tower after the adjusting window, moving in the direction of rotation. The other wires are then installed in a clockwise direction according to the firing order.

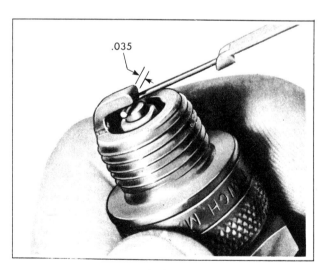

Setting spark plug cap

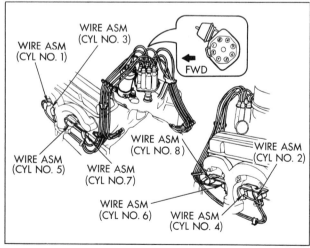

Spark plug wire installation

COOLING SYSTEM 5

INDEX

DESCRIPTION

The cooling system is sealed by a pressure type radiator filler cap, designed to operate the cooling system at higher than atmospheric pressure. This raises the boiling point of the coolant to increase the radiator efficiency. The filler cap contains a pressure relief valve and a vacuum relief valve. The pressure relief valve is held against its seat by a spring which when compressed, allows excessive pressure to be relieved out the radiator overflow.

Pressure radiator cap

The vacuum valve is also held against its seat by a spring which when compressed, opens the valve relieving the vacuum created when the system cools off.

The cooling systems water pump is of the centrifugal vane impeller type. The bearings are permanently lubricated during manufacture and are seated to prevent the loss of lubricant or the entry of dirt and water. The pump requires no care other than to make certain the air vent at the top of the housing and the drain holes in the bottom do not become plugged with dirt or grease.

Coolant level

The radiator coolant level should only be checked when the engine is cool, particularly on cars equipped with air conditioning. If the radiator cap is removed from a hot cooling system, serious personal injury may result.

Coolant level should be maintained three inches below the bottom of the filler neck when the system is cold. When a surge radiator supply tank is used the coolant level should be maintained at the one-half full level in the supply tank.

Upon repeated coolant loss, the pressure radiator cap and seat should be checked for sealing ability. Also, the cooling system should be checked for loose hose connections, defective hoses, gasket leaks, etc.

Coolant system checks

Test for restriction in the radiator, by warming the engine up and then turning the engine off and feeling the radiator. The radiator should be hot at the top and warm at the bottom, with an even temperature rise from bottom to top. Cold spots in the radiator indicate clogged sections.

Water pump operation may be checked by running the engine while squeezing the upper radiator hose. A pressure surge should be felt. Check for a plugged vent-hole in pump.

A defective head gasket may allow exhaust gases to leak into the cooling system. This is particularly damaging to the cooling system as the gases combine with the water to form acids which are harmful to the radiator and engine. To check for exhaust leaks into the cooling system, drain the system until the water level stands just above the top of the cylinder head, then disconnect the upper radiator hose and remove the thermostat and fan belt. Start the engine and quickly accelerate several times. At the same time note any appreciable water rise or the appearance of bubbles which are indicative of exhaust gases leaking into the cooling system.

Periodic maintenance

Every two years the cooling system should be serviced by flushing with plain water, then completely refilled with a fresh solution of water and a high-quality, inhibited (permanent-type) glycol base coolant and providing freezing protection at least to 0°F. Alcohol or methanol base coolants or plain water are not recommended at any time.

Pressure checking radiator cap

Cleaning
A good cleaning solution should be used to loosen the rust and scale before reverse flushing the cooling system. There are a number of cleaning solutions available and the manufacturer's instructions with the particular cleaner being used should always be followed.

An excellent preparation to use for this purpose is GM Cooling System Cleaner. The following directions for cleaning the system applies only when this type cleaner is used.

Drain the cooling system including the cylinder block and then close drain plugs. Remove thermostat and replace thermostat housing. Add the liquid portion (No. 1) of the cooling system cleaner. Fill the cooling system with water to a level of about 3 inches below the top of the overflow pipe. Cover the radiator and run the engine at moderate speed until engine coolant temperature reaches 180 degrees. Remove cover from radiator and continue to run the engine for 20 minutes. Avoid boiling. While the engine is still running, add the powder portion (No. 2) of the cooling system cleaner and continue to run the engine for 10 minutes. At the end of this time, stop the engine, wait a few minutes and then open the drain cocks. Also remove lower hose connection. *Be careful not to scald your hands.* Dirt and bugs may be cleaned out of the radiator air passages by blowing out with air pressure from the back of the core. Do not bend radiator fins.

Radiator
Remove the radiator upper and lower hoses and replace the radiator cap. Attach a lead-away hose at the top of the radiator. Attach a new piece of hose to the radiator outlet connection and insert the flushing gun in this hose. Connect the water hose of the flushing gun to a water outlet and the air hose to an air line. Turn on the water and when the radiator is full, turn on the air in short blasts, allowing the radiator to fill between blasts of air. *Apply air gradually as a clogged radiator will stand only 20 lbs. per in.* Continue this flushing until the water from the lead-away hose runs clear.

Cylinder block and cylinder head
With the thermostat removed, attach a lead-away hose to the water pump inlet and a length of new hose to the water outlet connection at the top of the engine. Disconnect the heater hose and cap connections at engine when reverse flushing engine. Insert the flushing gun in the new hose. Turn on the water and when the engine water jacket is full, turn on the air in short blasts. Continue this flushing until the water from the lead-away hose runs clear.

Heater core
Remove water outlet hose from heater core pipe. Remove inlet hose from engine connection. Insert flushing gun and flush heater core. Care must be taken when applying air pressure to prevent damage to the core.

FAN BELT
Adjustment
Loosen bolts at alternator mounting. Pull alternator away from engine until desired tension reading is obtained with a strand tension gauge, or approx. 1/2 play along longest straight surface. Tighten all bolts securely.

Replacing thermostat

THERMOSTAT
Replacement
Remove radiator to water outlet hose. Remove thermostat housing bolts and remove water outlet and gasket from thermostat housing. Inspect thermostat valve to make sure it is in good condition. Place thermostat in hot water 25° above the temperature stamped on the thermostat valve. Submerge the valve completely and agitate the water thoroughly. Under this condition the valve should open fully.

Remove the thermostat and place in water 10° below temperature indicated on the valve. With valve completely submerged and water agitated thoroughly, the valve should close completely. If thermostat checks satisfactorily, re-install, using a new housing gasket. Refill cooling system.

WATER PUMP
Removal
Drain radiator and break loose the fan pulley bolts. Disconnect heater hose, lower radiator hose and by pass hose (as required) at water pump.

Remove alternator upper brace, loosen swivel bolt and remove fan belt. Remove fan blade assembly attaching bolts, fan and pulley. Thermostatic fan clutches must be kept in an "in-car" position. When removed from the car the assembly should be supported so that the clutch disc remains in a vertical plane to prevent silicone fluid leakage. Remove pump to cylinder block bolts and remove pump and old gasket from engine.

bolt with the head removed) installed in one hole of the fan hub will aid in aligning hub, pulley and fan. Remove stud after starting the remaining three bolts. Connect hoses and fill cooling system. Install Delcotron upper brace and install fan belt. Adjust fan belt to specifications as previously outlined. Start engine and check for leaks.

Disassenbly
Remove bolts and fan and pulley. Remove back plate screws, plate and gasket. Support fan hub in an arbor press and press pump shaft out of hub. A 1/2" dia. x 2" rod will allow the shaft to be pushed through the hub. Support in an arbor press as shown. Press shaft and impeller assembly out of pump, applying pressure on the outer race of the shaft bearing only.

Removing fan hub

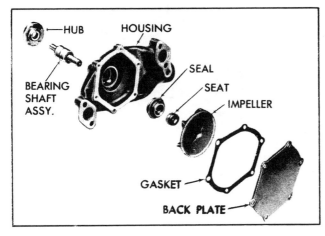

Pump disassembled

Installation
Install pump assembly on cylinder block then, using a new sealer-coated pump-to-block gasket, tighten bolts to specifications. Install pump pulley and fan on pump hub and tighten bolts to specifications. A guide stud (5/16"-24

Removing impeller

73

Shaft and bearing assembly must not be pushed out of housing by applying force on shaft, or bearings will be damaged. Use a 7/8" deep socket or piece of tubing 1-1/8" O.D. Shaft and bearing assembly should be pressed out of rear of pump only. Use a 1-7/16" O.D. tubing on Corvettes with 427 engine. Support impeller on seal surface, using Tool J-7028 in an arbor press, and using a 1/2" x 1" pin, press shaft out of impeller. Discard seal.

Installing fan hub

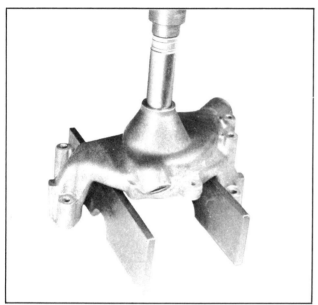

Removing shaft and bearing assembly

Assembly

Install pump shaft and bearing assembly into pump body bearing bore, applying pressure to outer race until it is flush with front of pump body with a socket. Apply pressure to outer race only. Lightly coat O.D. of new seal with suitable sealing compound and press into place with 1-1/4" socket applying pressure to the seal outer flange. The seal should bottom with the outer flange against the pump body. Press on fan hub. Check fan hub location using J-22162 fan hub locating gauge. See specification section for specific engine fan hub locating dimension.

Wet I.D. of rubber insert in ceramic seat assembly before slipping assembly onto bearing shaft. Unmarked white ceramic surface must face bearing seal. Support pump on front (hub end) of shaft and press on impeller. Press to obtain .010" to .035" clearance between impeller vanes and pump body. It will be necessary to put a bind about 1/2" from end of feeler gauge to enable clearance checking. Impeller must be flush to .018" below pump body back plate gasket surface. Install pump back plate and gasket with a light coat of gasket sealer and install screws, then tighten all six diagonally.

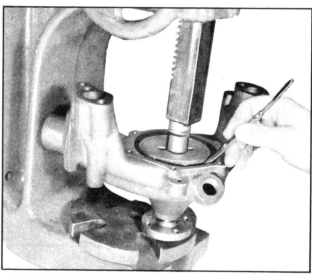

Installing impeller

RADIATOR
Description

All radiators incorporate a drain cock except the copper radiator which has a drain plug. Radiator assemblies are attached to the radiator support or radiator mounting box by cap screws and/or brackets.

Fan shrouds are attached to the radiator, radiator support panel or radiator support brackets by cap screws and/or clips. With the conventional downflow design radiator, coolant level should be one inch below the bottom of filler neck when cold. With the crossflow design, coolant level should be three inches below the bottom of the filler neck when cold.

Removal

Raise the hood and drain radiator. (Install a bolt in hole of hood support to prevent the hood from being released and falling.) Remove radiator inlet and outlet hoses and supply tank hoses at radiator. Remove transmission cooler lines if so equipped. Remove shroud to radiator support screws. Rest shroud on engine fan. Remove radiator upper support bracket screws and carefully lift radiator from vehicle.

If equipped with air conditioning: Remove the hood. Remove the bolts holding the receiver dehydrator bottle and let it rest in place. Back out the left front hinge to body bolt. Remove radiator upper support bolts (2 each side) and loosen the lower bolts (1 each side). Tilt the radiator support rearward and carefully remove the radiator.

Installation

Follow removal procedures in reverse order. Fill cooling system and check for leaks.

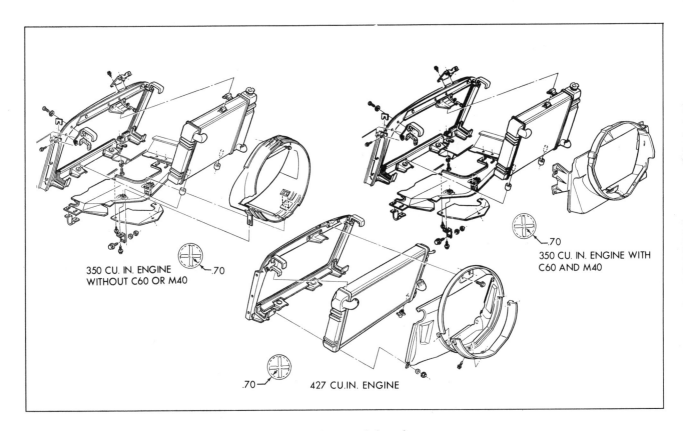

350 CU. IN. ENGINE WITHOUT C60 OR M40 — .70

350 CU. IN. ENGINE WITH C60 AND M40 — .70

427 CU.IN. ENGINE — .70

Radiator and shroud

ENGINE 6

INDEX

6

DESCRIPTION

The V8 engines covered in this section are the 327, 350, 427, and 454 cu. in. engines. In order to avoid repetition and to identify the engines involved in a particular procedure, the 327 and 350 cu. in. V8 engines are identified as "Small V8's." The 427-, and 454- cu. in. V8 engines are identified as "Mark IV V8's."

Removal

Drain cooling system and engine oil. Remove air cleaner and disconnect battery cables at battery. Remove hood hinge front bolts, loosen the rear bolts. Swing hood up and install Tool J-22382-1. Remove radiator and shroud. Remove fan blade and pulley.

Disconnect wires at starter solenoid, delcontron, temperature switch, oil pressure switch, and coil.

Disconnect accelerator linkage at pedal lever, exhaust pipes at manifold flanges, vacuum line to power brake unit at manifold (if so equipped), fuel line (front tank) at fuel pump, engine cooler lines (if so equipped), oil pressure gauge line (if so equipped). Remove power steering pump with hoses attached and lay aside. Raise vehicle on hoist. Remove propeller shaft. If plug for propeller shaft opening in transmission is not available, drain transmission.

Disconnect shift linkage at transmission, speedometer cable at transmission, transmission cooler lines (if so equipped).

On synchromesh equipped cars, disconnect clutch linkage at cross-shaft then remove cross-shaft at frame bracket. Connect lifting device or chain to engine lift brackets. Raise engine to take weight off front mounts and remove front mount thru bolts. Remove rear mount to crossmember bolts. Raise engine to take weight off rear mount, then remove crossmember. Remove engine-transmission assembly from vehicle as a unit. Remove transmission (and clutch): with synchromesh transmission − Remove clutch housing cover plate screws. Remove bolts attaching the clutch housing to engine block then remove transmission and clutch housing as a unit. Support the transmission as the last mounting bolt is removed, and as it is being pulled away from the engine (to prevent damage to clutch disc). Remove starter and clutch housing rear cover plate. Loosen clutch mounting bolts a turn at a time (to prevent distortion of clutch cover) until the spring pressure is released. Remove all bolts, clutch disc and pressure plate assembly.

With automatic transmission − Lower engine, secured by the hoist, and support engine on blocks. Remove starter and converter housing underpan. Remove flywheel-to-converter attaching bolts. Support transmission on blocks. Remove transmission-to-engine mounting bolts. With the hoist attached, remove blocks from the engine only and slowly guide the engine from the transmission. Mount engine in stand.

Installation

Attach lifting device or chain to engine lift brackets and remove engine from engine stand.

With synchromesh transmission − Install the clutch on flywheel. Install clutch housing rear cover and starter.

Install transmission and clutch housing. Install clutch housing cover screws and tighten securely.

With automatic transmission − Position engine adjacent to the transmission and align the converter with the flywheel. Bolt transmission to engine, then raise engine and transmission assembly and install flywheel to converter attaching bolts. Install converter housing underpan and starter.

Tilt and lower engine and transmission assembly into the chassis as a unit, guiding engine to align front mounts with frame supports. Install front mount thru bolts. Raise engine enough to install rear crossmember, then install crossmember, install rear mount and lower engine. Remove lifting device and torque all mount bolts to specifications. On synchromesh equipped vehicles, install clutch cross-shaft engine bracket, then adjust and connect clutch. Connect speedometer cable, shift linkage at transmission, and transmission cooler lines (if so equipped).

Install propeller shaft. Lower hoist and remove vehicle. Connect vacuum line to power brake unit (if so equipped). Exhaust pipes at manifold flanges, accelerator linkage at pedal lever, fuel line at fuel pump, engine cooler lines (if so equipped), and oil pressure gauge lines (if so equipped).

Connect wires at: coil, oil pressure switch, temperature switch, delcotron, and starter solenoid.

Install power steering pump. Install pulley, fan blade and fan belt. Install radiator and shroud. Remove hood prop tool and install front hinge bolts. Adjust hood.

Connect battery cables. Fill with coolant, engine oil and transmission oil, then start engine and check for leaks. Perform necessary adjustments and install air cleaner.

INTAKE MANIFOLD
Removal

Drain radiator and remove air cleaner. Disconnect battery cables at battery, upper radiator and heater hose at manifold, accelerator linkage at pedal lever, fuel line at carburetor, wires at coil (both sides) and temperature sending switch, power brake hose at carburetor base or manifold, spark advance hose at distributor, and crankcase ventilation hoses (as required).

Remove distributor cap and mark rotor position with chalk. Remove distributor clamp and distributor, then position distributor cap rearward clear of manifold. Remove delcotron upper bracket. Remove coil and bracket. Remove manifold-to-head attaching bolts, then remove manifold (with carburetor on) from engine and discard gaskets and seals. If manifold is to be replaced, transfer carburetor and carburetor mounting studs, water outlet and thermostat (use new gasket), heater hose adapter, and choke coil.

Installation

Clean gasket and seal surfaces of manifold, cylinder heads and block. Install manifold end seals on block. Install side gaskets on cylinder heads using sealing compound around water passages. Install manifold bolts, and torque to specifications in the sequence outlined on the Torque

78

Sequence Chart. Install coil. Install distributor with the rotor pointing at the chalk mark, then install distributor cap. If the crankshaft has been rotated while the distributor was removed, time distributor to number 1 cylinder.

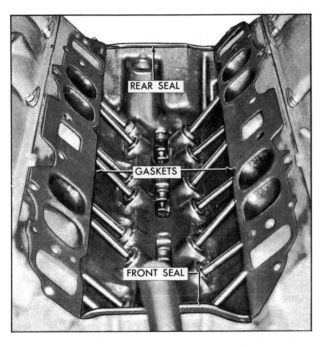

Intake manifold gasket and seal location

Install delcotron upper bracket. Adjust belt tension and torque attaching bolts to specifications. Connect battery cables at battery, upper radiator and heater hose at manifold, accelerator linkage at pedal lever, fuel line at carburetor, wires at ignition coil (both sides), power brake hose at carburetor base or manifold, spark advance hose at distributor, and crankcase ventilation hoses (as required). Fill with coolant, start engine, check for leaks and adjust timing and carburetor idle speed and mixture.

EXHAUST MANIFOLD
Removal
If equipped with "Air Injection Reactor System," remove air manifold and tubes. Disconnect battery ground cable. Remove air cleaner pre-heater air stove. Remove exhaust manifold flange nuts, then lower exhaust pipe assembly (hang exhaust pipe from frame with wire.) Remove end bolts then remove center bolts and remove manifold from engine.

Installation
Clean mating surfaces on manifold and head, then install manifold in position with center bolts. Install end bolts and snug all manifold bolts. Torque center bolts to specifications, then torque end bolts to specifications. Using a new flange gasket install exhaust pipe to manifold flange. Connect the battery ground cable. Install air cleaner preheater. Start engine and check for leaks.

ROCKER ARM COVER
Removal
Remove air cleaner. Disconnect crankcase ventilation hoses (as required). Disconnect temperature wire from rocker arm cover clips. Remove rocker arm cover.

Installation
Clean gasket surfaces on cylinder head and rocker arm cover with degreaser then, using a new gasket, install rocker arm cover and torque bolts to specifications. Connect crankcase ventilation hoses (if disconnected). Connect temperature wire at rocker arm cover clips. Install air cleaner.

VALVE MECHANISM
Removal
Remove rocker arm covers as outlined. Remove rocker arm nuts, rocker arm balls, rocker arms and push rods. Place rocker arms, rocker arm balls and push rods in a rack so they may be reinstalled in the same location.

Installation and adjustment
Whenever new rocker arms and/or rocker arm balls are being installed, coat bearing surfaces of rocker arms and rocker arm balls with "Molykote" or its equivalent. Install push rods. Be sure push rods seat in lifter socket. Install rocker arms, rocker arm balls and rocker arm nuts. Adjust valves when lifter is on base circle or cam-shaft lobe as follows: **with hydraulic valve lifters,** crank engine until mark on torsional damper lines up with center or "O" mark on the timing tab and the engine is in the number 1 firing position. This may be determined by placing fingers on the number 1 cylinder valve as the mark on the damper comes near the "O" mark on the front cover. If the valves are not moving, the engine is in the number 1 firing

Valve adjustment (hydraulic)

79

position. If the valves move as the mark comes up to the timing tab, the engine is in number 6 firing position and crankshaft should be rotated one more revolution to reach the number 1 position.

Valve adjustment is made by backing off the adjusting nut (rocker arm stud nut) until there is play in the push rod and then tighten nut to just remove all push rod to rocker arm clearance. This may be determined by rotating push rod with fingers as the nut is tightened. When push rod does not readily move in relation to the rocker arm, the clearance has been eliminated. The adjusting nut should then be tightened an additional 1 turn to place the hydraulic lifter plunger in the center of its travel. No other adjustment is required.

With the engine in the number 1 firing position as determined above, the following valves may be adjusted. Exhaust – 1, 3, 4, 8 Intake – 1, 2, 5, 7.

Crank the engine one revolution until the pointer "O" mark and torsional damper mark are again in alignment. This is number 6 firing position. With the engine in this position the following valve may be adjusted. Exhaust – 2, 5, 6, 7 Intake – 3, 4, 6, 8.

With mechanical valve lifters – Crank engine until mark on torsional damper lines up with center or "O" mark on

the timing tab and the engine is in the number 1 firing position. This may be determined by placing fingers on the number 1 cylinder valve as the mark on the damper comes near the "O" mark on the front cover. If the valves are not moving, the engine is in the number 1 firing position. If the valves move as the mark comes up to the timing tab, the engine is in number 6 firing position and crankshaft should be rotated one more revolution to reach the number 1 position.

With the engine in the number 1 firing position as determined above, adjust the following valves to specifications with a feeler gauge. Exhaust – 4, 8 Intake – 2, 7.

Turn crankshaft 1/2 revolution (180°) clockwise and adjust the following valve to specifications with a feeler gauge. Exhaust – 3, 6 Intake – 1, 8.

Turn crankshaft 1/2 revolution (180°) clockwise until the pointer "O" mark and torsional damper mark are again in alignment. This is number 6 firing position. With the engine in this position, adjust the following valves to specifications with a feeler gauge. Exhaust – 5, 7 Intake – 3, 4.

Turn crankshaft 1/2 revolution (180°) clockwise and adjust the following valves to specifications with a feeler gauge. Exhaust – 1, 2 Intake – 5, 6. Readjust valves (hot and running). Install rocker arm covers as outlined. Adjust carburetor idle speed and mixture.

Valve lifters

Hydraulic valve lifters very seldom require attention. The lifters are extremely simple in design, readjustments are not necessary, and servicing of the lifters requires only that care and cleanliness be exercised in the handling of parts.

Locating noisy hydraulic lifters

Locate a noisy valve lifter by using a piece of garden hose approximately four feet in length. Place one end of the hose near the end of each intake and exhaust valve with the other end of the hose to the ear. In this manner, the sound is localized making it easy to determine which lifter is at fault.

Another method is to place a finger on the face of the valve spring retainer. If the lifter is not functioning properly, a distinct shock will be felt when the valve returns to its seat.

The general types of valve lifter noise are, Hard Rapping Noise – Usually caused by the plunger becoming tight in the bore of the lifter body to such an extent that the return spring can no longer push the plunger back up to working position. Probable causes are: excessive varnish or carbon deposit causing abnormal stickiness, galling or "pick-up" between plunger and bore of lifter body, usually caused by an abrasive piece of dirt or metal wedging between plunger and lifter body.

Moderate Rapping Noise––Probable causes are excessively high leakdown rate, leaky check valve seat, improper adjustment.

General Noise Throughout the Valve Train – This will, in almost all cases, be a definite indication of insufficient oil

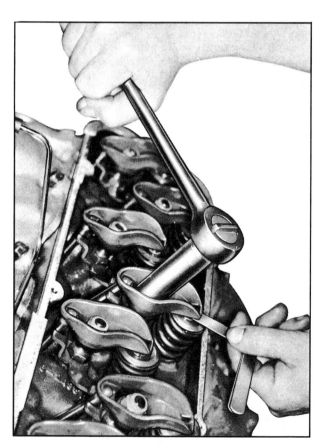

Valve adjustment (mechanical)

80

supply, or improper adjustment.

Intermittent Clicking——Probable cases are a microscopic piece of dirt momentarily caught between ball seat and check valve ball, in rare cases, the ball itself may be out-of-round or have a flat spot, and improper adjustment.

In most cases where noise exists in one or more lifters all lifter units should be removed, disassembled, cleaned in a solvent, reassembled, and reinstalled in the engine. If dirt, corrosion, carbon, etc., is shown to exist in one unit, it more than likely exists in all units, thus it would only be a matter of time before all lifters caused trouble.

Removal
Remove intake manifold as outlined. Remove valve mechanism as outlined. Remove valve lifters. Place valve lifters in a rack so they may be reinstalled in the same location.

Installation
Install valve lifters. Whenever new valve lifters are being installed coat foot of valve lifters with "Molykote" or its equivalent. Install intake manifold as outlined. Install and adjust valve mechanism as outlined.

VALVE STEM OIL SEAL AND/OR VALVE SPRING
Replacement
Remove rocker arm cover(s) as outlined. Remove spark plug, rocker arm and push rod on the cylinders to be serviced. Apply compressed air to the spark plug hole to hold the valves in place.

A tool to apply air to the cylinder is available through local jobbers or may be manufactured. In manufacturing this tool an AC-46N (or AC-43XL for aluminum heads) spark plug or its equivalent is recommended. Chisel the spark plug as shown, then drive the porcelain out of the plug by tapping the center electrode against a hard block. Using a 3/8" pipe tap, cut threads in the remaining portion of the spark plug and assemble.

Compress the valve spring, remove the valve locks, valve cap, and valve spring and damper. Remove valve stem oil seal. Remove as follows:

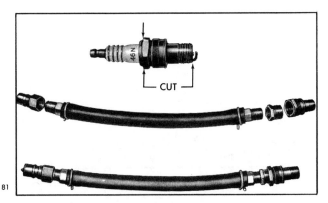

Air adapter tool

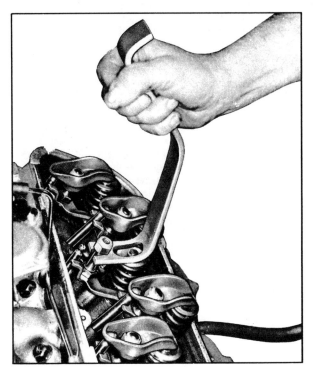

Compressing valve spring

Small V8 engines — to replace, set the valve spring and damper, valve shield and valve cap in place. The close coiled end of the spring is installed against the cylinder head. Compress the spring and install oil seal in the lower groove of the stem, making sure the seal is flat and not twisted. A light coat of oil on the seal will help prevent twisting.

Install the valve locks and release the compressor tool making sure the locks seat properly in the upper groove of the valve stem. Grease may be used to hold the locks in place while releasing the compressor tool.

Mark IV V8 engines — Install new valve stem oil seal (coated with oil) in position over valve guide. Set the valve spring and damper, and cap (or cap and seal assembly) in place. Compress the spring and install the valve locks, then release the compressor tool, making sure the locks seat properly in the groove of the valve stem. Grease may be used to hold the locks in place while releasing the compressor tool. Install spark plug and torque to specifications. Install and adjust valve mechanism as outlined.

CYLINDER HEAD ASSEMBLY
Removal
Removal intake manifold as outlined. Remove exhaust manifolds as outlined. Remove valve mechanism as outlined. Drain cooling system (block). Remove cylinder head bolts, cylinder head and gasket. Place cylinder head on two blocks of wood to prevent damage.

Installations

The gasket surfaces of head and block must be clean and free of nicks or heavy scratches. Cylinger bolt threads in the block and threads on the cylinder head bolt must be cleaned. (Dirt will affect bolt torque.) On engines using a STEEL gasket, coat both sides of a new gasket with a good sealer, spread the sealer thin and even. One method of applying the sealer that will assure the proper coat is with the use of a paint roller. Too much sealer may hold the beads of the gasket away from the head or block. *Use no sealer on engines using a composition STEEL ASBES-TOS gasket.*

Place the gasket in position over the dowel pins with the bead up. Carefully guide cylinder head into place over dowel pins and gasket. Coat threads of cylinder head bolts with sealing compound and install finger tight. Tighten cylinder head bolts a little at a time in the sequence shown on the torque sequence chart until the specified torque is reached. Install the exhaust manifold as outlined. Install the intake manifold as outlined. Install and adjust the valve mechanism as outlined.

OIL PAN
Removal

Disconnect battery positive cable and remove engine oil dipstick and tube. Raise vehicle on hoist and drain engine oil. Disconnect steering linkage idler at frame and lower the linkage. Remove oil pan bolts and remove oil pan. Discard used gaskets and seals.

Installation

Thoroughly clean all gasket sealing surfaces. If crankshaft was rotated while oil pan was off, place timing mark at 6:00 o'clock position. Using new gaskets, install side gasket on pan rails using gasket sealer as a retainer. Install front and rear pan seals, making sure their ends butt with side gaskets. Install oil pan and torque retaining bolts to specifications. Raise engine, remove wood blocks and lower engine onto front mounts. Connect steering linkage idler and torque attaching bolts to specifications. Lower vehicle and install engine oil dipstick and tube. Fill engine with oil and connect battery positive cable.

OIL SEAL (REAR MAIN)
Replacement

The rear main bearing oil seal can be replaced (both halves) without removal of the crankshaft. Always replace the upper and lower seal as a unit. Install with the lip facing toward the front of the engine.

With the oil pan and oil pump removed, remove the rear main bearing cap. Remove oil seal from the annular tang

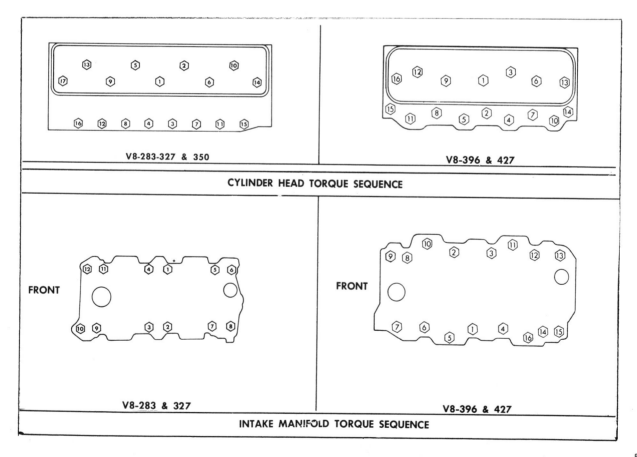

V8-283-327 & 350

V8-396 & 427

CYLINDER HEAD TORQUE SEQUENCE

FRONT

V8-283 & 327

FRONT

V8-396 & 427

INTAKE MANIFOLD TORQUE SEQUENCE

Torque sequence

by prying from the bottom with a small screw driver. Lubricate the lip of a new seal with engine oil. Keep oil off the parting line surface as this is treated with glue. Insert seal in cap and roll it into place with finger and thumb, using light pressure so seal tangs at parting line do not cut bead on back of seal. To remove the upper half of the seal, use a small hammer to tap a brass pin punch on one end of seal until it protrudes far enough to be removed with pliers. Always wipe crankshaft surface clean

before installing a new seal. Lubricate the lip of a new seal with engine oil. Keep oil off the parting line surface. Gradually push with a hammer handle, while turning crankshaft, until seal is rolled into place. (Similar to installing a main bearing.) Be careful that seal tangs at parting line do not cut bear on back of seal. Use sealer at parting line on cap half of seal and seal area of cylinder block. Install the rear main bearing cap (with new seal) and torque to specifications.

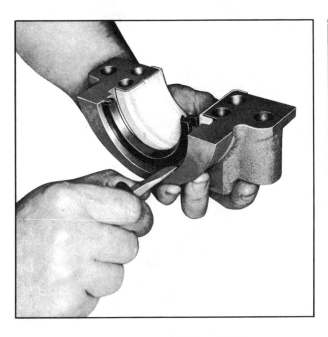

Removing oil seal (lower half)

Sealing bearing cap

TORSIONAL DAMPERS
Removal

Remove fan belt, fan and pulley. Remove the radiator and shroud. If additional operations such as camshaft removal are not being performed, the radiator will not have to be removed. On cars equipped with Mark IV engines, remove engine front mount through-bolts and raise front of engine enough for torsional damper to clear frame crossmember.

Remove accessory drive pulley, then remove torsional

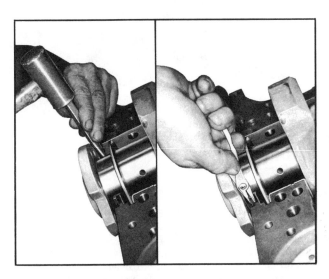

Removing oil seal (upper half)

Removing torsional damper

damper retaining bolt. Install Tool J-6978 to torsional damper and turn puller screw to remove damper from crankshaft. Remove tool.

Installation

It is necessary to use installer tool to prevent the inertia weight section from walking off the hub during installation of damper. Coat front cover seal contact area (on damper) with engine oil. Place damper in position over key on crankshaft. Pull damper onto crankshaft as follows: on Mark IV V8 engines use Tool J-21058. If engine or radiator has not been removed from the vehicle a 1/2"-20 x 5" bolt and a 1/2"-20 nut may be used in place of the bolt and nut of Tool J-21058. On small V8 engines install damper using a 7/16"-20 x 5" bolt and nut in place of bolt and nut furnished with J-21058. Install bolt in crankshaft with sufficient thread engagement (min. 1/2"). Remove tool from crankshaft. Install damper retaining bolt and torque to specifications, then install accessory drive pulley.

Install radiator and shroud. Install fan pulley and fan. Install fan belt and adjust to specifications using strand tension gauge. Lower engine and install front mount through-bolts. Fill cooling system, start engine and check for leaks.

CRANKCASE FRONT COVER
Removal

Remove oil pan as outlined. Remove torsional damper as outlined. Remove water pump. Remove crankcase front cover attaching bolts and remove front cover and gasket, then discard gasket.

Installation

Make certain that cover mounting face and cylinder block front face are clean and flat. Coat the oil seal with engine

oil and using a new cover gasket, coated with gasket sealer, install cover and gasket over dowel pins and cylinder block. Install cover screws and torque bolts to specifications. Install water pump. Install torsional damper as outlined. Install oil pan as outlined.

OIL SEAL (FRONT COVER)
Replacement
With cover removed

With cover removed, pry old seal out of cover from the front with a large screw driver. Install new seal so that open end of the seal is toward the inside of cover and drive it into position with Tool J-995 on Small V8 engines or Tool J-22102 on Mark IV V8 engines. *Support cover at sealing area to avoid distorting cover.*

Without cover removed

With torsional damper removed, pry old seal out of cover from the front with a large screw driver, being careful not to damage the surface on the crankshaft. Install new seal so that open end of seal is toward the inside of cover and drive it into position with Tool J-23042 on Small V8 engines or Tool J-22102 on Mark IV V8 engines.

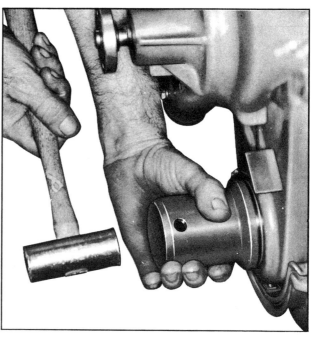

Installing oil seal (cover installed)

TIMING CHAIN AND/OR SPROCKETS
Replacement

Remove torsional damper and crankcase front cover as outlined. Crank engine until marks on camshaft and crankshaft sprockets are in alignment. Remove three camshaft sprocket to camshaft bolts. Remove camshaft sprocket and timing chain together. Sprocket is a light

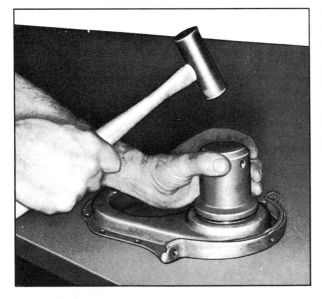

Installing oil seal (cover removed)

84

press fit on camshaft. To dislodge, tap lightly on lower edge of camshaft sprocket with a plastic hammer. If crankshaft sprocket is to be replaced, remove sprocket using Tool J-5825 on Small V8 engines or Tool J-1619 on Mark IV V8 engines. Install new sprocket using a 7/16"-20 x 5" bolt and nut on Small V8 engines or Tool J-21058 on Mark IV V8 engines. Install timing chain on camshaft sprocket. Hold the sprocket vertical with the chain hanging below, and align marks on camshaft and crankshaft sprockets. Align dowel in camshaft with dowel hole in camshaft sprocket and install sprocket on camshaft. Do not attempt to drive cam sprocket on shaft as welsh plug at rear of engine can be dislodged. Draw camshaft sprocket onto camshaft, using three mounting bolts. Torque bolt to specifications. Lubricate timing chain with engine oil. Install crankcase front cover and torsional damper as outlined.

Timing sprocket marks

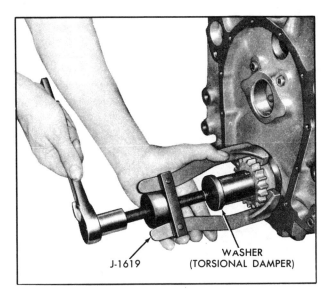

J-1619   WASHER
(TORSIONAL DAMPER)

Removing crankshaft sprocket (Mark IV V8)

Removing crankshaft sprocket (small V8)

Installing timing chain

Installing crankshaft sprocket   (396 and 427)

CAMSHAFT
Removal
Remove valve lifters as outlined. Remove crankshaft front cover as outlined. Remove grille. Remove fuel pump and push rod. All camshaft journals are the same diameter and care must be used in removing camshaft to avoid damage to bearings. Remove camshaft sprocket to camshaft bolts then remove sprocket and timing chain together. Sprocket is a tight fit on camshaft. If sprocket does not come off easily, a light blow on the lower edge of the sprocket (with a plastic mallet) should dislodge the sprocket. Install two 5/16" – 18 x 4" bolts in camshaft bolt holes then remove camshaft.

Installation
Whenever a new camshaft is installed coat camshaft lobes with "Molykote" or its equivalent. Lubricate camshaft journals with engine oil and install camshaft. Install timing

chain on camshaft sprocket. Hold the sprocket vertical with the chain hanging down, and align marks on camshaft and crankshaft sprockets. Align dowel in camshaft with dowel in hole in camshaft sprocket then install sprocket on camshaft. Draw the camshaft sprocket onto camshaft using the mounting bolts. Torque to specifications. Lubricate timing chain with engine oil. Install fuel pump push rod. Install grille. Install crankcase front cover as outlined. Install valve lifters as outlined.

FLYWHEEL
Removal
With transmission and/or clutch housing and clutch removed from engine, remove the flywheel.

Installation
Clean the mating surfaces of flywheel and crankshaft to make certain there are no burrs. Install flywheel on crankshaft and position to align dowel hole of crankshaft flange and flywheel. On Automatic Transmission equipped engines, the flywheel must be installed with the flange collar to transmission side.

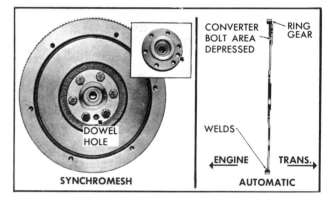

Flywheel installation (typical)

ENGINE MOUNTS
Front mount replacement
Remove nut, washer and engine mount through-bolt. Raise engine to release weight from mount. Remove mount from engine. Install new mount on engine. Lower engine, install through-bolt and tighten all mount bolts to specified torques.

Rear mount replacement
Remove crossmember-to-mount bolts. Raise transmission to release weight from mount. Remove mount-to-transmission bolts, then remove mount. Install new mount on transmission. While lowering transmission, align and start crossmember-to-mount bolts. Lower transmission and tighten all mounting bolts to specified torques.

ENGINE
Disassembly
With the engine mounted on an engine stand (transmission and/or clutch housing removed from the engine), remove the following sub-assemblies (if applicable):

Removing camshaft

Air Conditioning Compressor (with brackets). Air Compressor (with hoses and brackets). Power steering Pump (with brackets). Air Injection Reactor System (with brackets). Delcotron (with brackets). Accessory Drive Pulley(s) (and belts). Water Pump (and by pass hose). Fuel Pump and fuel pump push rod. Distributor Cap (with spark plug wires). Carburetor (and fuel lines). Oil Filter. Starter. Clutch Pressure Plate and Disc. Ground Strap. Oil Dip Stick and Oil Dip Stick Tube.

Remove intake and exhaust manifolds. Loosen valve rocker arm nuts until rocker arms can be pivoted, then remove push rods and valve lifters. Place push rods and valve lifters in a rack so they may be installed in the same location during engine assembly. Remove the cylinder head(s). Remove the torsional damper. Remove the oil pan. Remove crankcase front cover. On the performance engines, remove the oil baffle.

Remove oil pump and screen assembly and the extension shaft. Check connection rods and caps for cylinder number identification and if necessary, mark them. Check cylinder bores for ridge and if necessary, remove ridge. Remove connecting rod caps and using connecting rod guide set, Tool J-5239 (3/8") or J-6305 (11/32"), push connecting rod and piston assemblies out of block. It will be necessary to turn the crankshaft to disconnect and remove some of the connecting rod and piston assemblies.

Remove the camshaft. Use care in removing camshaft to avoid damaging bearings. Remove camshaft sprocket bolts then remove camshaft sprocket and timing chain. Sprocket is a light press fit on camshaft. To dislodge, tap lightly on lower edge of camshaft sprocket with a plastic hammer. Install two 5/16-18 bolts in camshaft sprocket bolt holes and carefully remove camshaft.

Remove the flywheel. Remove main bearing caps and lift crankshaft out of cylinder block. Remove rear main bearing oil seal from cylinder block and rear main bearing cap. Discard all gaskets and seals removed during engine disassembly.

Assembly

Install rear main bearing oil seal in cylinder block and rear main bearing cap grooves. Install with lip of seal toward front of engine. Where seal has two lips — isntall lip with helix towards front of engine. Lubricate lips of seal with engine oil. Keep oil off parting line surface. Install main bearings in cylinder block and main bearing caps then lubricate bearing surface with engine oil. Install crankshaft, being careful not to damage bearing surfaces.

Apply a thin coat of brush-on type oil sealing compound to block mating surface and corresponding surface of cap only. Do not allow sealer on crankshaft or seal lip. Install main bearing caps with arrows pointing toward front of engine. Torque all except rear main bearing cap bolts to specifications. Torque rear main bearing cap bolts to 10-12 ft. lbs. then tap end of crankshaft, first rearward then forward with a lead hammer. This will line up rear main bearing and crankshaft thrust surfaces. Retorque all main bearing cap bolts to specifications. Measure crankshaft end play with a feeler gauge. Force crankshaft forward and measure clearance between the front of the rear main bearing and the crankshaft thrust surface.

87

Sealing bearing cap and block

Install flywheel and torque to specifications. A wood block placed between the crankshaft and cylinder block will prevent crankshaft from rotating. Align dowel hole in flywheel with dowel hole in crankshaft. On vehicles equipped with automatic transmissions, install flywheel with the converter attaching pads towards transmission.

Timing sprocket marks

Whenever a new camshaft is installed, lubricate camshaft lobes with "Molykote" or its equivalent. Whenever new lifters are installed, polish lifter feet with No. 600 wet/dry emery paper. Install two 5/16-18 bolts in camshaft bolt holes, then lubricate camshaft journals with engine oil and install camshaft, being careful not to damage bearings. Remove the two 5/16-18 bolts. Install timing chain on camshaft sprocket then align marks on camshaft and crankshaft sprockets and connect chain to crankshaft sprocket. Align dowel on camshaft with dowel hole in camshaft sprocket and install sprocket on camshaft. Do not hammer camshaft sprocket onto camshaft. This may loosen camshaft rear welsh plug. Draw camshaft sprocket onto camshaft, using the mounting bolts. Torque bolts to specifications. Lubricate timing chain with engine oil.

Install connecting rod bearings in connecting rods and connecting rod caps then lubricate bearings, pistons, piston rings, connecting rod bolts and cylinder walls lightly with engine oil. Install connecting rod guide set J-5239 (3/8") or Tool J-6305 (11/32") on connecting rod bolt and using Tool J-8307 to compress the rings, install number one connecting rod and piston assembly in its respective bore. Use a hammer handle and light blows. Hold the ring compressor firmly against the cylinder block until all piston rings have entered the bore. Be sure ring gaps are properly positioned on piston and piston is properly positioned in cylinder block.

Installing camshaft

Remove connecting rod guide set then install connecting rod cap and torque nuts to specifications. Repeat for the remaining connecting rod and piston assemblies. Measure connecting rod side clearance (see specifications). Measure between connecting rod caps.

Install oil pump and screen assembly and extension shaft. On hi-performance engines, install the oil baffle. Install the crankcase front cover.

Place crankcase front cover gasket in position over dowel pins on cylinder block. Lubricate seal lip with engine oil then place crankcase front cover in position over dowel pins and torque bolts to specifications.

Install the oil pan. Install side gaskets on cylinder block. Do not use sealer. Install rear oil pan seal, in groove in rear main bearing cap, with ends butting side gaskets. Install oil pan front seal, in crankcase front cover, with ends butting side gaskets. Install oil pan and torque bolts to specifications.

TORSIONAL DAMPER
Installing
The inertia weight section of the torsional damper is assembled to the hub with a rubber type material. The installation procedures (with proper tool) must be followed or movement of the inertia weight section on the hub will destroy the tuning of the torsional damper.

Drive on type (without retaining bolt)
Coat front cover seal area (on damper) with engine oil. Attach damper Installer Tool J-22197 to damper. Tighten fingers of tool to prevent weight from moving. Position damper on crankshaft and drive into position using J-5590 until it bottoms against crankshaft gear or sprocket. Remove installer tool.

Pull on type (with retaining bolt)
Coat front cover seal contact area (on damper) with engine oil. Place damper in position over key on crankshaft. Using Tool J-21058 pull damper onto crankshaft. Install bolt in crankshaft with sufficient thread engagement (min. 1/2"). Remove tool from crankshaft. Install damper retaining bolt and torque to specification. On engines with a 7/16"-20 retaining bolt use a 7/16"-20 x 4" bolt and a 7/16"-20 nut in place of the bolt and nut furnished with Tool J-21058.

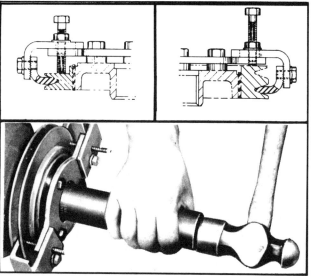

88

Installing torsional damper (drive on type)

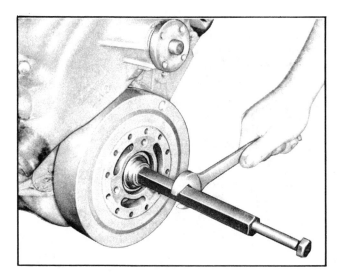

Installing torsional damper (pull on type)

CYLINDER HEAD
Installing
The gasket surfaces on both the head and the block must be clean of any foreign matter and free of nicks or heavy scratches. Cylinder bolt threads in the block and threads on the cylinder head bolts must be cleaned. (Dirt will effect bolt torque).

On engines using a STEEL gasket, coat both sides of gasket with sealer, spread the sealer thin and even. Too much sealer may hold the beads of the gasket away from the head or block. One method of applying the sealer that will assure the proper coat is with the use of a paint roller. Use no sealer on engines using a composition STEEL ASBESTOS gasket.

Place the gasket in position over the dowel pins, with the bead up. Carefully guide cylinder head into place over dowel pins and gasket. Coat threads of cylinder head bolts with sealing compound and install finger tight. Mark IV V8 engines with aluminum cylinder heads have hardened steel washers on the cylinder head bolts. These bolts require a special torque. Tighten cylinder head bolts a little at a time in torque sequence until the specified torque is reached. All engines using composition STEEL ASBESTOS gaskets and/or aluminum heads must have the cylinder head retorqued after engine warm-up.

Install valve lifters and push rods. Install in the same location from which removed during engine disassembly. Whenever new valve lifters and/or rocker arms and balls are being installed coat foot of valve lifters and surfaces of rocker arms and balls with "MOLYKOTE" or its equivalent. Polish feet of lifters with No. 600 wet/dry emery paper.

Install valve rocker arms, rocker arm balls and nuts and tighten rocker arm nuts until all push rod end play is taken up. Install intake and exhaust manifolds. Torque to specifications in the sequence outlined.

Install the following sub-assemblies (if applicable). Oil Dip Stick Tube and Oil Dip Stick. Ground Strap. Clutch Pressure Plate and Disc. Starter. Oil Filter (new). Carburetor and Fuel Lines. Distributor Cap (with spark plug wires). Fuel Pump (and fuel pump push rod). Water Pump (and by pass hose). Accessory Drive Pulley(s) (and belts). Delcotron (with brackets). Air Injection Reactor System (with brackets). Power Steering Pump (with brackets). Air Compressor (with hose and brackets). Air Conditioning Compressor (with brackets). Adjust all belts as necessary. Adjust valves. Attach engine lift at appropriate cylinder head bolt location.

CYLINDER HEAD ASSEMBLIES
Disassembly
With cylinder head removed, remove valve rocker arm nuts, balls and rocker arms (if not previously done). Using Tool J-8062, compress the valve springs and remove valve keys. Release the compressor tool and remove spring caps, spring shields (if so equipped) springs and spring damper, then remove oil seals and valve spring shims. Remove valves from cylinder head and place them in a rack in their proper sequence so that they can be assembled in their original positions.

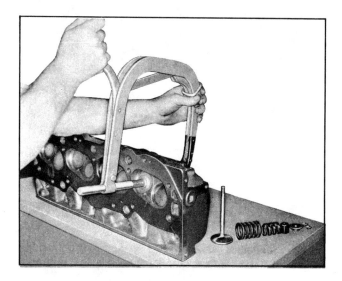

Compressing valve spring

Clean all carbon from combustion chambers and valve ports using Tool J-8089. Thoroughly clean the valve guides using Tool J-8101. Clean all carbon and sludge from push rods, rocker arms and push rod guides. Clean valve stems and heads on a buffing wheel. Clean carbon deposits from head gasket mating surface.

Inspection and repairs
Inspect the cylinder head for cracks in the exhaust ports, combustion chambers, or external cracks to the water chamber. Inspect the valves for burned heads, cracked faces or damaged stems. Excessive valve stem to bore

clearance will cause excessive oil consumption and may cause valve breakage. Insufficient clearance will result in noisy and sticky functioning of the valve and disturb engine smoothness. Measure valve stem clearance. Clamp a dial indicator on one side of the cylinder head rocker arm cover gasket rail, locating the indicator so that movement of the valve stem from side to side (crosswise to the head) will cause a direct movement of the indicator stem. The indicator stem must contact the side of the valve stem just above the valve guide. With the valve head dropped about 1/16" off the valve seat; move the stem of the valve from side to side using light pressure to obtain a clearance reading. If clearance exceeds specifications it will be necessary to ream valve guides for oversize valves as outlined.

Measuring valve stem clearance

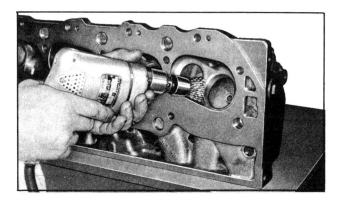

Cleaning combustion chambers

Check valve spring tension with Tool J-8056 spring tester. Springs should be compressed to the specified height and checked against the specifications chart. Springs should be replaced if not within 10 lbs. of the specified load (without dampers).

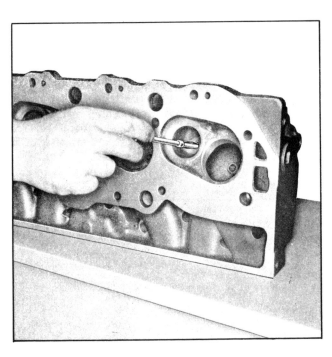

Cleaning valve guides

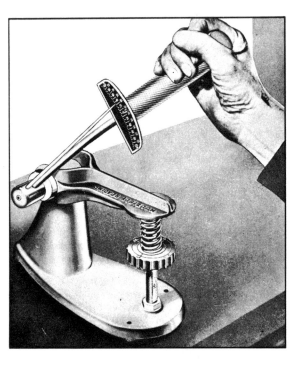

Checking valve spring tension

90

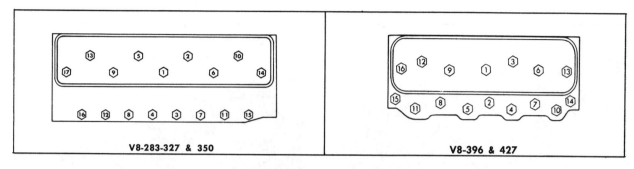

CYLINDER HEAD TORQUE SEQUENCE

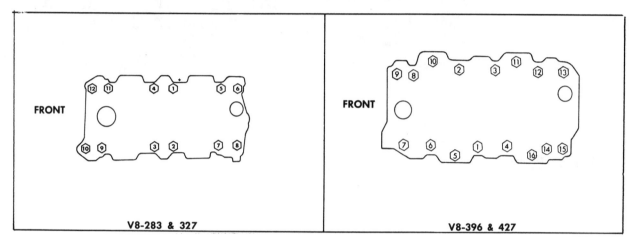

INTAKE MANIFOLD TORQUE SEQUENCE

Torque specifications

Inspect rocker arm studs for wear or damage. Inspect push rod guides on Mark IV V8s for wear or damage.

Rocker arm studs and push rod guides (Mark IV V8s)

The push rod guides are related to the cylinder head by the rocker arm studs. Replace where necessary and torque rocker arm studs to specifications. Coat Threads on cylinder head end of rocker arm studs with sealer before assembling to cylinder head.

Rocker arm studs ("Small V8")

Rocker arm studs that have damaged threads or are loose in cylinder heads should be replaced with new studs available in .003" and .013" oversize. Studs may be installed after reaming the holes. Remove old stud by placing Tool J-5802-1 over the stud, installing nut and flat washer and removing stud by turning nut. Ream hole for oversize stud using Tool J-5715 for .003" oversize or Tool J-6036 for .013" oversize. Do not attempt to install an oversize stud without reaming stud hole. Coat press-fit area of stud with hypoid axle lubricant. Install new stud, using Tool J-6880 as a guide. Gauge should bottom on head.

Valve guide bores

Valves with oversize stems are available. To ream the valve guide bores for oversize valves use Tool Set J-5830 ("Small V8") or J-7049 (Mark IV V8).

Valve seats

Reconditioning the valve seats is very important, because the seating of the valves must be perfect for the engine to deliver the power and performance built into it. Another important factor is the cooling of the valve heads. Good contact between each valve and its seat in the head is imperative to insure that the heat in the valve head will be properly carried away. Several different types of equipment are available for reseating valves seats. The recommendations of the manufacturer of the equipment being used should be carefully followed to attain proper results.

Regardless of what type of equipment is used, however, it is essential that valve guide bores be free from carbon or dirt to insure proper centering of pilot in the guide. Install expanding pilot in the valve guide bore and expand pilot.

Place roughing stone or forming stone over pilot and just clean up the valve seat. Use a stone cut to specifications. Remove roughing stone or forming stone from pilot, place finishing stone, cut to specifications, over pilot and cut just enough metal from the seat to provide a smooth finish. Narrow down the valve seat to the specified width.

91

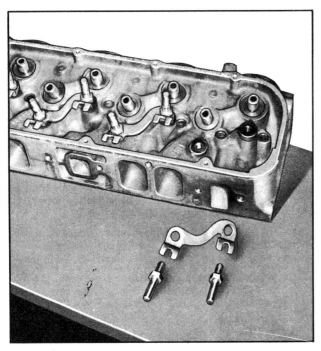

Rocker arm stud and push rod guide (396 and 427 V8)

Reaming rocker arm stud bore ("small V8")

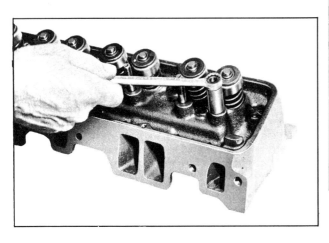

Removing rocker arm stud (in line and 283, 327, 350 V8)

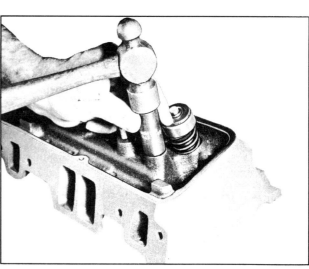

Installing rocker arm stud (in line and 283, 327, 350 V8)

This operation is done by grinding the port side with a 30° stone to lower seat and a 60° stone to raise seat. Remove expanding pilot and clean cylinder head carefully to remove all chips and grindings from above operations. Measure valve concentricity. Valve seats should be concentric to within .002" total indicator reading.

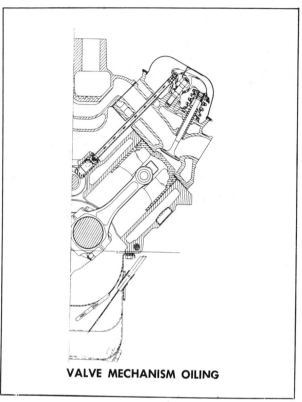

VALVE MECHANISM OILING

Reaming valve guide

Valves

Valves that are pitted can be refaced to the proper angle, insuring correct relation between the head and stem on a valve refacing mechanism. Valve stems which show excessive wear, or valves that are warped excessively should be replaced. When a valve head which is warped excessively is refaced, a knife edge will be ground on part or all of the valve head due to the amount of metal that must be removed to completely reface. Knife edges lead to breakage, burning or pre-ignition due to head localizing on this knife edge. If the edge of the valve head is less than 1/32" thick after grinding, replace the valve.

Assembly

Insert a valve in the proper port. Assemble the valve spring and related parts.

"Small V8"

Set the valve spring shim, valve spring (with damper if used), valve shield and valve cap or rotator in place. The close coiled end of the spring is installed against the cylinder head. Compress the spring with Tool J-8062. Install oil seal in the lower groove of the stem, making sure the seal is flat and not twisted. Install the valve locks and release the compressor tool, making sure the locks seat properly in the upper groove of the valve stem.

93

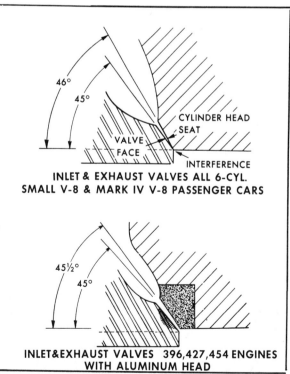

Relation of valve and seat angles

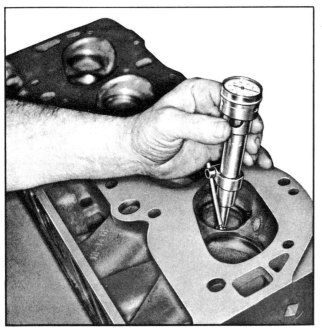

Measuring valve seat concentricity

the seal. A vacuum cup can be made from a small syringe and a high voltage shield. Check the installed height of the valve springs, using a narrow thin scale. Measure from the top of the shim or the spring seat to the top of the valve spring or valve spring sheild. If this is found to exceed the specified height, install a valve spring seat shim approximately 1/16" thick. At no time should the spring be shimmed to give an installed height under the minimum specified.

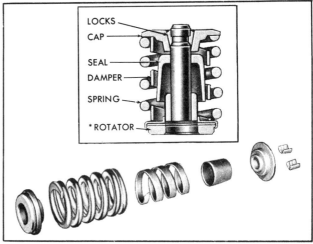

Exhaust valve spring installation ("Mark IV V8")

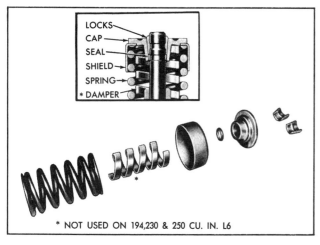

Valve spring installation ("small V8")

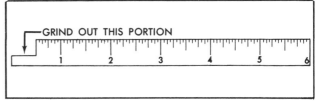

Cutaway scale

Mark IV

Install valve spring shim and/or exhaust valve rotator on valve spring seat then install a new valve stem oil seal over valve and valve guide. Set the valve spring (with damper), and valve cap in place. Compress the spring with Tool J-8062. Install the valve locks and release the compressor tool, making sure the locks seat properly in the groove of the valve stem. Grease may be used to hold the locks in place while releasing the compressor tool. Install the remaining valves. On "Small V8" engines check each valve stem oil seal by placing a vacuum cup or similar device over the end of the valve stem and against the cap. Operate the vacuum cup and make sure no air leaks past

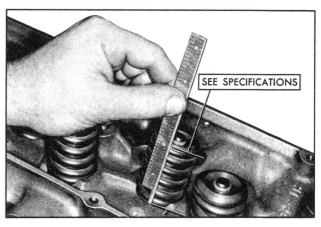

Measuring valve spring installed height

HYDRAULIC VALVE LIFTERS

Description

Two types of hydraulic lifters are used. Both types of lifters operate on the same principle and are serviced basically in the same manner. The complete lifter assemblies are interchangeable but parts from one lifter are not interchangeable with another. Both lifters are easily identified by the outside configuration of the lifter body. For purposes of identification we refer to them as lifter "A" and lifter "B".

Disassembly

Hold the plunger down with a push rod, and using the blade of a small screw driver, remove the push rod seat retainer. Remove the push rod seat and metering valve (lifter "A") or the push rod seat and inertia valve assembly (lifter "B"). Remove the plunger, ball check valve assembly and the plunger spring. Remove the ball check valve and spring by prying the ball retainer loose from the plunger with the blade of a small screw driver.

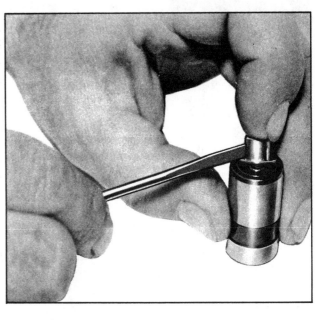

Removing ball check valve

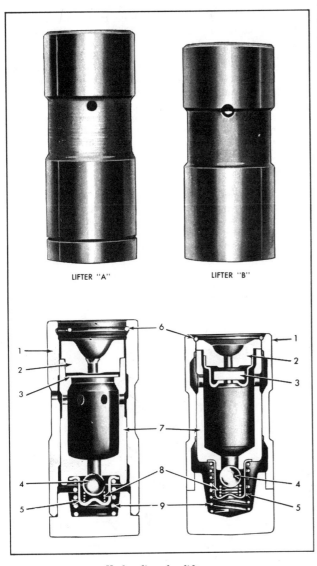

Hydraulic valve lifters

1. Lifter body	5. Check ball retainer
2. Push rod seat	6. Push rod seat retainer
3. Metering valve (lifter A)	7. Plunger
inertia valve (lifter B)	8. Check ball spring
4. Check ball	9. Plunger spring

Cleaning and inspection

Thoroughly clean all parts in cleaning solvent, and inspect them carefully. If any parts are damaged or worn the entire lifter assembly should be replaced. If the lifter body wall is scuffed or worn, inspect the cylinder block lifter bore, if the bottom of the lifter is scuffed or worn inspect the camshaft lobe, if the push rod seat is scuffed or worn inspect the push rod. Inertia valve and retainer (lifter "B") should not be removed from the push rod seat. To check the valve, shake the push rod seat and inertia valve assembly and the valve should move.

Assembly

Place the check ball on small hole in bottom of the plunger. Insert check ball spring on seat in ball retainer and place retainer over ball so that spring rests on the ball. Carefully press the retainer into position in plunger with the blade of a small screw driver. Place the plunger spring over the ball retainer and slide the lifter body over the spring and plunger, being careful to line up the oil feed holes in the lifter body and plunger. Fill the assembly with SAE 10 oil, then insert the end of a 1/8" drift pin into the plunger and press down solid. At this point oil holes in the lifter body and plunger assembly will be aligned. Do not attempt to force or pump the plunger. Insert a 1/16" drift pin through both oil holes to hold the plunger down against the lifter spring tension. On lifter "B" the drift pin must not extend inside the plunger. Remove the 1/8" drift pin, refill assembly with SAE 10 oil. Install the metering valve and push rod seat (lifter "A") or the push rod seat and inertia valve assembly (lifter "B").

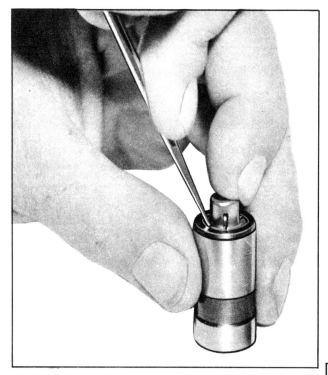

Installing ball check valve

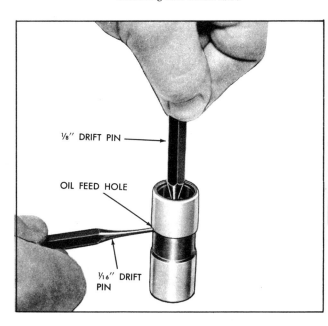

Assembling hydraulic lifter

Install the push rod seat retainer, press down on the push rod seat and remove the 1/16" drift pin from the oil holes. The lifter is now completely assembled, filled with oil and ready for installation. Before installing lifters, coat the bottom of the lifter with "Molykote" or its equivalent.

MECHANICAL VALVE LIFTERS
Cleaning and inspection
The mechanical valve lifter should never be disassembled. Wash lifter assembly in cleaning solvent and dry with compressed air. Blow out oil holes in lifter body and push rod seat with compressed air. Inspect push rod seat retainer. Inspect lifter for scuffed or worn lifter body or push rod seat. If the lifter body wall is scuffed or worn inspect the cylinder block lifter bore. If the bottom of the lifter is scuffed or worn inspect the camshaft lobe, if the push rod seat is scuffed or worn inspect the push rod. When using new lifters on used camshaft bottom of lifter should be polished with very fine emery cloth (such as No. 600 wet/dry) to reduce height of grinding marks. Additive containing EP lube such as EOS should always be added to crankcase oil for run-in when any new camshaft or lifters are installed. All damaged or worn lifters should be replaced. Before installing lifter, coat the bottom of the lifter with "Molykote" or its equivalent.

OIL PUMP
Description
The oil pump consists of two gears and a pressure regulator valve enclosed in a two-piece housing. The oil pump is driven by the distributor shaft which is driven by a helical gear on the camshaft. A baffle is incorporated on the pickup screen to eliminate pressure loss due to sudden or surging stops.

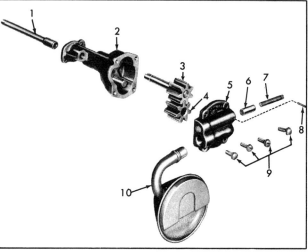

Oil pump ("small V8")

1. Shaft extension
2. Pump body
3. Drive gear and shaft
4. Idler gear
5. Pump cover
6. Pressure regulator valve
7. Pressure regulator spring
8. Retaining pin
9. Screws
10. Pickup screen and pipe

96

Removal

Remove oil pan as outlined. Remove pump to rear main bearing cap bolt and remove pump and extension shaft.

Installation

Assemble pump and extension shaft to rear main bearing cap, aligning slot on top end of extension shaft with drive tang on lower end of distributor drive shaft. Install pump to rear bearing cap bolt and torque to specifications. Installed position of oil pump screen is with bottom edge parallel to oil pan rails. Install oil pan as outlined.

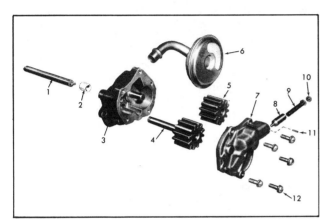

Oil pump (396 and 427 V8)

1. Shaft extension	7. Pump cover
2. Shaft coupling	8. Pressure regulator valve
3. Pump body	9. Pressure regulator spring
4. Drive gear and shaft	10. Washer
5. Idler gear	11. Retaining pin
6. Pickup screen and pipe	12. Screws

Dissassembly

Remove the pump cover attaching screws and the pump cover. Mark gear teeth so they may be reassembled with the same teeth indexing.

Remove the idler gear and the drive gear and shaft from the pump body. Remove the pressure regulator valve retaining pin, pressure regulator valve and related parts. If the pickup screen and pipe assembly need replacing, mount the pump in a soft-jawed vise and extract pipe from pump. Do not disturb the pickup screen on the pipe. This is serviced as an assembly.

Cleaning and inspection

Wash all parts in cleaning solvent and dry with compressed air. Inspect the pump body and cover for cracks or excessive wear. Inspect pump gears for damage or excessive wear. Check the drive gear shaft for looseness in the pump body. Inspect inside of pump cover for wear that would permit oil to leak past the ends of the gears. Inspect the pickup screen and pipe assembly for damage to screen, pipe or relief grommet. Check the pressure regulator valve for fit. The pump gears and body are not serviced separately. If the pump gears or body are

damaged or worn, replacement of the entire oil pump assembly is necessary.

Assembly

If the pickup screen and pipe assembly was removed, it should be replaced with a new part. Loss of press fit condition could result in an air leak and loss of oil pressure. Mount the pump in a soft-jawed vise, apply sealer to end of pipe, and using Tool J-8369 for "Small V8" or J-22144 for "Mark IV V8" tap the pipe in place with a plastic hammer. Be careful of twisting, shearing or collapsing pipe while installing in pump.

Install the pressure regulator valve and related parts. Install the drive gear and shaft in the pump body. Install the idler gear in the pump body with the smooth side of gear towards pump cover opening. Install the pump cover and torque attaching screws to specifications. Turn drive shaft by hand to check for smooth operation.

Installing screen (327, 350 V8)

Installing screen ("Mark IV V8")

MAIN BEARINGS
Description
Main bearings are of the precision insert type and do not utilize shim for adjustment. If clearances are found to be excessive a new bearing, both upper and lower halves will be required. Bearings are available in standard size and .001", .002", .009", .010", .020" and .030" undersize.

Selective fitting of both rod and main bearing inserts is necessary in production in order to obtain close tolerances. For this reason you may find one half of a standard insert with one half of a .001 undersize insert which will decrease the clearance .0005 from using a full standard bearing.

When a production crankshaft cannot be precision fitted by this method, it is then ground .009" undersize *on main journals only*. A .009 undersize bearing and a .010 undersize bearing may be used for precision fitting in the same manner as previously described. Any engine fitted with a .009 undersize crankshaft will be identified by the following markings: ".009" will be stamped on the crankshaft counterweight forward of the center main journal, a figure "9" will be stamped on the block at the left front oil pan rail. If, for any reason, main bearing caps are replaced, shimming may be necessary. Laminated shims for each cap are available for service. Shim requirement will be determined by bearing clearance.

Inspection
In general, (except No. 1 bearings) the lower half of the bearing shows a greater wear and the most distress from fatigue. If upon inspection the lower half is suitable for use, it can be assumed that the upper half is also satisfactory. If the lower half shows evidence of wear or damage, both upper and lower halves should be replaced. Never replace one half without replacing the other half.

Measuring crankshaft end play

Checking clearance
To obtain the most accurate results with "plastigage", (or its equivalent) a wax-like plastic material which will compress evenly between the bearing and journal surfaces without damaging either surface, certain precautions should be observed. If the engine is out of the vehicle and upside down, the crankshaft will rest on the upper bearings and the total clearance can be measured between the lower bearing and journal. If the engine is to remain in the vehicle, the crankshaft should be supported both front and rear (damper and flywheel) to remove the clearance from the upper bearing. The total clearance can then be measured between the lower bearing and journal. To assure the proper seating of the crankshaft all bearing cap bolts should be at their specified torque. In addition, preparatory to checking fit of bearings, the surface of the crankshaft journal and bearing should be wiped clean of oil.

With the oil pan and oil pump removed, and starting with the rear main bearing, remove bearing cap and wipe oil from journal and bearing cap. Place a piece of gauging plastic the full width of the bearing (parallel to the crankshaft) on the journal. Do not rotate the crankshaft while the gauging plastic is between the bearing and journal.

Install the bearing cap and evenly torque the retaining bolts to specifications. Remove bearing cap. The flattened gauging plastic will be found adhering to either the bearing shell or journal.

On the edge of gauging plastic envelope there is a graduated scale which is correlated in thousandths of an inch. Without removing the gauging plastic, measure its compressed width (at the widest point) with the graduations on the gauging plastic envelope. Normally, main bearing journals wear evenly and are not out-of-round. However, if a bearing is being fitted to an out-of-round journal (.001" max.), be sure to fit to the maximum diameter of the journal: If the bearing is fitted to the minimum diameter and the journal is out-of-round .001", interference between the bearing and journal will result in rapid bearing failure. If the flattened gauging plastic tapers toward the middle or ends, there is a difference in clearance indicating taper, low spot or other irregularity of the bearing of journal. Be sure to measure the journal with a micrometer if the flattened gauging plastic indicates more than .001" difference.

If the bearing clearance is within specifications, the bearing insert is satisfactory. If the clearance is not within specifications, replace the insert. Always replace both upper and lower insert as a unit. If a new bearing cap is being installed and clearance is less than .001", inspect for burrs or nicks; if none are found then install shims as required.

A standard .001" or .002" undersize bearing may produce the proper clearance. If not, it will be necessary to regrind the crankshaft journal for use with the next undersize bearing. Proceed to the next bearing. After all bearings have been checked rotate the crankshaft to see that there is no excessive drag. Measure crankshaft end play (see specifications) by forcing the crankshaft to the extreme front position. Measure at the front end of the rear main bearing with a feeler gauge. Install a new rear main bearing oil seal in the cylinder block and main bearing cap.

Replacement with crankshaft removal

Remove and inspect the crankshaft. Remove the main bearings from the cylinder block and main bearing caps. Coat bearing surfaces of new, correct size, main bearings with oil and install in the cylinder block and main bearing caps. Install the crankshaft.

Replacement without crankshaft removal

With oil pan, oil pump and spark plugs removed, remove cap on main bearing requiring replacement and remove bearing from cap.

Install a main bearing removing and installing tool in oil hole in crankshaft journal. If such a tool is not available, a cotter pin may be bent as required to do the job. Rotate the crankshaft clockwise as viewed from the front of engine. This will roll upper bearing out of block. Oil new selected size upper bearing and insert plain (unnotched) end between crankshaft and indented or notched side of block. Rotate the bearing into place and remove tool from oil hole in crankshaft journal.

Oil new lower bearing and install in bearing cap. Install main bearing cap with arrows pointing toward front of engine. Torque main bearing cap bolts to specifications.

CONNECTING ROD BEARINGS

Description

Connecting rod bearings are of the precision insert type and do not utilize shims for adjustment. DO NOT FILE RODS OR ROD CAPS. If clearances are found to be excessive a new bearing will be required. Bearings are available in standard size and .001" and .002" undersize for use with new and used standard size crankshafts, and in .010" and .020" undersize for use with reconditioned crankshafts.

Inspection and replacement

With oil pan and oil pump removed, remove the connecting rod cap and bearing. Inspect the bearing for evidence of wear or damage. (Bearings showing the above should not be installed). Wipe the bearings and crankpin clean of oil. Measure the crankpin for out-of-round or taper with a micrometer. If not within specifications replace or recondition the crankshaft. If within specifications and a new bearing is to be installed, measure the maximum diameter of the crankpin to determine new bearing size required.

If within specifications measure new or used bearing clearances with Plastigage or its equivalent. If a bearing is being fitted to an out-of-round crankpin, be sure to fit to the maximum diameter of the crankpin. If the bearing is fitted to the minimum diameter and the crankpin is out-of-round .001", interference between the bearing and crankpin will result in rapid bearing failure. Place a piece of gauging plastic the full width of the crankpin (parallel to the crankshaft). Install the bearing in the connecting rod and cap. Install the bearing cap and evenly torque nuts to specifications. Do not turn the crankshaft with the gauging plastic installed. Remove the bearing cap and using the scale on the gauging plastic envelope, measure the gauging plastic width at the widest point. If the clearance exceeds specifications select a new, correct size, bearing and remeasure the clearance. Coat the bearing

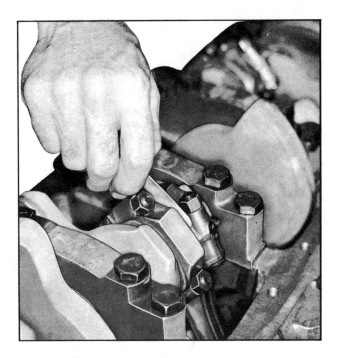

Measuring connecting rod side clearance

surface with oil, install the rod cap and torque nuts to specifications. When all connecting rod bearings have been installed tap each rod lightly (parallel to the crankpin) to make sure they have clearance. Measure all connecting rod side clearances (see specifications), between connecting rod caps on V8 engines.

CONNECTING ROD AND PISTON ASSEMBLIES

Removal

With oil pan, oil pump and cylinder head removed, use a ridge reamer to remove any ridge and/or deposits from the upper end of the cylinder bore. Before ridge and/or deposits are removed, turn crankshaft until piston is at the bottom of stroke and place a cloth on top of piston to collect the cuttings. After ridge and/or deposits are removed, turn crankshaft until piston is at top of stroke and remove cloth and cuttings.

Inspect connecting rods and connecting rod caps for cylinder identification. If necessary mark them. Remove connecting rod cap and install Tool J-5239 (3/8") or J-6305 (11/32") on studs. Push connecting rod and piston assembly out of top of cylinder block. It will be necessary to turn the crankshaft slightly to disconnect some of the connecting rod and piston assemblies and push them out of the cylinder.

Disassembly

Remove connecting rod bearings from connecting rods and caps. If connecting rod bearings are being reused, place them in a rack so they may be reinstalled in their original rod and cap.

Remove piston rings by expanding and sliding them off the pistons. Tools J-8020 (3-9/16"), J-8021 (3-7/8"),

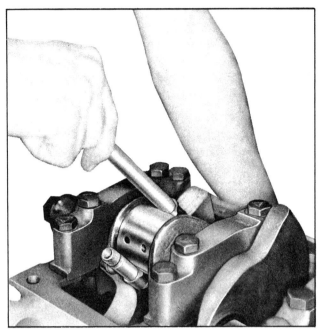

Removing connecting rod and piston assemblies

J-8032 (4"), J-22249 (3-15/16"), J-22147 (4-3/32"), and J-22250 (4-1/4") are available for this purpose.

Using Tool J-9510 for "Small V8" or Tool J-6994 for "Mark IV V8", place connecting rod and piston assembly in an arbor press with piston on Support then using Remover, press piston pin out of connecting rod and piston. Remove assembly from arbor press and remove tools, piston pin, connecting rod and piston.

Cleaning and inspection
Wash connecting rods in cleaning solvent and dry with compressed air. Check for twisted or bent rods and inspect for nicks or cracks. Replace connecting rods that are damaged.

Clean varnish from piston skirts and pins with a cleaning solvent. DO NOT WIRE BRUSH ANY PART OF THE PISTON. Clean the ring grooves with a groove cleaner and make sure oil ring holes and slots are clean. Inspect the piston for cracked ring lands, skirts or pin bosses, wavy or worn ring lands, scuffed or damaged skirts, eroded areas at top of the piston. Replace pistons that are damaged or show signs of excessive wear. Inspect the grooves for nicks or burrs that might cause the rings to hang up.

The piston pin clearance is designed to maintain adequate clearance under all engine operating conditions. Because of this, the piston and piston pin are a matched set and not serviced separately. Inspect piston pin bores and piston pins for wear. Piston pin bores and piston pins must be free of varnish or scuffing when being measured. The piston pin should be measured with a micrometer and the piston pin bore should be measured with a dial bore gauge or an inside micrometer. If clearance is in excess of the .001" wear limit, the piston and piston pin assembly should be replaced.

Assembly
Lubricate piston pin holes in piston and connecting rod to facilitate installation of pin. Using Tool J-9510 for "Small VE" or Tool J-6994 for "Mark IV V8" place Support with spring and pilot in place on an arbor press. Position piston on connecting rod with appropriate side of piston and connecting rod bearing tangs aligned. Place piston on support, indexing pilot through piston and rod. Place Installer on piston pin, start piston pin into piston and press on installer until pilot bottoms in support. Remove installer from connecting rod and piston assembly and check piston for freedom of movement on piston pin.

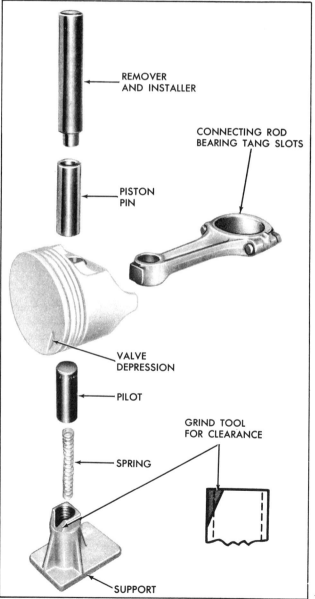

Piston pin and tool layout ("Mark IV V8")

ring into the respective piston ring groove and roll the ring entirely around the groove to make sure that the ring is free. If binding occurs at any point the cause should be determined, and if caused by ring groove, remove by dressing with a fine cut file. If the binding is caused by a distorted ring, check a new ring.

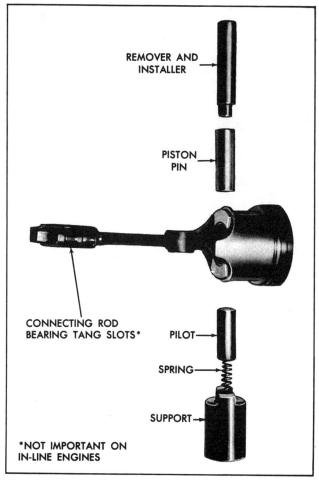

REMOVER AND INSTALLER

PISTON PIN

CONNECTING ROD BEARING TANG SLOTS*

PILOT

SPRING

SUPPORT

*NOT IMPORTANT ON IN-LINE ENGINES

Piston pin and tool layout (in-line and "small V8")

Measuring ring gap

PISTON RINGS

All compression rings are marked on the upper side of the ring. When installing compression rings, make sure the marked side is toward the top of the piston. The top ring is chrome faced, or treated with molybdenum for maximum life.

The oil control rings (except for MK-IV V8) are of the three piece type, consisting of two segments (rails) and a spacer. The MK-IV V8 engines use a chrome plated cast iron ring and an expander.

Select rings comparable in size to the piston being used. Slip the compression ring in the cylinder bore; then press the ring down into the cylinder bore about ¼ inch (above ring travel). Be sure ring is square with cylinder wall. Measure the space or gap between the ends of the ring with a feeler gauge. If the gap between the ends of the ring is below specifications, remove the ring and try another for fit. Fit each compression ring to the cylinder in which it is going to be used. If the pistons have not been cleaned and inspected as previously outlined, do so. Slip the outer surface of the top and second compression

101

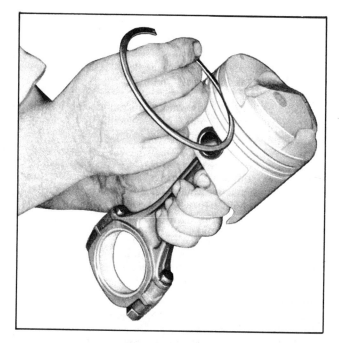

Checking ring in groove

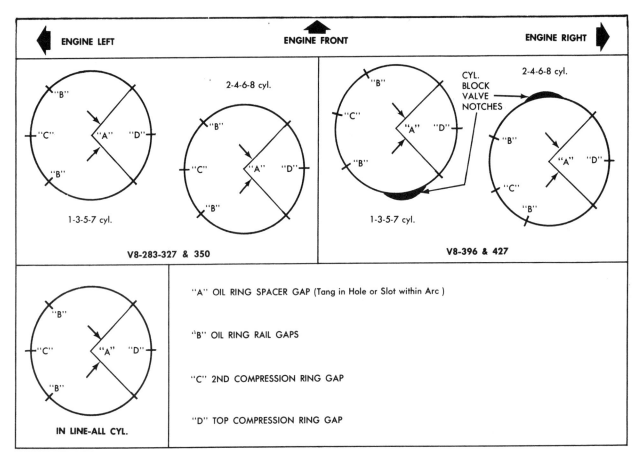

ENGINE LEFT   ENGINE FRONT   ENGINE RIGHT

2-4-6-8 cyl.

"B"

"C"   "A"   "D"

"B"

1-3-5-7 cyl.

"B"

"C"   "A"   "D"

"B"

V8-283-327 & 350

2-4-6-8 cyl.

CYL.
BLOCK
VALVE
NOTCHES

"B"

"C"   "A"   "D"

"B"

"B"

"A"   "D"

"C"

"B"

1-3-5-7 cyl.

V8-396 & 427

"A" OIL RING SPACER GAP (Tang in Hole or Slot within Arc)

"B"

"C"   "A"   "D"

"B"

IN LINE-ALL CYL.

"B" OIL RING RAIL GAPS

"C" 2ND COMPRESSION RING GAP

"D" TOP COMPRESSION RING GAP

Ring gap location

Install piston rings as follows: Tools J-8020 (3-9/16"), J-8021 (3-7/8"), J-8032 (4"), J-22249 (3-15/16"), J-22147 (4-3/32"), and J-22250 (4¼") are available for this purpose. Install oil ring spacer in groove and, except on 250 cu. in. engine, insert anti-rotation tang in oil hole. Hold spacer ends butted and install lower steel oil ring rail with gap properly located. Install upper steel oil ring rail with gap properly located. Flex the oil ring assembly to make sure ring is free. If binding occurs at any point the cause should be determined, and if caused by ring groove, remove by dressing groove with a fine cut file. If binding is caused by a distroted ring, check a new ring. Install second compression ring expander then ring with gaps properly located. Install top compression ring with gap properly located.

Proper clearance of the piston ring in its piston ring groove is very important to provide proper ring action and reduce wear. Therefore, when fitting new rings, the clearances between the surfaces of the ring and groove should be measured. (See specifications.)

Installation
Cylinder bores must be clean before piston installation. This may be accomplished with a hot water and detergent wash or with a light honing as necessary. After cleaning, the bores should be swabbed several times with light

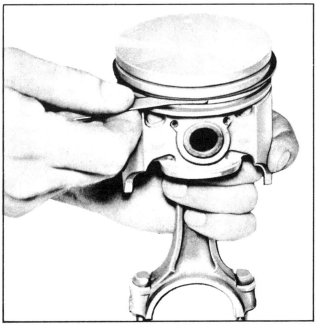

102

Measuring ring groove clearance

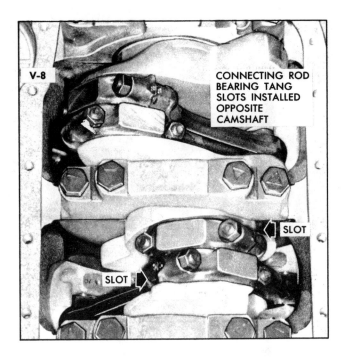

Connecting rods-installed position

engine oil and a clean dry cloth. Lubricate connecting rod bearings and install in rods and rod caps. Lightly coat pistons, rings and cylinder walls with light engine oil. With bearing caps removed, install Tool J-5239 (3/8") or J-6305 (11/32") on connecting rod bolts. Be sure ring gaps are properly positioned as previously outlined.

Install each connecting rod and piston assembly in its respective bore. Install with connecting rod bearing tang slots on side opposite camshaft. Use Tool J-8037 to compress the rings. Guide the connecting rod into place on the crankshaft journal with Tool J-5239 (3/8") or J-6305 (11/32"). Use a hammer handle and light blows to install the piston into the bore. Hold the ring compressor firmly against the cylinder block until all piston rings have entered the cylinder bore. Remove Tool J-5239 or J-6305. Install the bearing caps and torque nuts to specification. If bearing replacement is required refer to "Connecting Rod Bearings". Be sure to install new pistons in the same cylinders for which they were fitted, and used pistons in the same cylinder from which they were removed. Each connecting rod and bearing cap should be marked, beginning at the front of the engine. 1, 3, 5, and 7 in the left bank and, 2, 4, 6, and 8 in the right bank. The numbers on the connecting rod and bearing cap must be on the same side when installed in the cylinder bore. If a connecting rod is ever transposed from one block or cylinder to another, new bearings should be fitted and the connecting rod should be numbered to correspond with the new cylinder number.

CRANKSHAFT

103   **Removal**

The crankshaft can be removed while the engine is disassembled for overhaul, as previously outlined or without complete disassembly.

With the engine removed from the vehicle and the transmission and/or clutch housing removed from the engine, mount engine in overhaul stand and clamp securely. Remove the oil dip stick and oil dip stick tube, (if applicable). Remove the starting motor, clutch assembly (if equipped) and flywheel. Remove the spark plugs. Remove crankshaft pulley and torsional damper. Remove oil pan and oil pump. Remove crankcase front cover, and if so equipped timing chain and camshaft sprocket.

Check the connecting rod caps for cylinder number identification. If necessary mark them. Remove the connecting rod caps and push the pistons to top of bores. Remove main bearing caps and lift crankshaft out of cylinder block. Remove rear main bearing oil seal and main bearings from cylinder block and main bearing caps.

Cleaning and Inspection

Wash crankshaft in solvent and dry with compressed air. Measure dimensions of main bearing journals and crankpins with a micrometer for out-of-round, taper or undersize (see specifications). Check crankshaft for run-out by supporting at the front and rear main bearings journals in "V" blocks and check at the front and rear intermediate journals with a dial indicator (see specifications). Replace or recondition the crankshaft if out of specifications.

SPROCKET OR GEAR
Replacement

On "Small V8" engines, remove crankshaft sprocket using Tool J-5825, install using Tool J-5590. On Mark IV V8 engines, remove crankshaft sprocket using Tool J-1619, install using Tool J-21058.

Installation

For installation of crankshaft, refer to the applicable steps under "Engine Assembly" previously outlined.

CAMSHAFT
Inspection

The camshaft bearing journals should be measured with a micrometer for an out-of-round condition. If the journals exceed .001" out-of-round, the camshaft should be replaced. The camshaft should also be checked for alignment. The best method is by use of "V" blocks and a dial indicator. The dial indicator will indicate the exact amount the camshaft is out of true. If it is out more than .002" dial indicator reading, the camshaft should be replaced.

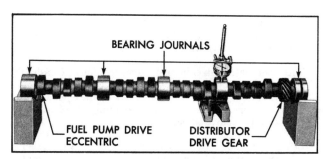

Checking camshaft alignment

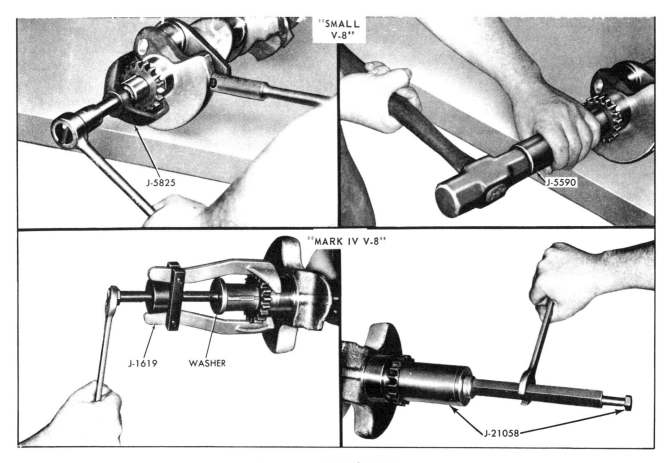

Sprocket or gear replacement

CAMSHAFT BEARINGS

Inspection

With the camshaft removed, inspect the bearings for evidence of wear or damage. (Bearings showing the above should be replaced).

Removal

Camshaft bearings can be replaced while engine is disassembled for overhaul, or without complete disassembly of the engine. To replace bearings without complete disassembly remove the camshaft and crankshaft leaving cylinder heads attached and pistons in place. Before removing crankshaft, tape threads of connecting rod bolts to prevent damage to crankshaft. Fasten connecting rods against sides of engine so they will not be in the way while replacing camshaft bearings.

With camshaft and crankshaft removed, drive camshaft rear plug from cylinder block. This procedure is based on removal of the bearings nearest center of the engine first. With this method a minimum amount of turns are necessary to remove all bearings.

Using Tool Set J-6098, with nut then thrust washer installed to end of threads, index pilot in camshaft front bearing and install puller screw through pilot. Install

remover and installer tool with shoulder toward bearing, making sure a sufficient amount of threads are engaged. Using two wrenches, hold puller screw while turning nut. When bearing has been pulled from bore, remove remover and installer tool and bearing from puller screw. Remove remaining bearings (except front and rear) in the same manner. It will be necessary to index pilot in camshaft rear bearing to remove the rear intermediate bearing. Assemble remover and installer tool on driver handle and remove camshaft front and rear bearings by driving towards center of cylinder block.

Installation

The camshaft front and rear bearings should be installed first. These bearings will act as guides for the pilot and center the remaining bearings being pulled into place. Assemble remover and installer tool on driver handle and install camshaft front and rear bearings by driving towards center of cylinder block. Oil holes in bearings must line up with oil holes in cylinder block.

Using Tool Set J-6098, with nut then thrust washer installed to end of threads, index pilot in camshaft front bearing and install puller screw through pilot. Index camshaft bearing in bore, then install remover and installer tool on puller screw with shoulder toward bearing.

104

Replacing camshaft center bearing

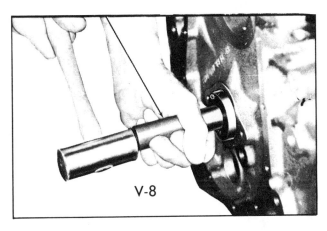

V-8

Replacing front camshaft bearing

Using two wrenches, hold puller screw while turning nut. After bearing has been pulled into bore, remove the remover and installer tool from puller screw and check alignment of oil hole in camshaft bearing. Install remaining bearings in the same manner. It will be necessary to index pilot in the camshaft rear bearing to install the rear intermediate bearing. Install a new camshaft rear plug. Plug should be installed flush to 1/32" deep and be parallel with rear surface of cylinder block.

CYLINDER BLOCK

Cleaning and inspection
Wash cylinder block thoroughly in cleaning solvent and clean all gasket surfaces. Remove oil gallery plugs and clean all oil passages. These plugs may be removed with a sharp punch or they may be drilled and pried out. Clean and inspect water passages in the cylinder block. Inspect the cylinder block for cracks in the cylinder walls, water jacket valve lifter bores and main bearing webs.

105

Measure the cylinder walls for taper, out-of-round or excessive ridge at top of ring travel. This should be done with a dial indicator. Set the gauge so that the thrust pin must be forced in about ¼" to enter gauge in cylinder bore. Center gauge in cylinder and turn dial to "O". Carefully work gauge up and down cylinder to determine taper and turn it to different points around cylinder wall to determine the out-of-round condition. If cylinders were found to exceed specifications, honing or boring will be necessary.

Measuring cylinder bore

OIL FILTER BYPASS VALVE

Inspection and replacement
With the oil filter removed, check the spring and fibre valve for operation. Inspect for a cracked or broken valve. If replacement is necessary, the oil filter adapter and bypass valve assembly must be replaced as an assembly. Clean valve chamber in cylinder block thoroughly. Torque retaining screws to specifications.

Piston selection.
Check used piston to cylinder bore clearance. Measure the "Cylinder Bore diameter" with a telescope gauge (2½" from top of cylinder bore). Measure the piston diameter (at skirt across center line of piston pin).

Subtract piston diameter from cylinder bore diameter to determine piston to bore clearance. Locate piston to bore clearance of piston selection chart and determine if piston to bore clearance is in the acceptable range.

If used piston is not acceptable, check and determine if a new piston can be selected to fit cylinder bore within the acceptable range. If cylinder bore must be reconditioned, measure new piston diameter (across center line of piston pin) then hone cylinder bore to correct clearance (preferable range). Mark the piston to identify the cylinder for which it was fitted.

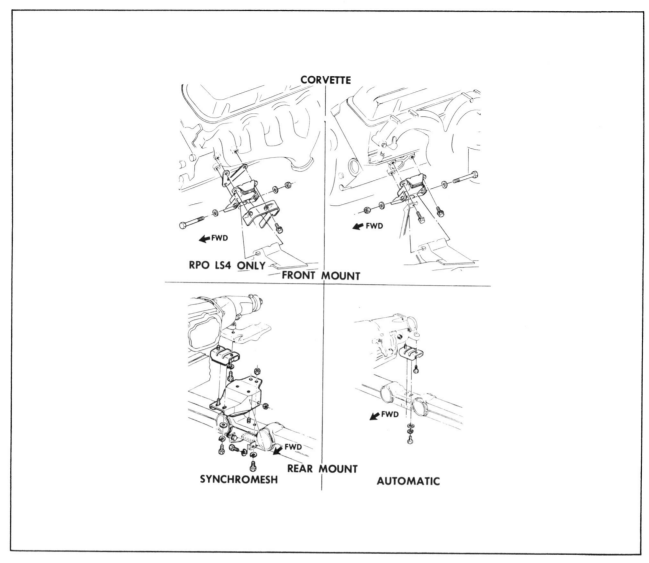

Engine mounts

CLUTCH 7

INDEX

DESCRIPTION

The diaphragm spring type clutch assemblies used with the different engine-transmission combinations consist of single plate units with two designs of diaphragm springs, flat and bent finger, and a dual plate bent finger spring type clutch. The latter unit is an optional heavy duty clutch assembly used with Mark IV engines having a 14 inch ring gear diameter.

The integral release fingers in the bent finger design are bent back to gain a centrifugal boost and to insure quick reengagement at high engine speeds. This type of clutch has the advantages of increasing pressure plate load as the driven plate wears, and of low pedal effort with high plate load without requiring over center booster springs on the clutch linkage.

The clutch release bearing used with the bent finger design has an overall length of 1-1/4" and is shorter than the release bearing used with the flat finger design clutch. Do not interchange the two bearings. The longer bearing, if used with the bent finger spring clutch, will cause inability to obtain free pedal travel, especially as the clutch wears, resulting in slippage and rapid wear.

Linkage inspection

Check the clutch linkage to be sure the clutch releases fully. With engine running, hold the clutch pedal approximately 1/2" from floor mat and move shift lever between first and reverse several times. If this can be done smoothly, the clutch is fully releasing. If shift is not smooth, clutch is not fully releasing and adjustment is necessary.

Check clutch pedal bushings for sticking or excessive wear. Check fork for proper installation on ball stud. Lack of lubrication on fork can cause fork to be pulled off the ball. Check for bent, cracked or damaged cross shaft levers or support bracket. Loose or damaged engine mounts may allow the engine to shift its position causing a bind on clutch linkage at the cross shaft. Check to be sure there is some clearance between cross shaft and both mount-bracket. Check throw out bearing end clearance between spring fingers and front bearing retainer on the transmission. If no clearance exists, fork may be improperly installed on ball stud or clutch disc may be worn out.

Adjustment

There is one linkage adjustment (clutch fork push rod or pedal push rod) to compensate for all normal clutch wear. The clutch pedal should have free travel (measured at clutch pedal pad) before the throwout bearing engages the clutch diaphragm spring levers. Free travel is required to prevent clutch slippage which would occur if the bearing was held against the fingers or to prevent the bearing from running continually until failure. A clutch that has been slipping prior to free play adjustment may still slip right after the new adjustment due to previous heat damage. The car should be allowed to cool at least 12 hours to give clutch time to cool to normal temperatures. Any slippage should then be evaluated.

Disconnect spring between clutch push rod and cross shaft lever. With clutch pedal against stop: loosen jam nuts sufficiently to allow the adjusting rod to move against the

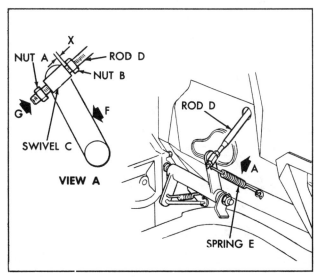

Clutch pedal free travel adjustment

clutch fork until the release bearing contacts the pressure plate fingers lightly. Rotate upper nut against swivel and back of 4-1/2 turns. Tighten lower nut to lock swivel against nut. Install return spring and check clutch pedal free travel. Pedal free travel should be Corvette (STD) – 1-1/4" to 2"; Corvette (HVY DTY) – 2" to 2-1/2".

Removal

Support engine and remove transmission as outlined in transmission section. Disconnect clutch fork push rod and spring. Remove flywheel housing. Slide clutch fork from ball stud and remove fork from dust boot. Ball stud is threaded into clutch housing and is easily replaced, if necessary. Install Tool J-5824 to support the clutch assembly during removal. Look for "X" mark on flywheel and on clutch cover. If "X" mark is not evident, prick punch marks on flywheel and clutch cover for indexing purposes during installation. Loosen the clutch-to-flywheel attaching bolts evenly 1 turn at a time until spring pressure is released, then remove the bolts, and remove clutch assembly.

Installation

Clean pressure plate and flywheel face. (They should be free of oil, grease, metal deposits or burned spots). Position the clutch disc and pressure plate in relative installed position and support them with alignment Tool J-5824. The driven disc is installed with damper springs to the transmission side. The grease slinger is always on the transmission side.

Turn clutch assembly until "X" mark on cover lines up with "X" mark on flywheel, then align cover bolt holes to nearest flywheel holes. Install a bolt in every hole and

tighten down evenly and gradually until tight (to avoid possible clutch distortion). Cover loads are as high as 1-1/4 tons. Remove pilot tool. Unhook clutch fork and lubricate ball socket and fork fingers at release bearing end with a high melting point grease and reinstall fork on ball stud. Lubricate the recess on the inside of throwout bearing collar and the throwout fork groove with a light coat of graphite grease (Moly Grease). Install clutch fork and dust boot into clutch housing and install throwout bearing to the throwout fork, then install flywheel housing. Install transmission. Connect fork push rod and spring. Lubricate spring and push rod ends. Adjust shift linkage. Perform linkage adjustment for pedal free play and check clutch release position.

CLUTCH PEDAL
Description

The clutch pedal is the pendant-type hung from a support brace common to the brake pedal and must be removed to remove brake pedal.

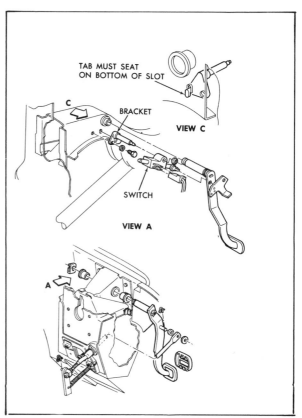

Clutch pedal linkage composite

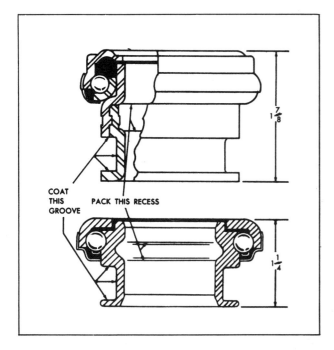

Release bearing lubrication and comparison

CROSS SHAFT
Removal

Remove linkage return and lower linkage springs and disconnect clutch pedal and fork push rods from respective cross shaft levers. Loosen outboard ball stud nut and slide stud out of bracket slot. Move cross shaft outboard, and as required to clear inboard ball stud, then merely lift out to remove from vehicle. The cross shaft has nylon ball stud seats which should be inspected for wear or damage. Also check condition of engine bracket ball stud assembly and special anti-rattle "O" ring. Figures show component parts of cross shaft. Replace parts as necessary based on wear or damage. Lubricate ball studs and seats with graphite grease before reassembly.

Installation

Reverse removal procedure to install. Adjust clutch linkage as previously outlined.

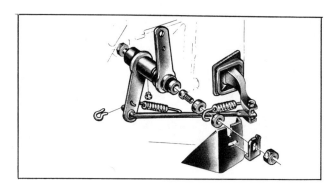

lower linkage details

TRANSMISSION, MANUAL 8

INDEX

THREE-SPEED (SAGINAW) TRANSMISSIONS
Description

The Saginaw three-speed manual transmission is fully synchronized in all forward gears and incorporates helical drive gears throughout. The main drive gear is supported by a ball bearing at the front end of the transmission and is piloted at the front end in an oil impregnated bushing mounted in the engine crankshaft. The front end of the mainshaft is piloted in a row of roller bearings set into the hollow end of the main drive gear and is carried at the rear end by a bearing mounted in the front of the extension housing.

The countergear is carried on rollers at both ends and thrust is taken on thrust washers located between the ends of the gear and the thrust bosses in the case. An anti-rattle plate assembly at the front of the countergear provides a constant spring tension between the counter and clutch gears to reduce torsional vibrations. The reverse idler gear is carried on a bushing finish bored in place and thrust is taken on the thrust bosses of the case.

Gear shifting is manual through shift control rods from the shifter tube in the mast jacket to the rearward shift lever of the side cover assembly for first and reverse gear; and through a cross shaft assembly attached to the forward side cover lever for second and third gear. All three forward gears are fully synchronized. The synchronizer assemblies consist of a clutch hub, clutch sleeve, two clutch key springs and three energizer clutch keys and are retained as an assembly on the main shaft by a snap ring. The transmission may be used as an aid in deceleration by downshifting in sequence without double-clutching or any gear clashing. Reverse is not synchronized, however, it is a helical gear to insure quiet operation.

Linkage Adjustment

Turn ignition switch to the "LOCK" position and raise vehicle on hoist. Loosen swivel lock nuts on both shift rods. Rods should pass freely through swivels. Disconnect back drive cable from column lock tube lever. Place shift lever control (in car) into Neutral. Insert locking gauge (.664 in.) in notch of lever and bracket assembly. Place transmission levers in the neutral position. Hold 1st /REVERSE rod and lever against locating gauge, tighten

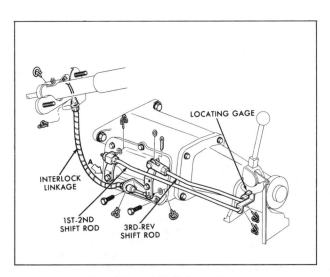

Speed floor shift linkage adjustment

locknut against swivel. Hold 2nd/3rd rod and lever against locating gauge, tighten the forward nut against the swivel and then the aft. Remove locating gauge. Check transmission shifting to ensure proper operation. Readjust linkage, if required. *Shift lever must be properly centered in console to prevent fore and aft and side contact with console. Shim lever support bracket as required but be sure to maintain support to extension clearance.*

From inside vehicle, loosen two nuts at the steering column to dash panel bracket. Place transmission shift lever in "Reverse". Rotate the lock tube lever counter-clockwise (viewed from front of column) to remove any free play. Reposition the cable bracket until the cable eye passes over the retaining pin on the bracket.

Hold the bracket in position and have an assistant tighten the steering column to dash panel bracket retaining nuts. Reinstall the cotter pin and washer retaining the back drive cable to the lever pin.

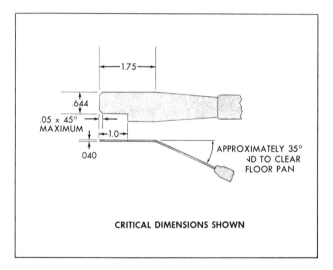

CRITICAL DIMENSIONS SHOWN

Locating gage

Back drive linkage adjustment

Place shift lever in reverse position and ignition switch in "LOCK" position. Raise vehicle on hoist. Remove the back drive cable cotter pin and washer at the column lock tube lever. Disconnect cable from the lever retaining pin. From inside the vehicle, loosen the two nuts at the steering column to dash panel bracket. Rotate the lock tube lever counter-clockwise (viewed from front of column) to remove any free play from the column. Reposition the cable bracket until the cable eye passes freely over the retaining pin on the bracket. Holding the bracket in position, have an assistant tighten the bracket retaining nuts inside the vehicle. Reinstall the cotter pin and washer retaining the cable to the lever retaining pin. Check interlock. Key must move freely to and from the "LOCK" position with the shift control in the "Reverse" position. Lower and remove vehicle from hoist.

Shift control lever and bracket assembly replacement

This procedure is for 3 or 4 speed lever assembly. Remove shift control lever knob inside vehicle. Remove console trim plate and seat assembly. Raise and support vehicle on hoist. Disconnect transmission control rods from shift control levers. Remove shift control to support bracket retaining bolts. Remove support bracket to crossmember retaining bolts. Remove support assembly from vehicle, then pull down on control assembly to remove from console boot and vehicle.

To install, reverse removal procedure. *Shift lever must be positioned in console opening to prevent fore and aft and side contact with the console. Shim support bracket as required but be sure to maintain support to extension clearance.*

Speedometer driven gear replacement

Raise vehicle on hoist. Disconnect speedometer cable, remove lock plate to extension bolt and lock washer and remove lock plate. Insert screwdriver in lock plate slot in fitting and pry fitting gear and shaft from extension. Pry "O" ring from groove in fitting. Install new "O" ring in groove in fitting. Coat "O" ring and driven gear shaft with transmission lubricant and insert shaft. Hold the assembly so slot in fitting is toward lock plate boss on extension and install in extension. Push fitting into extension until lock plate can be inserted in groove and attach to extension. Lower and remove vehicle from hoist.

EXTENSION OIL SEAL
Replacement

Raise vehicle on hoist then remove propeller shaft and disconnect any necessary items to obtain clearance. Pry seal out of extension or use Tool J-5859 and J-2619. Wash counterbore with cleaning solvent and inspect for damage. Prelubricate between sealing lips and coat new seal O.D. with Permatex or equivalent and start straight in bore in case extension. Using Tool J-5154, tap seal into counterbore until flange bottoms against extension. Reinstall propeller shaft and any items removed to obtain clearance.

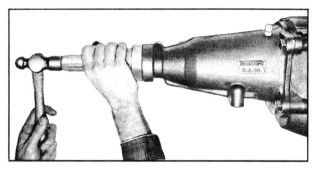

Installing extension oil seal using J-5154

TRANSMISSION SIDE COVER—SAGINAW
Removal

Raise vehicle on hoist. Disconnect TCS switch (where

112

used) and control rods from levers. Shift transmission into neutral detent positions before removing cover. Remove cover assembly from transmission case carefully and allow oil to drain.

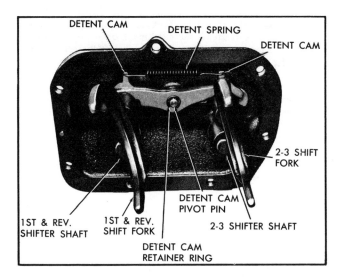

Transmission side cover – saginaw

Disassembly

Remove the outer shifter levers and TCS switch (where used) from side cover. Remove both shift forks from shifter shaft assemblies. Remove both shifter shaft assemblies from cover. Pry out 2-3 shifter shaft lip seal from side cover bore and remove "O" ring seal from 1st-rev. shaft if replacement is required.

Remove detent cam spring and pivot retainer "C" ring. Remove both detent cams. Replace damaged parts.

Assembly

With detent spring tang projecting up over the 2nd and 3rd shifter shaft cover opening install the first and reverse detent cam onto the detent cam pivot pin. With the detent spring tang projecting up over the first and reverse shifter shaft cover hole install the 2nd and 3rd detent cam. Install detent cam retaining "C" ring to pivot shaft, and hook spring into detent cam notches. Install both shifter shaft assemblies in cover being careful not to damage seals. Install both shift forks to shifter shaft assemblies, lifting up on detent cam to allow forks to fully seat into position. Install outer shifter levers, flat washers, lock washers and bolts. Install TCS switch (where used).

Installation

Shift shifter levers into neutral detent (center position.) Position cover gasket on case. Carefully position side cover into place making sure the shift forks are aligned with their respective mainshaft clutch sliding sleeves. Install cover attaching bolts and tighten evenly to specified torque. Connect TCS switch wiring. Remove filler

plug and add lubricant to level of filler plug hole. Lower vehicle and remove from hoist.

TRANSMISSION
Removal and installation

Disconnect battery ground cable. From inside vehicle, remove shifter ball. Remove console trim plate. Raise vehicle on hoist.

Completely remove right and left exhaust pipes from vehicle. In order to remove the exhaust pipes on vehicles equipped with MK IV engines, it is necessary to remove the forward stud on each manifold. Disconnect propshaft at transmission slip yoke. Lower front of propshaft. Remove slip yoke from transmission. Remove bolts attaching rear mount to rear mount bracket. Raise engine lifting transmission off mount bracket. Remove the bolts retaining transmission linkage mechanism mounting bracket to frame. Remove bolts attaching gearshift assembly to mounting brackets and remove mounting bracket. Remove shifter mechanism with rods attached. Disconnect shifter levers at transmission. Disconnect speedometer cable. Remove transmission mount bracket to crossmember bolts and remove mount bracket. Remove bolts retaining rear mount cushion and exhaust pipe yoke.

Remove transmission-to-clutch housing retaining bolts and extension bottom bolt. Pull transmission rearward and rotate clockwise until it is clear of clutch housing. To allow room for transmission removal slowly lower the rear of engine until the tachometer drive cable at the distributor just clears the horizontal ledge across the front of dash. The tachometer cable can be easily damaged by heavy contact with the dash. Slide transmission rearward out of the clutch, then tip front end of transmission downward and lower the assembly from vehicle. Reinstall transmission assembly by performing above steps in reverse order. Lower and remove vehicle from hoist.

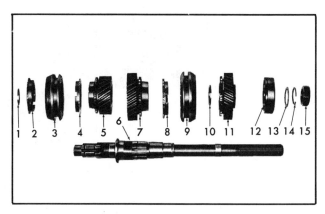

Mainshaft assembly – layout

1. Snap ring	9. 1st reverse clutch assembly
2. 3rd speed blocker ring	10. Snap ring
3. 2-3 clutch assembly	11. Reverse gear
4. 2nd speed blocker ring	12. Rear bearing
5. 2nd speed gear	13. Special washer
6. Mainshaft	14. Snap ring
7. 1st speed gear	15. Speedo drive gear
8. 1st speed blocker ring	

FOUR-SPEED (MUNCIE AND SAGINAW) TRANSMISSION

Description

The Muncie four-speed synchromesh transmission incorporates helical gears specially designed to provide high torque capacity without additional weight, and gear teeth proportioned to operate at high speeds with neither excessive heat generation nor excessive frictional losses. Shafts, bearings, high capacity clutches and other precision parts are held to close limits, providing proper clearances necessary for durability during extended heavy usage.

The main drive gear is supported by a heavy-duty bearing at the front end of the transmission case and is piloted at its front end in an oil impregnated bushing mounted in the engine crankshaft. The front end of the mainshaft is piloted in a row of roller bearings set into the hollow end of the main drive gear and the rear end is carried by a heavy-duty ball bearing mounted at the rear end of the transmission case in a retainer casing. The counter gear is carried on a double row of rollers at both ends while thrust is taken on thrust washers located between the ends of the gear and the thrust bosses in the case.

The two-piece reverse idler gear is carried on bronze bushings while thrust is taken on thrust washers located between the front of the gear and the back of the reverse idler thrust boss and between the rear of the gear and the reverse idler shaft boss in the case extension.

Gearshifting is manual through shift control rods to the transmission cover shifter levers for first through fourth gears, and to the reverse lever located in the case

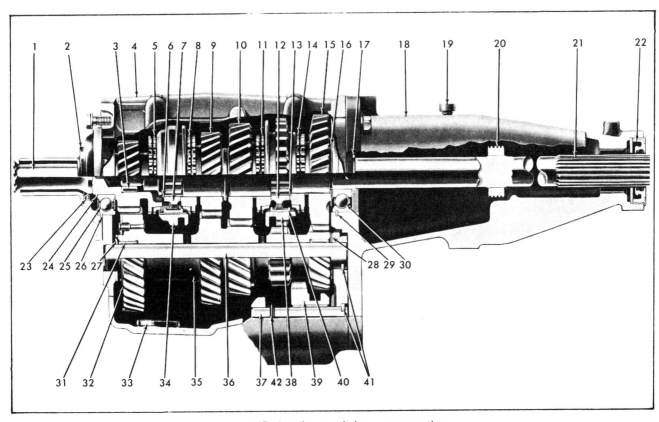

4-speed (Saginaw) transmission — cross section

1. Clutch gear	12. 1-2 speed synch. hub	22. Rear oil seal	33. Magnet
2. Bearing retainer	13. 1-2 speed synch. snap ring	23. Retainer oil seal	34. 4-3 synch. sleeve
3. Pilot bearings	14. 1st speed blocker ring	24. Snap ring – bearing to gear	35. Countergear assembly
4. Case	15. First gear	25. Clutch gear bearing	36. Counter shaft
5. 4th speed blocker ring	16. Reverse gear thrust and spring washers	26. Snap ring – bearing to case	37. Reverse idler shaft
6. 4-3 synch. snap ring		27. Thrust washer – front	38. 1-2 speed synch. sleeve and reverse gear
7. 4-3 synch. hub	17. Snap ring – bearing to mainshaft	28. Thrust washer – rear	
8. 3rd speed blocker ring	18. Extension	29. Snap ring – bearing to extension	39. Reverse idler gear (sliding)
9. 3rd speed gear	19. Vent	30. Rear bearing	40. Clutch key
10. 2nd speed gear	20. Speedometer drive gear	31. Countergear roller bearings	41. Woodruff key
11. 2nd speed blocker ring	21. Mainshaft	32. Anti-lash plate assembly	42. Ring – reverse idler gear stop

114

extension. The shifter lever to the rear of the transmission cover controls first and second gears while the lever to the front controls third and fourth gears. All four forward gears are fully synchronized. The transmission may be used as an aid in deceleration by down-shifting in sequence without double clutching. Reverse is not synchronized, however, it is a helical gear to insure quiet operation.

The Saginaw four-speed fully synchronized (all forward gears) transmission incorporates helical drive gears. The main drive gear is supported by a ball bearing at the front end of the transmission case and is piloted at its front end in an oil impregnated bushing mounted in the engine crankshaft. The front end of the mainshaft is piloted in a row of roller bearings set into the hollow end of the main drive gear and the rear end is carried by a ball bearing mounted in the front of the extension housing. The countergear is carried on a single row of rollers at both ends while thrust is taken on thrust washers located between the ends of the gear and the thrust bosses in the case. An anti-rattle plate assembly at the front of the countergear provides a constant spring tension between the counter and clutch gears to reduce torsional vibrations.

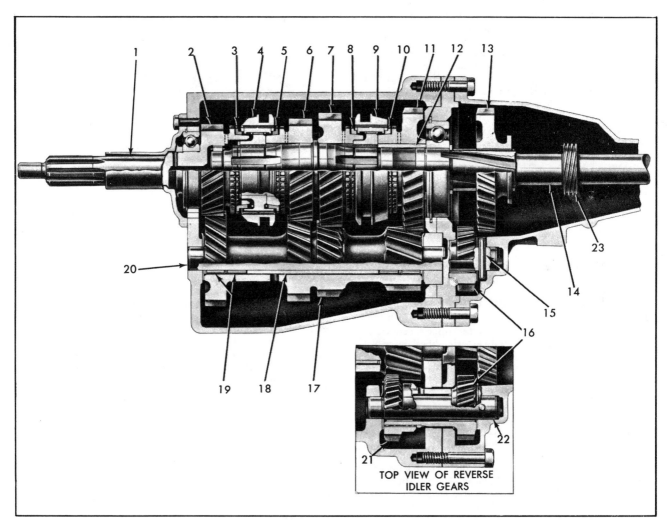

Four-speed transmission cross section (close ratio shown) (muncie)

1. Bearing retainer
2. Main drive gear
3. Fourth speed synchronizing ring
4. Third and fourth speed clutch assembly
5. Third speed synchronizing ring
6. Third speed gear
7. Second speed gear
8. Second speed synchronizing ring
9. First and second speed clutch assembly
10. First speed synchronizing ring
11. First speed gear
12. First speed gear sleeve
13. Reverse gear
14. Mainshaft
15. Reverse idler shaft roll pin
16. Reverse idler gear (rear)
17. Countergear
18. Countershaft bearing roller spacer
19. Countershaft bearing roller
20. Countershaft
21. Reverse idler gear (front)
22. Reverse idler shaft
23. Speedo drive gear

The reverse idler gear is held in constant mesh with the 1st and 2nd synchronizer gear regardless of its position. The idler gear is carried on a bushing finish bored in place and rotates on a short idler shaft retainer by a Woodruff key.

The synchronizer assemblies consist of a clutch hub, clutch sleeve, two clutch key springs and three energizer clutch keys and are retained as an assembly on the main shaft by a snap ring.

A great deal of similarity and interchangeability now exists between the new 3 and 4-speed Saginaw transmissions.

The synchronizer assembly at the front of the mainshaft is used for the third and fourth, rather than the second and third shift. The synchronizer assembly at the rear of the mainshaft is used for the first and second, rather than the first and reverse shift. Gear teeth cut in the first and second synchronizer sleeve (reverse gear) distinguish it from the third and fourth synchronizer sleeve.

Starting from the front, gears on the mainshaft are third, second and first rather than second, first and reverse. A fourth blocker ring is used between the 1-2 synchronizer assembly and the first gear on the four-speed transmissions.

The cover on the new four-speed transmission is located on the left-hand side of the case. It is similar to the three-speed cover with the addition of a reverse shifter shaft assembly, detent ball and detent spring.

Disassembly and assembly procedures are covered in this section.

SHIFT CONTROL LEVER AND BRACKET ASSEMBLY
Maintanance and adjustments
Refer to replacement procedures for three-speed transmission control lever and bracket assemblies described under 3-Speed Transmissions.

SPEEDOMETER DRIVEN GEAR AND OIL SEAL
Replacement
Raise vehicle on hoist. Disconnect speedometer cable, remove retainer to housing bolt and lock washer and remove retainer. Insert screw driver in slot in fitting and pry fitting, gear and shaft from housing. Pry "O" ring in groove in fitting. Install new "O" ring in groove and insert shaft. Hold the assembly so slot in fitting is toward boss on housing and install in housing. Push fitting into housing until retainer can be inserted in groove and install retainer lock washer and bolt. Lower and remove vehicle from hoist.

Shift linkage adjustments
Turn ignition switch to the "LOCK" position and raise vehicle on hoist. Loosen swivel lock nuts on both shift rods. Rods should pass freely through swivels. Disconnect back drive cable from column lock tube lever. Place shift lever control (in car) into Neutral. Insert locking gauge (.644 in.) in notch of lever and bracket assembly. Place transmission levers in the neutral position. Hold Reverse rod and lever against locating gauge, tighten locknut against swivel. Hold 1st/2nd rod and lever against locating gauge, tighten the forward nut against the swivel and then

the aft. Repeat step with 3rd/4th rod and lever. Remove locating gauge. Check transmission shifting to ensure proper operation. Readjust linkage, if required. *Shift lever must be centered in console opening to prevent fore and aft, and side contact with console. Shim bracket support as required but be sure to maintain support to extension clearance.*

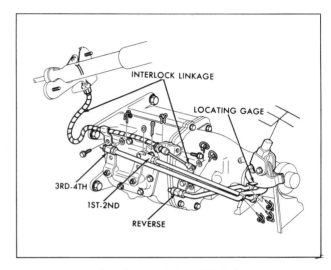

4 Speed transmission shift linkage

From inside vehicle, loosen two nuts at the steering column to dash panel bracket. Place transmission shift lever in "Reverse" and rotate ignition switch to the "LOCK" position (if not previously done). Rotate the lock tube lever counter-clockwise (viewed from front of column) to remove any free play. Reposition the cable bracket until the cable eye passes over the retaining pin on the bracket.

Hold the bracket in position and have an assistant tighten the steering column to dash panel bracket retaining nuts. Reinstall the cotter pin and washer retaining the back drive cable to the lever pin. Lower and remove vehicle from hoist.

BACK DRIVE CONTROL LINKAGE
Adjustment
Adjustment procedures of the back drive control linkage on vehicles equipped with floor shift four-speed transmissions is basically the same as on vehicles having a floor shift Three-speed transmission. Refer to the adjustment procedures described under three-speed transmissions.

TRANSMISSION SIDE COVER (MUNCIE)
Removal
Raise vehicle on hoist. Disconnect control rods from levers. Disconnect TCS switch (where used) and back-up lamp at side cover. Shift transmission into second speed before removing cover, by moving 1-2 (Rear Cover) shifter lever into forward detent position. Remove cover assem-

bly from transmission case carefully and allow oil to drain.

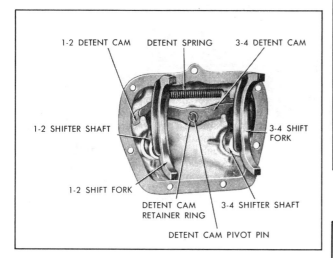

Transmission side cover, shift fork and detent assembly

Removing reverse shifter shaft lock pin

Disassembly

Remove TCS switch (where used), the outer shifter lever nuts, lock washers and flat washers. Pull levers from shafts. Remove both shift forks from shifter shaft and detent plate assemblies. Remove both shifter shaft assemblies from cover. Seal in side cover and "O" ring on 1st/2nd shaft may now be removed if replacement is required because of damage. *Components should be marked so they may be reinstalled with same relationship.* Remove detent cam spring and pivot retainer "C" ring. Remove both detent cams. Replace necessary parts.

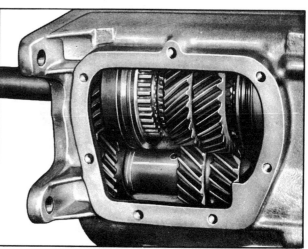

Third and fourth speed synchronizer clutch sleeve in fourth gear position

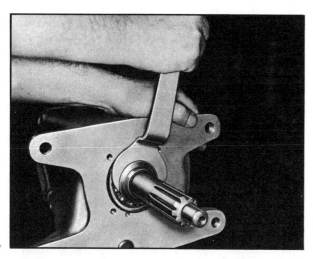

117

Removing main drive gear retaining nut

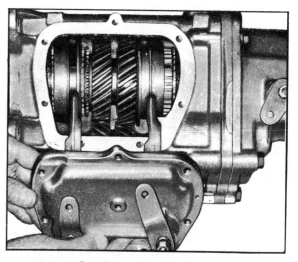

Installing side cover assembly

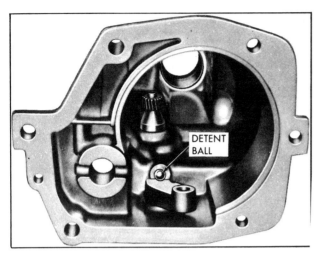

Installing reverse shifter shaft and detent ball

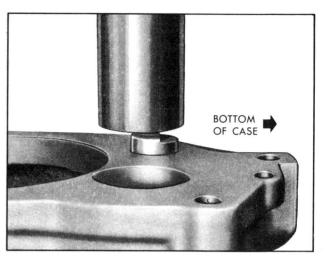

Installing countershaft

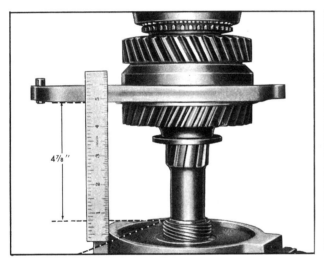

Installing speedometer drive gear

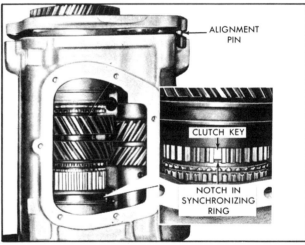

Installing mainshaft assembly

Assembly

Install 1-2 detent cam to cover pivot pin first, then install 3-4 detent cam so the detent spring notches are offset or opposite each other. Detent cam notches must be facing downward. Install detent cam retaining "C" ring to pivot shaft, and hook spring into detent cam notches. Install both shifter shaft assemblies in cover being careful not to damage seals. Install both shift forks to detent plates, lifting up on detent cam to allow forks to fully seat into position. Install outer shifter levers, flat washers, lock washers and nuts. Install TCS switch (where used).

Installation

Shift 1-2 shifter lever into second speed (forward) position. Position cover gasket on case. Carefully position side cover into place making sure the shift forks are aligned with their respective mainshaft clutch sliding sleeves. Install cover attaching bolts and tighten evenly to 15-20 ft. lbs. torque. Connect shift rods to levers at cover. Connect back-up lamp switch and TCS switch (where used) at cover. Remove filler plug and add lubricant to level of filler plug hole. Lower and remove vehicle from hoist.

TRANSMISSION SIDE COVER (SAGINAW)
Removal

Raise vehicle on hoist. Disconnect control rods from levers. Disconnect back-up lamp and TCS switch (where used) at cover. Shift transmission into neutral detent positions before removing cover. Remove cover assembly from transmission case carefully and allow to drain.

Removing clutch gear & mainshaft assembly

Removing countershaft

J-22246

COUNTER SHAFT

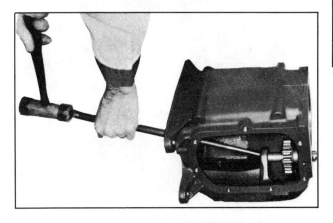

Removing reverse idler gear shaft

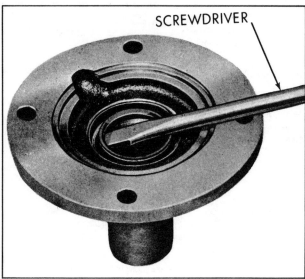

SCREWDRIVER

Removing bearing retainer oil seal

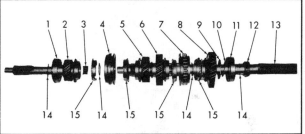

Clutch gear & mainshaft details

1. Clutch gear bearing
2. Clutch gear
3. Mainshaft pilot bearings
4. 3-4 synchronizer assembly
5. Third speed gear
6. Second speed gear
7. 1-2 synchronizer & reverse gear assy.
8. First speed gear
9. Thrust washer
10. Spring washer
11. Rear bearing
12. Speedo drive gear
13. mainshaft
14. Snap ring
15. Synchronizing "blocker" ring

Disassembly

Remove the outer shifter levers and TCS switch (where used). Remove both shift forks from shifter shaft assemblies. Remove all three shifter shaft assemblies from cover. The lip seal in the cover and "O" ring seal on the 1-2 and reverse shafts may now be pryed out if required because of damage. Remove reverse shifter shaft detent ball and spring. Remove detent cam spring and pivot retainer "C" ring. Mark to identify for reassembly, then remove both detent cams. Replace damaged parts.

Assembly

With detent spring tang projecting up over the 3rd and 4th shifter shaft cover opening install the first and second detent cam onto the detent cam pivot pin. With the detent spring tang projecting up over the first and second shifter shaft cover hole install the 3rd and 4th detent cam. The 1-2 detent cam has .090" greater contour on the inside detent notch.

Install detent cam retaining "C" ring to pivot shaft, and

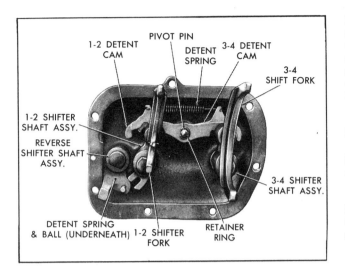

Transmission side cover assembly — saginaw

hook spring into detent cam notches. Install 1-2 and 3-4 shifter shaft assemblies in cover being careful not to damage seals. Install both shift forks to shifter shaft assemblies, lifting up on detent cam to allow forks to fully seat into position. Install reverse detent ball and spring to cover, then install reverse shifter shaft assembly to cover. Install TCS switch in cover. Install outer shifter levers, flat washers, lock washers and bolts.

Installation

Shift shifter levers into neutral detent (center) position. Position cover gasket on case. Carefully position side cover into place making sure the shift forks are aligned with their respective mainshaft clutch sliding sleeves. Install cover attaching bolts and tighten evenly to specified torque. Connect shift rods to levers at side cover. Connect back-up lamp and TCS switch (where used) at side cover. Remove filler plug and add lubricant to level of filler plug hole. Lower and remove vehicle from hoist.

Installing speedometer drive gear

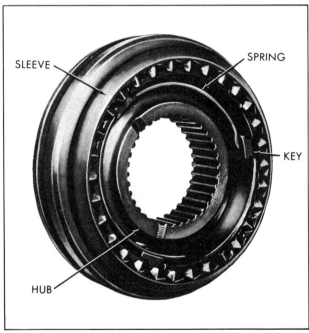

Three-four synchronizer assembly

120

TRANSMISSION, AUTOMATIC 8A

ALUMINUM POWERGLIDE
Description
The case and converter housing of the two speed aluminum Powerglide Transmission is a single case aluminum unit. When the manual control is placed in the drive position, the transmission automatically shifts to low gear for initial vehicle movement. As the car gains speed and depending on load and throttle position, an automatic shift is made to high gear. A forced down-shift feature provides a passing gear by returning the transmission to low range.

The front pump assembly is a conventional gear type and the oil pump housing is of the large diameter type acting as the front bulkhead of the transmission. The conventional gear type rear oil pump is bolted outside the rear bulkhead of the case and is completely enclosed by the one piece aluminum extension housing. The torque converter is a conventional three element welded design bolted to the engine flywheel which drives through a two-speed planetary gearset. The high clutch assembly is typical of the designs used in this type transmission.

The aluminum Powerglide uses an output shaft mounted governor which requires a hole through the output shaft. The reverse clutch assembly is a multiple disc type clutch. The steel plates are splined directly to the case while the face plates are splined to the internal or ring gear. The clutch piston operates within the rear portion of the case. The internal diameter of the piston is sealed to an integral hub portion of the case rear bulkhead. The outside diameter is sealed to a machined portion of the case. The piston is hydraulically applied and is released by separate coil springs. The valve body assembly is bolted to the bottom of the transmission case and is accessible for service by removing the oil pan assembly. The valve body consists of an upper and lower body located on either side of a transfer plate. The vacuum modulator is located on the left rear face of the transmission case. The modulator valve bore is located in the upper valve body.

MAINTENANCE AND ADJUSTMENTS
Oil level and requirements
The transmission oil level should be checked periodically. Oil should be added only when level is near the "ADD" mark on the dip stick with oil hot or at operating temperature. The oil level dip stick is located at the right rear of the engine compartment. In order to check oil level accurately, the engine should be idled with the transmission oil hot and the control lever in neutral (N) position.

It is important that the oil level be maintained no higher than the "FULL" mark on the transmission oil level gauge. DO NOT OVERFILL, for when the oil level is at the full mark on the dip stick, it is just slightly below the planetary gear unit. If additional oil is added, bringing the oil level above the full mark, the planetary unit will run in the oil, foaming and aerating the oil. This aerated oil carried through the various oil pressure passages (low servo, reverse servo, clutch apply, converter, etc.) may cause malfunction of the transmission assembly, resulting in cavitation noise in the converter and improper band or clutch application. Overheating may also occur..

If the transmission is found consistently low on oil, a thorough inspection should be made to find and correct all external oil leaks. Transmission oil leakage is now easily identified as all automatic transmission fluid used is dyed red. The mating surfaces of servo cover, converter housing, transmission case and transmission case extension should be carefully examined for signs of leakage.

The vacuum modulator must also be checked to insure that the diaphragm has not ruptured as this would allow transmission oil to be drawn into the intake manifold. Usually, the exhaust will be excessively smokey if the diaphragm ruptures due to the transmission oil added to the combustion. The transmission case extension rear oil seal should also be checked. All test plugs should be checked to make sure that they are tight and that there is no sign of leakage at these points. The converter underpan should also be removed. Any appreciable quantity of oil in this area would indicate leakage at the front pump square seal ring, front pump seal assembly, or front pump bolt sealing washers.

Draining and refilling
To drain the transmission, remove the oil pan drain plug. Position a pan or can to catch the draining oil. If the transmission is to be removed from the vehicle for repairs, the draining operation may be performed after removal if desired.

To refill the transmission, remove dip stick from oil filler tube and refill transmission with oil using a filler tube and funnel. Then, after shifting into all ranges at idle speed to fill all oil passages, the engine should be run at 800-1000 rpm with the transmission in Neutral until the oil warms up, then add oil as required to raise the fluid level to the full mark on the dip stick. Total capacity including coverter is 17.7 pts. for air cooled and 20.1 pts. for water cooled models.

Service adjustments
Four service adjustments are required for Aluminum Powerglide equipped cars: Shift linkage, throttle valve linkage, neutral safety switch and throttle return check valve (Dashpot) adjustment.

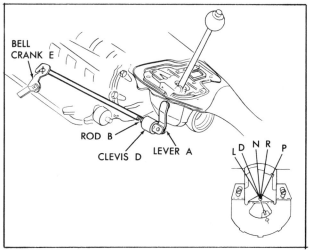

gear shift control adjustment

linkage adjustment

Set transmission Bell Crank (E) in "PARK" position. Set Control Shaft Lever (A) in "PARK" position. Install Clevis (D) on Rod (B) loosely, attach Rod to Bell Crank (E) and secure with retainer. With Lever (A) in "PARK" position, adjust Clevis at Lever (A) until clevis pin passes freely through holes and secure with washer and cotter pin. Check shifts to insure proper operation. Readjust clevis as necessary.

powerglide control lever and bracket assembly

Powerglide control lever and bracket assembly is a straight line shift type, mounted to the floor of the body rather than the transmission extension as in the past. Shifting from one detent position to another is done by depressing a round button located in the center of the spherical shift lever knob. If disassembly of this control lever and bracket assembly is necessary, refer to the Figure for parts breakdown and relative positioning for assembly.

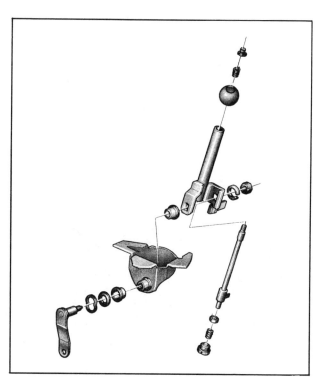

control lever and bracket assembly – exploded

Throttle valve linkage adjustment

With accelerator pedal depressed, Carburetor Lever on V-8 models must be at wide open throttle position. Dash Lever must be 1/64"-1/16" off Lever Stop and Transmission Lever must be against transmission internal stop.

POWERGLIDE TRANSMISSION
Replacement

Disconnect battery ground cable. Remove ball end from

123

transmission shift control lever. Raise front and rear of vehicle. Insert a block of wood between the top of the differential carrier housing and the underbody (to prevent upward travel of the carrier when the carrier front support is disconnected).

Disconnect the differential carrier front support from its frame bracket, by removing the nut on the underside of the biscuit mount. Pry the carrier downward to relieve load while removing the two center mounting bolts from the carrier front support. (To pry carrier downward insert crowfoot end of a pry bar through the opening in the carrier front support, hooking end of bar over top of the center mounting bolt pad cast in the underside of the carrier).

Pivot carrier support downward for access to prop shaft "U"-joint. Disconnect prop shaft front and then rear "U" bolts. Disconnect parking brake cable from ball socket at idler lever located near center of underbody. Remove prop shaft by moving shaft forward. Remove left bank exhaust pipe. Remove right bank exhaust pipe and heat riser.

Disassemble the transmission mount by removing the two bolts that attach rear mount cushion to the rear mount bracket. Support engine under oil pan and raise engine to remove load from rear mount cushion. To avoid damage to oil pan, a suitable wide base, heavy wood platform should be placed between the jack pad and the oil pan.

Remove the three transmission mount bracket-to-crossmember bolts, then remove mount bracket. Remove the two bolts from mount pad to transmission case and remove rubber mount cushion and exhaust pipe "yoke".

Disconnect oil cooler lines at transmission and swing lines clear. Remove connector underpan. Remove converter to flywheel attaching bolts. At transmission disconnect vacuum modulator line and speedometer cable. Disconnect transmission shift linkage from transmission gearshift control lever. Remove shift lever. At transmission, remove throttle valve linkage and disconnect neutral safety switch linkage, also remove the gearshift control linkage.

Remove neutral safety switch from transmission control bracket. Remove transmission output shaft slip yoke and insert a plastic shipping plug in end of extension to prevent spillage of transmission fluid. The yoke is removed to avoid tearing the heat reflecting pad on the underbody, when the transmission is being removed. Remove bright metal ignition shielding from distributor area.

Remove the transmission dip stick and tube assembly. Disconnect transmission vacuum modulator line at distributor advance line tee. Position transmission hoist under transmission and attach safety chain to transmission. Remove transmission converter housing-to-engine attaching bolts and slide transmission rearward. Observe converter when moving transmission rearward. If converter does not move with the transmission, pry it free of flywheel before proceeding.

Install converter retaining strap. Lower and remove transmission from vehicle by tilting the front down and to the right while intermittently lowering the transmission to facilitate its removal. Reinstall transmission assembly by performing the above steps in reverse order.

Bolt torques

Transmission Case to Flywheel Housing Bolts ....35 ft. lbs.
Converter to Flywheel Bolts ....35 ft. lbs.

Diagnosis

Proper operation of the Powerglide transmission may be affected by a number of factors, all of which must be considered when trouble in the unit is diagnosed. Proper trouble diagnosis can only be accomplished when performed in a thorough step by step procedure. The following procedure has been devised and tested and is recommended for all trouble diagnosis complaints and if the serviceman will follow this checking procedure, accurate and dependable diagnosis may be accomplished. This will result in a savings of time, not only to the serviceman, but to the customer as well.

Warming up transmission

Before attempting to check and/or correct any complaints on the Powerglide transmission it is absolutely essential that the oil level be checked and corrected if necessary. An oil level which is either too high or too low can be the cause of a number of abnormal conditions from excessive noise to slippage in all ranges.

It must be remembered that cold oil will slow up the action of the hydraulic controls in the transmission. For this reason a trouble or oil leak diagnosis should not be attempted until the transmission has been warmed.

Shop warm up

Connect tachometer to engine. Set parking brake tight and start engine. Place selector light in "D" (drive) range. Adjust carburetor idle speed adjusting screw to run engine at approximately 750 rpm and operate in this manner for two minutes. At the end of two minutes of operation, the transmission will be sufficiently warmed up for diagnosis purposes. At this point, readjust the engine idle speed to 450-475 rpm in "D" range.

Road warm up

Drive the car approximately 5 miles with frequent starts and stops. At this point, make sure the engine idle speed is set to 450-475 rpm in "D" range.

Checking fluid level

After transmission has been warmed up, check the fluid level with the engine idling, parking brake set and control lever in "N" (neutral). If the fluid level is low, add fluid to bring level up to the full mark on gauge rod.

If fluid level is too high, fluid may be aerated by the planet carrier. Aerated fluid will cause turbulence in the converter which will result in lost power, lower stall speed and lower pressures in control circuits. Lower fluid level to full mark, then shut off engine to allow air bubbles to work out of fluid.

Basic pressure checks

Four basic pressure checks are used for diagnosis and operational checks for the Aluminum Powerglide transmission. All checks should be made only after thoroughly warming up the transmission.

1. Transmission case
2. Welded converter
3. Front oil pump seal assembly
4. Front oil pump body
5. Front oil pump body square ring seal
6. Front oil pump cover
7. Front oil pump body cover
8. Clutch relief valve ball
9. Clutch piston inner and outer seal
10. Clutch piston
11. Clutch drum
12. Clutch hub
13. Clutch hub thrust washer
14. Clutch flange
15. Low gear and clutch flange assembly
16. Planet short pinion
17. Planet input sun gear
18. Planet carrier
19. Planet input sun gear thrust washer
20. Ring gear
21. Reverse piston
22. Reverse piston outer seal
23. Reverse piston inner seal
24. Rear pump wear plate
25. Extension seal ring
26. Rear pump
27. Extension
28. Governor hub
29. Governor hub drive screw
30. Governor body
31. Governor shaft retainer clip
32. Governor outer weight retainer ring
33. Governor inner weight retainer ring
34. Governor outer weight
35. Governor spring
36. Governor inner weight
37. Extension rear oil seal
38. Extension rear bushing
39. Output shaft
40. Speedometer drive and driven gear
41. Governor shaft belleville springs
42. Governor shaft
43. Governor valve
44. Governor valve retaining clip
45. Governor hub seal rings
46. Rear pump drive pin
47. Rear pump bushing
48. Rear pump priming valve
49. Rear pump drive gear
50. Rear pump driven gear
51. Reverse piston return springs, retainer and retainer ring
52. Transmission rear case bushing
53. Output shaft thrust bearing
54. Reverse clutch pack
54A. Reverse clutch cushion spring (waved)
55. Pinion thrust washer
56. Planet long pinion
57. Low sun gear needle thrust bearing
58. Low sun gear bushing (Splined)
59. Pinion thrust washer
60. Parking lock gear
61. Transmission oil pan
62. Valve body
63. High clutch pack
64. Clutch piston return spring, retainer and retainer ring
65. Clutch drum bushing
66. Low brake band
67. High clutch seal rings
68. Clutch drum thrust washer (selective)
69. Turbine shaft seal rings
70. Front pump driven gear
71. Front pump drive gear
72. Stator shaft
73. Input shaft

124

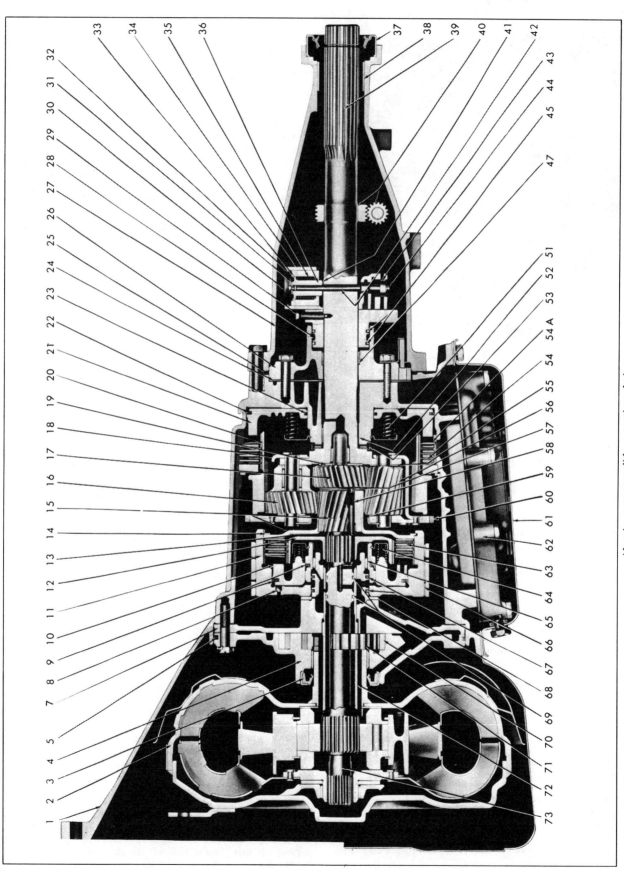

Aluminum powerglide – sectioned view

Pressure test plug

It is not recommended that stall tests be conducted which would result in engine vacuum falling below 10" Hg. Pressure gauge hose connections should be made at the low servo apply (main line) test point. Run the gauge line into the driving compartment by pushing aside the mast jacket seal. Tie line out of the way of the drivers feet and connect to pressure gauge J-21867.

Wide Open Throttle Upshift Pressure Check — Refer to the pressure check chart for upshift pressure points as indicated on the low servo apply (main line) gauge.

Idle Pressure in "Drive" Range — In addition to the oil pressure gauges, a vacuum gauge is needed for this check. With the parking brake applied and the shift selector lever in "Drive", low servo apply (main line) pressure should be as shown on the pressure check chart.

If pressures are not within these ranges, the following items should be checked for oil circuit leakage: Pressure regulator valve stuck. Vacuum modulator valve stuck. Hydraulic modulator valve stuck. Leak at low servo piston ring (between ring and bore). Leak at low servo piston rod (between rod and bore). Leak at valve body to case gasket. Leak at valve body gaskets. Front pump clearances. Check passages in transmission case for porosity.

Manual "Low" Range Pressure Check — Connect a tachometer, apply the parking brake, place the selector lever in "Low" range, and adjust the engine speed to 1000 rpm with the car stationary. Low servo apply (main line) pressure should be as shown on the pressure check chart.

Pressures not within this range can indicate the following possibilities: Partially plugged oil suction screen. Broken or damaged ring low servo. Pressure regulator valve stuck. Leak at valve body to case gasket. Leak between valve body gaskets. Leak at servo center. Front pump clearances.

Drive Range Overrun (Coast) Pressure — With the vehicle coasting in "Drive" range at 20-25 mph with engine vacuum at approximately 20" Hg., low servo apply (main line) pressure should be as shown on the pressure check chart.

TRANSMISSION OVERHAUL OPERATIONS
Disassembly

Place transmission in Holding Fixture J-3289-01 and Adapters J-9506. Cleanliness is an important factor in the overhaul of the transmission. Before attemtping any disassembly operation, the exterior of the case should be thoroughly cleaned to prevent the possibility of dirt entering the transmission internal mechanism. During disassembly, all parts should be thoroughly cleaned in cleaning fluid and then air dried. Wiping cloths or rags should not be used to dry parts as lint may be deposited on the parts which may cause later trouble. Do not use solvents which could damage rubber seals or clutch plate facings. Remove converter holding tool previously installed and remove converter assembly.

Transmission mounted in fixture

Extensions, governor and rear oil pump

If replacement is necessary, remove speedometer driven gear. Loosen capscrew and retainer clip holding speedometer driven gear in extension and remove gear. Remove transmission extension by removing five bolts retaining extension to case. Note seal ring on rear pump body. Remove the speedometer drive gear from output shaft, using J-5814.

Remove the "C" clip from the governor shaft on the weight side of the governor, then remove the shaft and governor valve from the opposite side of the governor assembly and the two belleville springs. Loosen the governor drive screw and remove the governor assembly over the end of the output shaft.

Remove the four bolts retaining the rear oil pump to the transmission case and remove the pump body, and drain back baffle, extension seal ring, drive and driven gears. When the drive gear is removed, the drive pin may fall out if the hole is on the bottom of the shaft and the shaft is horizontal.

Removing rear oil pump drive pin

Remove the oil pump drive pin. This is of extreme importance. Do not fail to remove this drive pin. After removing the drive pin, remove the rear pump wear plate.

Transmission internal components

Rotate the holding fixture until the front of the transmission is pointing up and remove the seven front oil pump bolts. The bolt holes are offset to facilitate proper location upon installation.

Remove the front oil pump and stator shaft assembly and the selective fit thrust washer using J-9539 (or two 3/8"-16"x10" stove bolts) and the slide weights from Tool J-6585. Note the two threaded holes in the pump to mount the pullers. Remove the front pump ring seal and gasket. The front pump bolts have special sealing washers which must be in place upon installation.

Release the tension on the low band adjustment, then, with transmission horizontal, grasp the transmission input shaft and carefully work it and the clutch assembly out of the case. Use care so as not to lose the low sun gear (splined) bushing from the input shaft. The low sun gear thrust washer will probably remain in the planet carrier.

Use care so as not to damage the machined face on the front of the clutch drum. The low brake band and struts may now be removed. Make certain that the rear pump drive pin has been removed, then remove the planet carrier and the output shaft thrust caged bearing from the front of the transmission.

Remove the reverse ring gear if it did not come out with the planet carrier. Using a large screw driver remove the reverse clutch pack retainer ring and then lift out the reverse clutch plates and the (waved) cushion spring. If difficulty is experienced in getting the snap ring past the shoulder on the reverse pack pressure plate, a feeler gauge may be used as a guide.

Install Tool J-9542 through the rear bore of the case with the flat plate on the rear face of the case and turn down the wing nut to compress the rear piston spring retainer and springs, then remove the snap ring. Tool J-8039 may be used to remove the snap ring if desired. Remove Tool J-9542, the reverse piston spring retainer and the 17 piston return springs.

Remove the rear piston by applying air to the reverse port in the rear of the transmission case. Remove the inner and outer seals. Remove the three servo cover bolts, servo cover, piston and spring.

Removing clutch drum and input shaft

Applying air to remove rear piston

Oil pan and valve body

The oil pan and valve body may be serviced without the necessity of removing the extension and internal components covered in the preceding steps. Rotate the holding fixture until the transmission is upside down and the oil pan is at the top. Remove the oil pan attaching bolts, oil pan and gasket. Remove the vacuum modulator and gasket, and the vacuum modulator plunger, dampening spring and valve.

Remove the two bolts attaching the detent guide plate to the valve body and the transmission case. Remove the guide plate and the range selector detent roller spring.

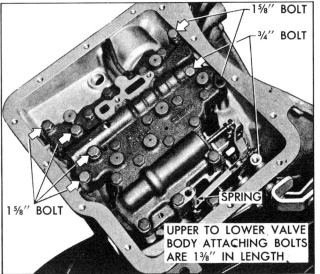

Valve body removal

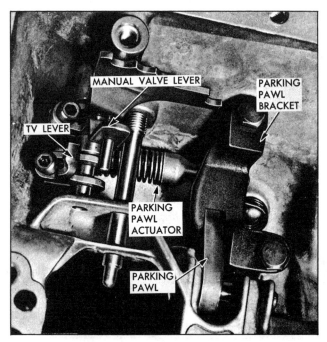

Inner control levers, parking pawl and bracket

Remove the remaining valve body-to-transmission case attaching bolts and carefully lift out the valve body and gasket, disengaging the servo apply tube from the transmission case as the valve body is removed. If necessary, the TV, shift and parking actuator assembly levers, and the parking pawl and bracket may be removed.

OVERHAULING UNIT ASSEMBLIES

Converter and stator

The converter is a welded assembly and no internal repairs are possible. Check the seams for stress or breaks and replace converter if necessary.

FRONT PUMP

Seal replacement

If the front pump seal requires replacement, remove the pump from the transmission, pry out and replace the seal. (Drive new seal into place, fully seated in counterbore, using J-6839). Then, if no further work is required on the front pump, reinstall it in the case. Outer diameter of the seal should be coated with non-hardening sealer prior to installation.

Disassembly

Remove bolts attaching pump cover to body and remove the cover. Remove pump gears from body. Do not drop or nick gears. These gears are not heat treated. Remove the rubber seal ring from the pump body.

Checking pump body bushing to converter
pump hub clearance

Inspection

Wash all parts in cleaning solvent and blow out all oil passages. DO NOT USE RAGS TO DRY PARTS. Some solvents may be harmful to rubber seals.

Inspect pump gears for nicks or damage. Inspect body and cover faces for nicks or scoring. Inspect cover hub O.D. for nicks or burrs which might damage clutch drum bushing journal. Check for free operation of the priming valve and replace if necessary.

Inspect body bushing for galling or scoring. Check clearance between body bushing and converter pump hub. Maximum clearance is .005". If the bushing is damaged, the front pump body should be replaced.

Inspect converter housing hub O.D. for nicks or burrs which might damage front pump seal or bushing. Repair or replace as necessary. If oil seal is damaged or is leaking (and the pump body is otherwise suitable for reuse), pry out and install a new seal, fully seated in counterbore, using Seal Driver J-6839. Outer diameter of seal should be coated with a non-hardening sealer prior to installation.

On water cooled models so equipped, check condition of oil cooler by-pass valve and replace if valve leaks excessively. For removal, an "Easy Out" or its equivalent may be used. For installation tap seat in place with soft hammer or brass drift so it is flush to .010" below the surface.

With parts clean and dry, install pump gears and check: Clearance between O.D. of driven gear and body should be .0035"-.0065". Clearance between I.D. of driven gear and crescent should be .003"-.009". Gear end clearance should be .0005"-.0015".

Assembly

With the transmission facing up, remove the input shaft, clutch drum, low band and struts as outlined under "Transmission — Disassembly." Install the downshift timing valve, conical end out, into place in the pump cover to a height of 17/32" measured from the shoulder of the valve assembly to face of pump cover. Oil drive and driven gears generously and install in the pump body. Assemble drive gear with recessed side of the drive lugs downward — facing the converter.

Carefully set the pump cover in place over the body and loosely install 2 attaching bolts. Place the pump assembly, less the rubber seal ring, upside down into the pump bore of the case (use guide pins if desired). Install remaining attaching bolts and torque to specifications.

Remove pump assembly from case bore. Replace the clutch drum and input shaft, low band and struts as outlined under "Transmission-Assembly." If necessary, remove two bolts and use J-6585 puller and J-6585-3 adapters to remove pump assembly. Replace and retorque bolts.

Replace rubber seal ring in its groove in the pump body and install the pump assembly properly in place in the case bore, using a new gasket, being sure that the selective fit thrust washer is in place. Install the attaching bolts, using new bolt sealing washers if necessary.

Rear pump

The rear pump is removed and disassembled as described in the "Transmission-Disassembly" procedures earlier in this section. General cleaning and clearance check information will remain the same as for the front pump. Assembly of the rear pump is described in the "Transmission-Assembly" procedure later in this section.

When reinstalling the rear pump priming valve, retain the washer, spring and valve in the bore with the retaining seat, install small hole first. Install retaining seat flush to

.02" below surface of pump face. When properly installed the tip of the valve must extend above the face of the pump to insure priming valve operation.

Rear pump bushing replacement

If the rear pump bushing must be replaced, it may be removed using Tool J-9557 (and Handle J-7079) and re-installed using Tool J-6582, pressing or driving the bushing in from the front of the pump.

CLUTCH DRUM
Disassembly

When working with the clutch drum, use extreme care that the machined face on the front of the drum not be scratched, scored, nicked or otherwise damaged during any of the following service operations. This machined face must be protected whenever it must be brought to bear on a press or tool of any sort.

Remove retainer ring and low sun gear and clutch flange assembly from the clutch drum. Remove the hub rear thrust washer. Lift out the clutch hub, then remove the clutch pack and the hub front thrust washer.

Remove the spring retainer using J-9542 or if using an arbor press, use J-5133 and J-7782 adapter ring. Compress the springs far enough to allow removal of the retainer snap ring; then, releasing pressure on the springs; remove the retainer and the springs. When using J-9542, place a piece of cloth or cardboard between the tool and the front side of the clutch drum as protection for the machined face. Lift up on the piston with a twisting motion to remove from the drum, then remove the inner and outer seals.

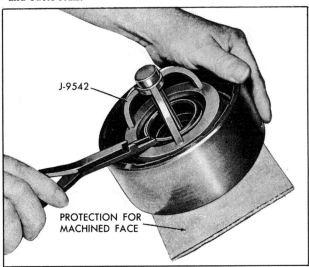

Removing clutch spring retainer snap ring

Inspection

Wash all parts in cleaning solvent (air dry). Do not use rags to dry parts. Check drum bushing for scoring or excessive wear. Check the steel ball in the clutch drum that acts as a relief valve. Be sure that it is free to move in the hole and that the oriface leading to the front of the drum is open. If the clutch relief valve check ball in the clutch drum is loose enough to come out or not loose enough to rattle, replace the clutch drum as an assembly. Replacement or restaking of the ball should not be attempted.

Check fit of clutch flange in drum slots. There should be no appreciable radial play between these two parts. Also check low sun gear for nicks or burrs and bushing for wear. Check clutch plates for burning and wear.

Bushing replacement

Remove the old bushing with Tool J-9546 using care not to damage the bushing bore or the machined face on the front of the clutch drum. Use the same tool to install the new bushing. Press (do not hammer) the bushing into the clutch drum from the machined face side of clutch drum. Press only far enough so that the tool meets the clutch drum. Do not force the tool against the clutch drum machined face.

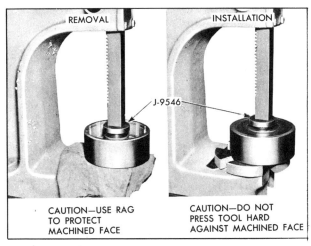

Removing and installing clutch drum bushing

Assembly

Install new piston inner seal in hub of clutch drum with seal lip downward (toward front of transmission). Install a new piston seal in clutch piston. Seal lips must be pointed toward the clutch drum (front of transmission). Lubricate seals generously and install piston in clutch drum with a twisting motion. Place the springs in position on the piston, then place the retainer in place on the springs. Using Tools J-5133 and J-7782 and a press, or J-9542 as a hand operation, depress the retainer plate and springs far enough to allow installation of the spring retainer snap ring in its groove on the clutch drum hub.

Install the hub front washer with its lip toward the clutch drum, then install the clutch hub. Install (wave) cushion spring wherever used, see chart. On 396 V-8 models only, the first driven plate to the rear of the cushion spring is a selective fit. Before installing plates to clutch drum, stack the (5) drive and (5) driven (exc. selective fit) plates and measure this stack height. New refer to chart and select the proper selective driven plate and install it into clutch drum on top of the cushion spring.

Install the steel reaction plates and faced drive plates alternately, beginning with a steel reaction plate. See clutch chart. Install the rear hub thrust washer with its flange toward the low sun gear, then install the low sun gear and flange assembly and secure with retainer ring. When installed, the openings in the retainer ring should be adjacent to one of the lands of the clutch drum. Check the assembly by turning the clutch hub to be sure it is free to rotate.

Low band

The brake band used in the Aluminim Powerglide transmission has bonded linings which, due to the transmission characteristics and band usage, should require very little attention. However, whenever a transmission is disassembled the band should be cleaned of metal particles and inspected.

Check lining for evidence of scoring or burning. Check band and lining for cracks. Check all band linkage for excessive wear.

PLANET ASSEMBLY AND INPUT SHAFT
Inspection

Wash planet carrier and input shaft in cleaning solvent, blow out all oil passages and air dry. Do not use rags to dry parts. Inspect planet pinions for nicks or other tooth damage. Check end clearance of planet gears. This clearance should be .006”-.030”.

Check input sun gear for tooth damage, also check input sun gear rear thrust washer for damage. Inspect output shaft bearing surface for nicks or scoring and inspect input pilot bushing. Inspect input shaft splines for nicks or damage and check fit in clutch hub and input sun gear. Also check fit of splines in turbine hub. Check oil seal rings for damage; rings must be free in input shaft ring grooves. Remove rings and insert in stator support bore and check to see that hooked ring ends have clearance. Replace rings on shaft.

Checking planet gear end clearance

Repairs

The large planet carrier assembly now has the pinion shafts flared at each end for retention into the carrier. No overhaul of the large planet carrier assembly should be attempted.

Small planet carrier assembly — overhaul

If during inspection, the planet pinions, pinion needle bearing, pinion thrust washers, input sun gear, and/or input sun gear thrust washer should show evidence of excessive wear or damage, they should be replaced.

Place the planet carrier assembly in a fixture or vise so that the front (parking lock gear, end) of the assembly faces up. Using prick punches or other similar means, mark each pinion shaft and also the planet carrier assembly so that reassembling, each pinion shaft will be reinstalled in the same location from which it was removed. The pinion shafts are not selectively fit but it is good practice to reinstall them in their original locations.

Remove the pinion shaft lock plate screws and rotate the lock plate counter-clockwise sufficiently to remove it. Starting with a short planet pinion, and using a soft steel drift, drive on the lower end of the pinion shaft until the pinion shaft is raised above the press fit area of the output shaft flange. Feed J-4599 into the short planet pinion from the lower end, pushing the planet pinion shaft ahead of it until the tool is centered in the pinion and the pinion shaft is removed from the assembly. Planet pinion remover and replacer Tool J-4599, comes in two pieces, both alike. Only one is used when removing the planet pinion; two, however, must be used when reassembling.

Remove the short planet pinion from the assembly. Remove J-4599, needle bearings and needle bearing spacers (3) from short planet pinion. Use care so as not to lose any of the planet pinion needle bearings. Twenty needle bearings are used in each end and are separated by a bearing spacer in the center.

By following the procedure as outlined, remove the adjacent long planet pinion that was paired by thrust washers to the short planet pinion now removed. Twenty needle bearings are used in each end of the long pinion, separated by a bearing spacer in the center.

Remove the upper and lower thrust washers. Remove and disassemble the remaining planet pinions, in pairs, by first removing a short planet pinion and then the adjacent long planet pinion. Remove low sun gear needle thrust bearing, input sun gear and input sun gear thrust washer. Wash all parts in cleaning solvent and air dry.

Recheck the planet pinion gears and input sun gear for nicks or other tooth damage; also check the planet pinion thrust washers and input sun gear thrust washer. Check low sun gear needle thrust bearing for spalled needles. Replace worn or damaged parts. Inspect the planet pinion needle bearings closely and if there is indication of excessive wear, all the needle bearings must be replaced. Also inspect pinion shafts closely and, if worn, replace the worn shafts.

Inspect the input shaft bushing installed in the base of the output shaft. If damaged, it may be removed by threading Tool J-9534 into the bushing and pulling the bushing out using Slide Hammer J-6585. New bearing can be installed by pressing in flush or below thrust surface with the pilot end of input shaft as press tool.

Using J-4599, assemble needle bearing spacer and needle bearings (20 in each path) in one of the long planet pinions. Use petroleum jelly to aid in assembling and holding the needle bearings in position. Position the long planet pinion with J-4599, centered in the pinion assembly and with thrust washers at each end, in the planet carrier. Oil grooves on thrust washers must be toward gears. The long planet pinions are located opposite the closed portions of the carrier, while the short planet pinions are located in the openings.

Feed the second J-4599 in from the top picking up the upper thrust washer and the planet pinion and pushing the already installed Tool J-4599, out of the lower end. As the first tool is pushed down, check that it picks up the lower thrust washer. Select the proper pinion shaft, as marked, lubricate the shaft and install it from the top, pushing the assembling tools ahead of it. Turn the pinion

shaft so that the slot or groove at the upper end faces the center of the assembly. With a brass or soft steel drift, drive the pinion shaft in until the lower end is flush with the lower face of the planet carrier.

Following the same general procedure as outlined, assemble and install a short planet pinion in the planet carrier adjacent to the long planet pinion now installed. The thrust washers already installed with the long planet pinion also suffice for this short planet pinion as the two pinions are paired together on one set of thrust washers.

Install the input sun gear thrust washer, the input sun gear and low sun gear needle thrust bearing. Assemble and install the remaining planet pinions, in pairs, by first installing the long planet pinion and then the adjacent short planet pinion. Check end clearance of planet gears. This clearance should be .005"-.030". Place the pinion shaft lock plate in position, then with the extended portions of the lock plate aligned with slots in the planet pinion shafts, rotate the lock plate clockwise until the three attaching screw holes are accesible. Install the pinion shaft lock plate attaching screws and tighten to specifications.

GOVERNOR
The governor assembly is a factory balanced unit. If body replacement is necessary, the two sections must be replaced as a unit. Remove the governor as outlined under "Transmission-Disassembly."

Disassembly
The governor valve and shaft were already disassembled from the assembly during the removal procedures. Remove the outer weight assembly by sliding toward center of body. Remove the smaller inner weight retaining snap ring and remove the inner weight and spring. If it is considered necessary, remove the four body assembly bolts and separate the body, hub and gasket. Remove the two seal rings.

Inspection
Clean all parts thoroughly in a solvent and air dry. Check conditions of all component parts of the assembly. Replace any bent, damaged or scored parts. Body and hub must be replaced as a unit.

Assembly
Reassemble governor weights and reinstall in body bore. Replace seal rings on hub. Slide hub into place on output shaft and lock into place with the drive screw. Install gasket and governor body over output shaft, install governor shaft, line up properly with output shaft and install body attaching bolts. Torque to specifications. Place transmission selector lever in PARK to keep shaft from turning while tightening these bolts. Check the governor weight for free fit in body after the four attaching bolts are torqued. If the weight sticks or binds, loosen the bolts and retorque.

VALVE BODY
Removal
Remove valve body as described under "Transmission-Disassembly." If performing the operation on the vehicle, the vacuum modulator and valve, oil pan and gasket, guide detent plate and range selector detent roller spring must be removed in order to remove the valve body from the transmission.

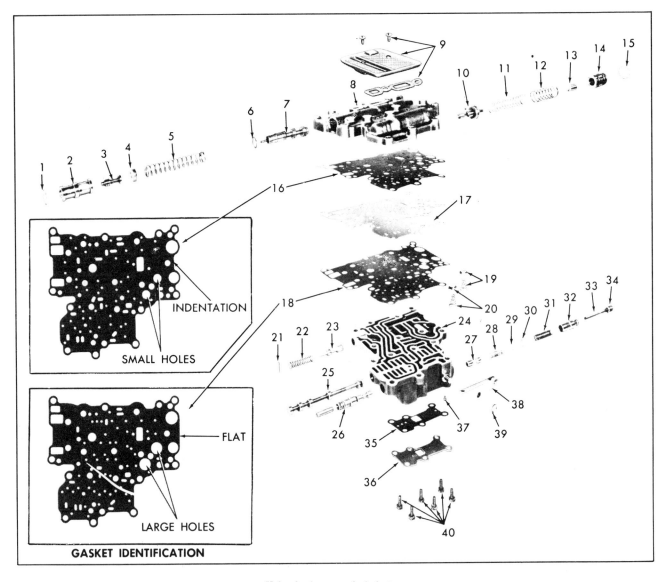

Valve body — exploded view

1. Snap ring
2. Hydraulic modulator valve sleeve
3. Hydraulic modulator valve
4. Pressure regulator spring retainer
5. Pressure regulator spring
6. Pressure regulator spring seat
7. Pressure regulator valve and
   dampener assembly
8. Lower valve body
9. Suction screen, gasket and attaching
   screws
10. Low and drive valve
11. Low and drive valve inner spring
12. Low and drive valve outer spring
13. Low and drive regulator valve
14. Low and drive regulator valve
    sleeve and cap

15. Snap ring
16. Transfer plate to lower valve
    body gasket
17. Transfer plate
18. Transfer plate to upper valve
    body gasket
19. Front pump check valve and spring
20. Rear pump check valve and spring
21. High speed down shift timing
    valve stop pin
22. High speed down shift timing
    valve spring
23. High speed down shift timing valve
24. Upper valve body
25. Manual control valve
26. Vacuum modulator valve,
    plunger and spring (exc. L-4 & L-194)

27. Throttle valve
28. Throttle valve spring
29. Throttle valve spring seat
30. Throttle valve spring regulator
    guide washer
31. Detent valve spring
32. Detent valve
33. Throttle valve spring regulator
34. Throttle valve spring regulator nut
35. Upper valve body plate gasket
36. Upper valve body plate
37. Detent valve and spring retaining stud
38. Range selector detent lever
39. Snap ring
40. Upper valve body plate to upper valve
    body attaching bolts and washers

Disassembly

Remove the manual valve, suction screen and gasket. Remove valve body bolts and carefully remove lower valve body and transfer plate from upper valve body. Discard gaskets. Remove the front and rear pump check valves and springs.

From the upper valve body, remove the TV and detent valves and the downshift timing valve. TV and Detent Valve — Remove the retaining pin by wedging a thin screw driver between its head and the valve body, then remove the detent valve assembly and throttle valve spring. Tilt the valve body to allow the throttle valve to fall out. If necessary, remove the "C" clip and disassemble the detent valve assembly. Do not disturb the setting of the adjustment hex nut on the detent valve assembly. This is a factory adjustment and should not normally be changed. However, some adjustment is possible if desired. See "Throttle Valve Adjustment." Downshift Timing Valve — Drive out the roll pin, remove the valve spring and the downshift timing valve.

From the lower valve body, remove the low-drive shift valve and the pressure regulator valve. Low-Drive Shift Valve — Remove the snap ring and tilt the valve body to remove the low-drive regulator valve sleeve and valve assembly, valve springs and the shifter valve. Pressure Regulator Valve — Remove the snap ring, then tilt valve body to remove the hydraulic modulator valve sleeve and valve, pressure regulator valve spring retainer, spring and pressure regulator valve assembly.

Inspection

Since most valve failures are caused initially by dirt or other foreign material preventing a valve from functioning properly, a thorough cleaning of all parts in clean solvent is mandatory. Check all valves and their bores for burrs or other deformities which could result in valve hang-up.

Assembly

Replace valve components in the proper bores, reversing the disassembly procedures given above and checking if necessary. Place front and rear pump check valves and springs into place in the upper valve body and install the gasket and transfer plate. Carefully install the lower valve body and gasket and install 15 1-3/8" attaching bolts. Torque to specifications.

Installation

Install the valve body onto the transmission as outlined under "Transmission-Assembly."

VACUUM MODULATOR

The vacuum modulator is mounted on the left rear of the transmission and can be serviced from beneath the vehicle.

Removal

Remove the vacuum line at the vacuum modulator. Unscrew the vacuum modulator from the transmission using J-9543, if available, or any thin 1" tappet type wrench. Remove the vacuum modulator plunger, dampening spring and valve from the transmission case.

Inspection and repairs

Check the vacuum modulator plunger and valve for nicks and burrs. If such cannot be repaired with a slip stone, replace the part. The vacuum modulator can be checked with a vacuum source for leakage. However, leakage normally results in transmission oil pull-over and results in oil smokey exhaust and continually low transmission oil. No vacuum modulator repairs are possible; replace as an assembly.

Installation

Install vacuum modulator valve, dampening spring and plunger in bore of transmission. Place a new gasket on vacuum modulator. The gasket has centering tabs to hold it centered during installation. Install vacuum modulator, tighten firmly, and install vacuum line. Rubber tubing "A" should bottom against modulator cam. Pipe assembly "B" should bottom against the modulator extension.

TRANSMISSION CASE

Inspection

Wash case thoroughly with cleaning solvent, air dry and blow out all oil passages. Do not use rags to dry parts. Inspect case for cracks which may contribute to leakage. Inspect case rear bushing for damage or excessive wear. This is a precision bushing and if damaged or worn excessively must be replaced. Check shifter shaft seal. If it shows signs of damage or leaking, pry it out and install a new seal. The new seal must be firmly seated in case counterbore.

Rear bushing — replacement

Transmission case rear bushing is a precision bushing which requires no reaming or finishing after assembly. Remove bushing by driving or pressing from within case using J-9557 and Handle J-7079. To install new bushing, drive or press bushing into place from rear of case using Tool J-9557 and Handle J-7079. Install bushing only until shoulder of J-9557 contacts the rear face of the case. Excessive force, either hammering or pressing may crack or otherwise damage the aluminum case.

TRANSMISSION EXTENSION

Inspection

Wash extension thoroughly with cleaning solvent and air dry. Do not use rags to dry parts. Inspect extension for cracks that may contribute to leakage. Inspect extension rear bushing for damage or excessive wear. Inspect rear oil seal and replace if damaged or worn.

Rear bushing — replacement

For service, the transmission extension rear bushing is of a precision type which requires no reaming or finishing after installation. Place transmission extension in an arbor press rear end up. Using J-5778, press old bushing from extension. Place new bushing on pilot end of J-5778 and press it into place. Replace extension rear oil seal, using Seal Installer J-5154. Prelubricate between lips of seal with cup grease.

TRANSMISSION

Assembly

Use only transmission oil or petroleum jelly as lubricants to retain bearings or races during assembly. Lubricate all bearings, seal rings and clutch plates prior to assembly.

Install the parking lock pawl and shaft and insert a new "E" ring retainer. Install the parking lock pawl pull-back spring over its boss to the rear of the pawl. The short leg of the spring should locate in the hole in the parking pawl. Install the parking lock pawl reaction bracket with its two bolts. Fit the actuator assembly between the parking lock pawl and the bracket.

Insert the outer shift lever into the case, being careful of the shaft seal, and pick up the inner shift lever and parking lock assembly and tighten allen head screw. Insert outer TV lever and shaft, special washer and "O" ring into case and pick up inner TV lever. Tighten allen head nut. To prevent possible binding between throttle and range selector controls; .010 to .020 clearance must exist between inner TV lever and inner shift lever after assembly. Thread the low band adjusting screw into case.

Transmission internal components

Install the inner and outer seals on the reverse piston and, lubricating the piston and case with transmission oil, install the piston into the case. If necessary, carefully slide a feeler gauge around the outer diameter of the piston to start the seal ring into the bore. With the support fixture turned so that the transmission case is facing up, install the 17 reverse piston return springs and their retainer ring.

Installing reverse piston

Carefully install Tool J-9542 over the retainer ring and through the rear bore of the case. With the flat plate on the rear face of the case, turn down on the wing nut to compress the return springs and allow the retaining ring snap ring to be installed. Remove tool J-9542. Use care when performing this operation that the spring retainer is correctly guided over the case internal hub and is not damaged by catching on the edge of the hub or in the snap ring groove. Install the large (waved) cushion spring.

Lubricate and install the reverse clutch pack beginning with a reaction (spacer) plate and alternating with the drive plates (faced) until all reaction plates and all drive plates are in place. The notched lug on each reaction plate is installed in the groove at 7 o'clock position in the case. Then install the thick pressure plate which has a "dimple" in one lug to align with the same slot in the case as the notched lugs on the other reaction plates. 327 and 396 cu. in. V-8 models have 6 reaction and 6 drive plates.

Install the clutch plate retaining ring. With the rear of the transmission case downward, align the internal lands and grooves of the reverse clutch pack faced plates, then engage the reverse ring gear with these plates. This engagement must be made by "feel" while jiggling and turning the ring gear.

Place the output shaft thrust bearing over the output shaft and install the planetary carrier and output shaft into the transmission case. Move the transmission into a horizontal position. The two input shaft seal rings should be in place on the shaft. Install the clutch drum (machined face first) onto the input shaft and install the low sun gear bushing (splined) against shoulder on shaft. Install clutch drum and input shaft assembly into case, aligning thrust needle bearing on input shaft and indexing low sun gear with the short pinions on the planet carrier.

Remove the rubber seal ring from the front pump body and, using guide studs from J-3387 set, install clutch drum selective thrust washer, front pump gasket and front pump to case. Install two pump-to-case bolts.

To check for correct thickness of the selective fit thrust washer, move transmission so that output shaft points down. Mount a dial indicator so that plunger of indicator is resting on end of the input shaft. J-5492 may be used to support the dial indicator. Zero the indicator. Push up on the transmission output shaft and observe the total indicator movement. The indicator should read .028" to .059". If the reading is within limits, the proper selective fit washer is being used. If the reading is not within limits, it will be necessary to remove the front pump, change to a thicker or thinner selective fit thrust washer, as required to obtain the specified clearance, and repeat the above checking procedure. Clutch drum selective thrust washers are available in thicknesses of .061", .078", .092" and .106".

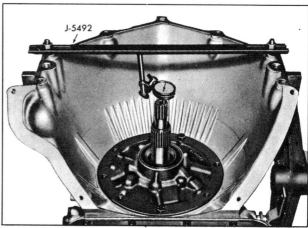

Checking end play for proper thrust washer selection

Install the servo piston, piston ring, and spring into the servo bore. Then, using a new gasket and "O" ring, install the servo cover. See that gasket is properly aligned with the three bolt holes and the drain back passage in the case. Remove the front pump and the selective fit washer from the case, and install the low brake band, anchor and apply struts into the case. Tighten the low band adjusting screw enough to prevent struts from falling out of case.

Place the seal ring in the groove around the front pump body and the two seal rings on the pump cover extension. Install clutch drum selective thrust washer, front pump gasket and front pump to case. Remove guide pins and install all pump bolts, replacing any damaged bolt sealing washers necessary and torque bolts to specifications.

Extension, governor and rear oil pump

Turn transmission so that shaft points upward, Install rear pump wear plate, drive pin, and drive gear, indexing gear to drive pin. Install rear pump body and driven gear, drain back baffle, and pump to case attaching bolts. Bolt holes are positioned so that the pump may be assembled only in the proper position.

Install governor over output shaft (See "Governor-Assembly" for body to hub installation.) Install governor shaft and valve, two Belleville washers (concave side of washers against output shaft), and retaining "C" clips. Center shaft in output shaft bore and tighten governor hub drive screw.

Using Tool J-5814, install speedometer gear into output shaft. Place extension seal ring over rear pump body and install transmission extension and five retaining bolts. If removed, replace speedometer driven gear.

Oil pan and valve body

With transmission upside down, and manual linkage installed as previously described, and the selector lever detent roller installed, install the valve body (servo apply tube installed) and a new gasket. Carefully guide the servo apply line into its boss in the case as the valve body is set into place. Install six mounting bolts and range selector detent roller spring. Position the manual valve actuating lever fully forward when installing valve body to more easily pick up the manual valve.

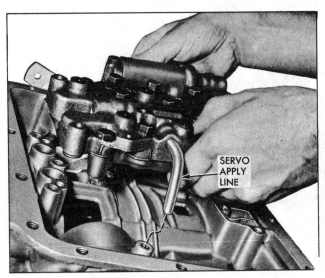

SERVO APPLY LINE

Installing valve body

Install the guide plate making sure that the inner lever properly picks up the manual valve. Install attaching bolts. Install the vacuum modulator valve and the vacuum modulator and gasket. Install the oil pan, using a new gasket, and the oil pan attaching bolts. Install converter and Safety Holding Strap J-9549 or a suitable substitute.

Low band adjustment

Tighten the low servo adjusting screw to 40 inch lbs. using torque wrench, J-5853 and a 7/32" hex head driver. The input and output shaft must be rotated simultaneously to properly center the low band on the clutch drum. Then back off four (4) complete turns for a band which has been in use for 1,000 miles or more, or 3 turns for a new band. Tighten the locknut to specified torque. The amount of back-off is not an approximate figure, it must be exact.

Installing detent guide plate

Throttle valve adjustment

No provision is made for checking TV pressures. However, if operation of the transmission is such that some adjustment of the TV is indicated, pressures may be raised or lowered by adjusting the position of the jam nut on the throttle valve assembly. To raise TV pressure 3 psi, backoff the jam nut one (1) full turn. This increases the dimension from the jam nut to the throttle valve assembly stop. Conversely, tightening the jam nut one (1) full turn lowers TV pressure 3 psi. A difference of 3 psi in TV pressure will cause a change of approximately 2 to 3 MPH in the wide open throttle upshift point. Smaller pressure adjustments can be made by partial turns of the jam nut. The end of TV adjusting screw has an allen head so the screw may be held stationary while the jam nut is moved. Use care when making this adjustment since no pressure tap is provided to check TV pressure.

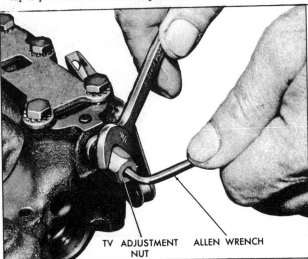

TV ADJUSTMENT NUT   ALLEN WRENCH

TV adjustment nut

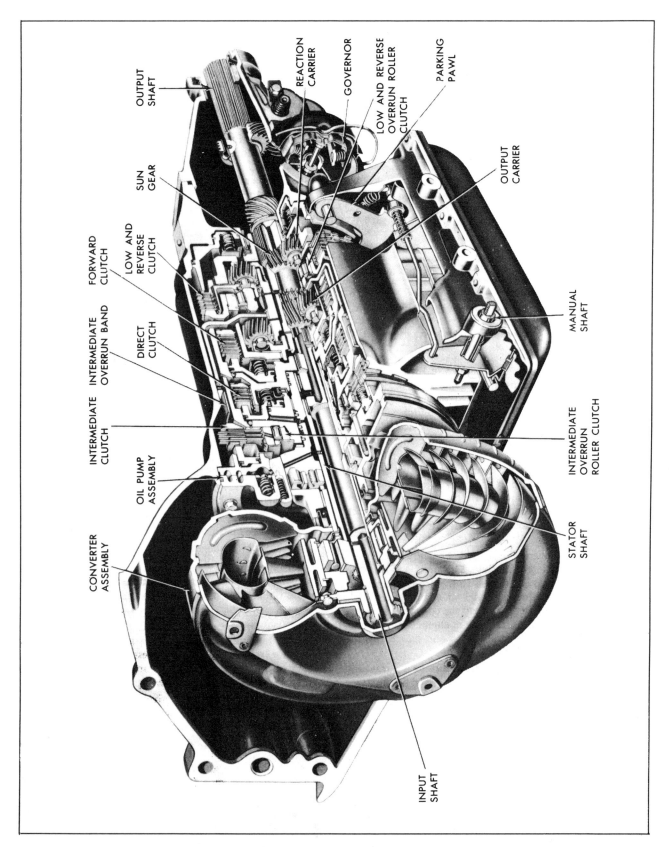

CBC 350 Transmission, Sectional View

CBC 350 AUTOMATIC TRANSMISSION
Description

This fully automatic transmission has a three-element torque converter and two planetary gear sets. There are four multiple-disc chutes, two roller clutches and an intermediate overrun band to properly operate the planetary gears.

The three elements of the torque converter are the pump, turbine and a stator. The stator runs on a clockwise-only roller clutch. The converter is serviced only as a complete assembly. It is fastened to the engine crankshaft by a flexplate and rotates at engine speed. The pump is part of the converter housing, and as it runs at engine speed it moves the fluid and therefore the turbine.

At low turbine speeds the converter multiplies engine torque, while at higher speeds it acts merely as a fluid coupling.

A gear pump is incorporated in the system to operate the clutches and automatic controls.

Manual linkage is provided to select the operating range. Engine vacuum operates the vacuum modulator and a cable control operates the detent valve.

The modulator valve senses changes in torque input to the transmission, transmits this signal to the pressure regulator which in turn allows smooth shifts at all throttle openings.

The detent valve is cable connected to the accelerator lever. At half throttle the valve is actuated causing a downshift at speeds below 50 mph. When the throttle is fully open the downshift occurs from 3 to 1 at speeds below 40 mph and from 3 to 2 below 75 mph.

MAINTENANCE AND ADJUSTMENTS
Oil Level

The dip stick is located at the right rear of the engine. One pint of fluid brings the level from the Add to Full marks. Check this level at every engine oil change. The level should be at the Full mark at normal operating temperature of 200°F. At 70°F the level should be ¼″ below the Add level. It requires 15 miles or more of highway driving to bring the oil to normal operating temperature.

6 pints of fluid are required to fill the transmission when the pan has been drained. The converter takes approximately 20 pints. Determine the proper level by the amount on the dipstick. Use only DEXRON or DEXRON II transmission fluid.

Raise the vehicle and support the transmission. If necessary, remove the transmission crossmember. Remove the oil pan bolts from the front and sides of the pan (use suitable oil receptacle). Loosen the rear bolts four turns and carefully pry the pan loose and allow the fluid to drain. Remove the remaining bolts and the oil pan. Discard the gasket, clean the pan with solvent and dry with compressed air.

Draining and refilling

Remove the two screws holding the strainer to valve body. Remove the strainer and gasket, discarding the gasket. Clean the strainer in solvent and dry with compressed air.

Reverse the above procedure for assembly, using new gaskets and tightening the pan bolts to 12 ft. lbs. Add 6 pints of fluid, start and idle the engine, move the selector through each position and check the fluid level with the selector in Park. Bring the fluid level to ¼″ below the add level.

Transmission Mount

While the vehicle is on a hoist, push up and down on the transmission extension while observing the mount. If the rubber separates from the metal plate or if the mount bottoms out, replace the mount. Check the mount attaching screws for tightness.

Console shift cable and linkage

To check the operation move the selector lever to each position and see that the transmission lever is in the detent position. With the key in "Run" and transmission in "Reverse", the key cannot be removed and the steering wheel is unlocked. With the key in "Lock" and the transmission in "Park" the key can be removed and the steering wheel is locked.

COOL
(65°-85°F.)
(18°-30°C.)

HOT
(190°-200°F.)
(88°-93°C.)

ADD .5 LITER
(1 PT.)

FULL HOT

WARM

NOTE: DO NOT OVERFILL. IT TAKES ONLY ONE PINT TO RAISE LEVEL FROM "ADD" TO "FULL" WITH A HOT TRANSMISSION.

Dipstick Markings

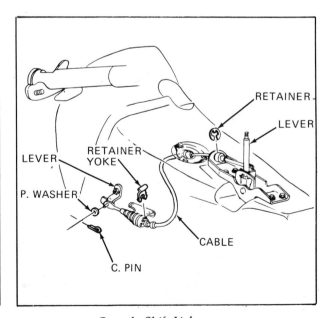

RETAINER

LEVER

LEVER

RETAINER
YOKE

P. WASHER

C. PIN

CABLE

Console Shift Linkage

When the console lever button is depressed and the lever is moved toward reverse there can be no rotation of the column shift bowl in relation to the column shroud.

With the selector lever in Drive, loosen the nut so the pin moves in the slot of the transmission lever. Place the lever in Drive by moving the lever counterclockwise to the detent and then clockwise three detent positions to Drive. Tighten the nut to 20 ft. lbs.

Place the selector lever in Park and ignition switch in Lock. Pull down lightly on the rod against the lock stop and tighten the screw to 20 ft. lbs.

DIAGNOSIS

The following sequence should be followed to properly diagnose transmission difficulties:

Check fluid level
Check detent downshift cable adjustment
Check vacuum line and fittings
Check manual linkage
Road test
Install oil pressure gauge and check pressures in each range during road test

DETERMINING SOURCE OF FLUID LEAK

Before trying to correct a leak, the actual source of the leak must be precisely pinpointed. Air flow around the transmission and engine can cause the oil from a leak to travel a relatively long distance from the source and deceive the trouble shooter. Therefore the suspected area should be thoroughly degreased then checked for leaks.

With the engine and transmission at normal operating temperature check very carefully to be sure where the leak is actually occuring. The following are some of the more common points of fluid leaks:

Transmission pan
Damaged pan gasket
Pan bolts not tight
Pan lip not flat
Extension housing to main body leak
Loose extension housing bolts
Rear seal leak
Porous castings
Filler pipe seal
Modulator "O" ring
Detent cable "O" ring
Loose governor cover
Speedometer "O" ring
Manual shaft seal
Loose line pressure tap plug
Front pump seal
Loose front pump bolts

Fluid pressure check

With the engine running and the parking brake on, and the engine speed set at 1200 rpm, the gauge connected at the line pressure tap and the vacuum modulator tube disconnected the gauge should read 162 psi in Drive, Neutral, Park, L1 and L2 and 255 psi in Reverse. These figures are for the V8 engine.

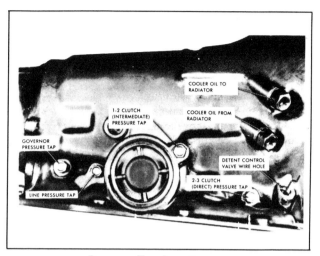

Pressure Tap Locations

Porous case repair

Porous cases can be easily repaired using epoxy. First be certain where the leak is occuring by thoroughly warming the engine and transmission and checking for leaks. The area to be repaired must be completely cleaned with a solvent such as acetone using a stiff brush and air drying. Apply the properly mixed epoxy to the area and let dry for at least four hours before checking for leaks.

Vacuum modulator diagnosis

An improperly operating vacuum modulator will cause harsh up-and-down-shifts, late upshifts, soft up-and downshifts, slip in low, drive and reverse, overheating, or engine burning transmission fluid.

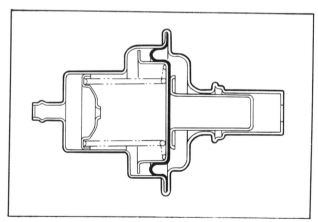

Vacuum Modulator Assembly

To check the vacuum diaphragm for leaks, insert a pipe cleaner into the connector pipe as far as possible. If any transmission fluid is found in the modulator it must be replaced. A trace of gasoline or water does not matter.

Check the modulator body by rolling it on a flat surface to see that the sleeve is concentric with the cam. If they are concentric and the plunger is free, the modulator is serviceable.

Burned clutch plates

These are some of the causes of burned clutch plates. Forward clutch: Check ball in clutch housing stuck, missing or damaged. Clutch piston cracked and/or seals damaged. Low pressure. Pump cover seal rings worn, broken or missing. Oversize ring groove in pump cover. Valve body face not flat.

Intermediate clutch: Piston seals missing or worn. Low pressure. Valve body face not flat.

Direct clutch: Vacuum line to modulator restricted. Check ball in clutch piston stuck, missing or damaged. Defective modulator bellows. Clutch piston seals damaged or missing. Valve body face not flat. Clutch installed backwards.

Governor pressure check

Install pressure gauge at line pressure tap. Disconnect vacuum line to modulator. With rear whells off the ground, selector in Drive and handbrake on, check pressure at 1000rpm. Increase rpm's to 3000 and if pressure does not drop 7 psi or more check the following: Stuck valve in governor, free weight in governor, restricted orifice in governor valve, check screen in control valve assembly, restriction in feed line, scored bore in governor.

Manual linkage

The manual linkage adjustment and starter safety switch should restrict starting to the Park and Neutral positions only. Adjust as necessary. With the selector in Park, the pawl should engage easily and prevent the car from moving. The pointer should line up with each position marker.

Road test

Drive range: Place selector in Drive and accelerate. 1-2 and 2-3 shifts should occur at all throttle openings. As vehicle slows to a stop 3-2 and 2-1 shifts should occur. In the L2 range, 1-2 and 2-1 shifts should occur. In the L1 range no shift should occur.

With the selector in Drive and the vehicle running at 35 mph move the selector to L2 at which point the transmission should downshift to 2nd. A braking effect should increase from 100 to downshift to 2nd. A braking effect should be noticed along with an increase in engine rpm. Line pressure should increase from 100 to 125 psi.

With the selector in L2 and the vehicle running at 30 mph place the selector in L1 at which point the downshift should occur. Braking should begin accompanied by an increase in engine rpm. Line pressure should increase to 150 psi.

With the selector in Reverse check for reverse operation.

Trouble diagnosis

No drive in Drive range: Install pressure gauge and check fluid level. Check for leaks and defective vacuum modulator. Manual linkage misadjusted.

Low oil pressure: Forward clutch does not engage - seals leaking, piston cracked or clutch plates burned. Pump feed to forward clutch oil seal rings missing or broken. Leak in feed circuits. Pump to case gasket leaking. Clutch ball check stuck or missing. Low - reverse clutch assembly spring broken, damaged cage or installed backwards.

High line pressure: Vacuum leak caused by disconnected vacuum line, line leak from engine to modulator, high or low engine vacuum, vacuum leak in accesory hoses. Modulator valve stuck, water in modulator or modulator damaged. Detent valve stuck in detent position. Pressure regulator or boost valve stuck. Boost valve sleeve broken.

Low line pressure: Low fluid level. Defective vacuum modulator. Blocked or restricted screen assembly or damaged gasket. Oil pump gears worn. Leaking pump to case gasket. Defective pump body or cover. Bottom seal ring on pump cover hub leaking. Valve body pressure regulator or boost valve stuck. Weak pressure regulator valve spring. Loose valve bolts. Stuck valve in reverse and modulator booster. Check pump oil seal rings and forward clutch seals for leaks. Check direct clutch outer seal. Check both 1-2 and 2-3 accumulator pistons and rings. Check for broken intermediate servo piston seal ring. Check ball missing from case face.

No line pressure: Blockage in suction cavity in case. Missing priming valve in pump. Broken front pump drive gear lugs. Missing vacuum modulator valve. Pump to case gasket improperly installed.

No upshift: Stuck governor valve. Stuck shift valve or stuck shift control. Broken gear in governor assembly. Loose governor locating pin in case. Broken piston ring in 1-2 accumulator. Broken intermediate piston seals. Jammed governor weights.

Low 3-2 or 2-1 downshifts: Stuck detent regulator valve. Missing detent regulator valve spring seat. Sticking shift control or shift valves. Disconnected detent linkage. Wrong vacuum setting. Defective governor. Sticking modulator valve.

High 3-2 or 2-1 downshifts: Sticking detent regulator valve. Defective governor. Sticking shift control or shift valves.

No drive in Drive: Damaged forward clutch piston seals. Relief ball in forward clutch drum not sealing. Excessive clearance in pressure plate. Slipping rear roller clutch. Damaged shifter shaft. Improperly installed manual valve.

No drive in Reverse: Broken direct clutch apply seal rings. Blocked reverse apply holes. Improperly installed manual vlave.

1-2 shift requires full throttle: Sticking detent valve. Leaking vacuum line or fittings. Leaking valve body gaskets. Stuck detent valve train. 1-2 valve stuck closed.

No 1-2 shift: Binding downshift detent cable. Sticking governor valve. Loose driven gear. Leaking valve body gaskets. Blocked governor feed channels. Shift valve train stuck. Defective clutch piston seals. Broken spring or damaged cage in intermediate clutch.

No 2-3 shift: Leaking valve body gaskets. Shift valve train stuck. Broken seals at direct clutch to pump hub. Defective clutch piston seals. Burned clutch plates.

Drive obtained in Neutral: Misadjusted manual linkage. Manual valve disconnected. Line pressure leaking into forward clutch apply passage. Burned forward clutch plates.

No Reverse or slip in Reverse: Low fluid level. Misadjusted manual linkage. Low line pressure. Leaking valve body gaskets. 2-3 shift valve train stuck open. Piston or pin in intermediate servo stuck and intermediate overrun band is applied. Damaged Low-Reverse clutch outer seal. Burned clutch plates. Forward clutch does not release.

Slip in all ranges: Low fluid level. Low line pressure. Burned clutch plates. Seal rings on pump cover damaged. Cross leaks in case.

1-2 shift slips: Low fluid level. Low line pressure. Either accumulator oil rings damaged. Damaged pump-to-case gasket. Damaged intermediate clutch piston seals. Burned clutch plates.

SERVICE OPERATIONS
Drive range detent downshift cable

To remove cable, push up bottom of snap lock and release lock and cable. Remove cable from the carburetor lever. Disconnect snap lock assembly by compressing the locking tabs. Remove clamp from filler tube, remove screw holding cable to transmission and remove detent cable.

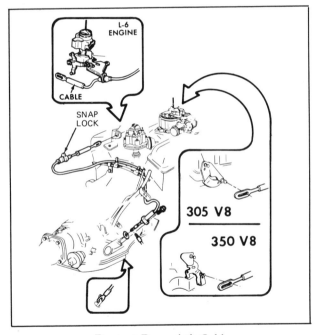

Dentent Downshift Cable

To install, place new seal on cable and lubricate seal with transmission fluid. Connect the transmission end of the cable and fasten cable to case tightening bolt to 20 ft. lbs. Route cable in front of filler tube and install clamp. Locate clamp 2 inches above filler tube bracket. Pass cable through bracket and engage locking tabs of snap lock. Fasten cable to carburetor lever.

To adjust, disengage the snap lock, place the carburetor in wide open throttle position and push snap lock downward until top is flush with cable.

Vacuum modulator and modulator valve assembly

Remove the vacuum hose from the modulator stem and remove the attaching screw and retainer. Remove the modulator assembly with its O-ring. Remove modulator valve from case. Reverse this procedure to install the modulator and modulator valve. Check fluid level.

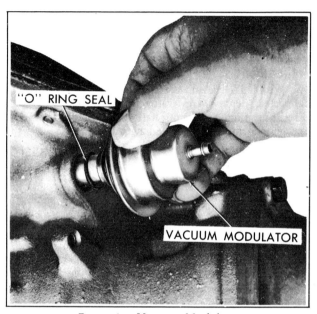

Removing Vacuum Modulator

Oil cooler pipes

If the oil cooler pipes need replacing use only steel tubing with double lap flared fittings.

Removing Governor Cover

Governor

To remove the governor, remove the speedometer cable at the transmission and remove the governor cover retainer and the governor cover. Remove the governor and inspect the weights and the valve for free travel.

To install the governor use a brass drift around the outside flange of the cover. Install the retainer and reconnect the speedometer cable. Check fluid level.

Speedometer driven gear

Disconnect the speedometer cable. Remove the retainer bolt, retainer, the driven gear and the O-ring. Reverse this procedure to install. Check fluid level.

Speedometer drive gear

To remove the drive gear first raise the vehicle and support the transmission with a proper jack. Remove the propeller shaft and disconnect the speedometer cable. Disconnect the transmission rear mount from the crossmember. Remove the crossmember and the extension housing. Using special tools J-21427-01 and J-8105 and output shaft, remove the speedometer drive gear. Remove the retaining clip.

To install, place the drive gear retaining clip in the hole in the output shaft. Align the slot in the drive gear with the retaining clip and install. Install extension housing and tighten bolts to 25 ft. lbs. Connect speedometer cable and install crossmember to frame and transmission. Install propeller shaft.

Extension housing oil seal

Remove the propeller shaft and pry out the oil seal with a small screwdriver. Drive a new seal in place with tool J-21426. Install propeller shaft and check fluid level.

Accumulator

To remove 1-2 accumulator remove the two oil pan bolts below the accumulator cover. Install tool J-23069 in place of the bolts removed. Remove retaining ring by pressing in on cover. Remove the cover O-ring seal, spring and accumulator.

To install, first replace the accumulator piston. Place the spring, O-ring seal and cover in position. Press in on cover with the special tool J-23069 and install the retaining ring. Install the oil pan bolts after removing the tool.

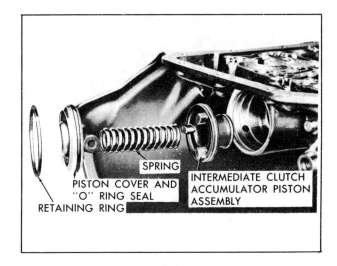

SPRING

PISTON COVER AND "O" RING SEAL

RETAINING RING

INTERMEDIATE CLUTCH ACCUMULATOR PISTON ASSEMBLY

Removing 1-2 Accumulator Valve Body

Valve body assembly

To remove the valve body assembly, first drain the oil pan. Remove the oil pan and strainer and discard their gaskets. Remove the detent spring and roller assembly from the valve body and remove the valve body bolts. Disconnect the manual control valve link from the range selector inner lever. Remove the detent control valve link from its actuating lever. Then remove the valve body assembly. Remove the manual valve and link assembly from the valve body assembly.

Installation of the valve body assembly is the reverse of the removal procedure. Use new gaskets and check fluid level.

Manual shaft, range selector inner lever and parking linkage

First drain the oil pan then remove the valve body assembly, discarding the gaskets. Remove the manual shaft to case retainer and unthread the jam nut holding the range selector inner lever to the manual shaft. Remove the jam nut, manual shaft from range selector inner lever and case. Remove the parking pawl actuating rod and range selector inner lever from the case. Remove the parking lock bracket bolts and bracket. Remove the parking pawl spring.

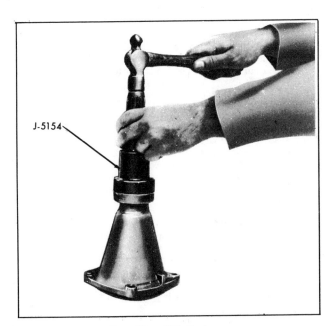

J-5154

Installing Oil Seal

Installation is the reverse of removal, using new gaskets and check the fluid level.

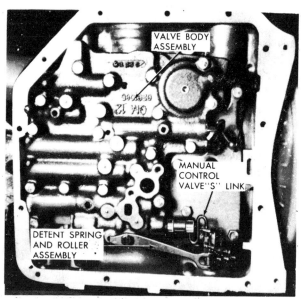

valve body

Transmission replacement
Disconnect the battery negative cable, the detent downshift cable at the carburetor and release the parking brake. Raise the vehicle and remove the propeller shaft. Disconnect the speedometer cable, detent downshift cable, modulator vacuum line, and oil cooler tubing at the transmission. Disconnect the shift control linkage and properly support the transmission. Disconnect the rear mount from the frame crossmember, then remove the crossmember. Remove the converter under pan. Remove the converter to flexplate bolts. Loosen the exhaust pipe to manifold bolts and ¼". Remove the transmission to engine bolts and the oil filler tube at the transmission. Raise the transmission to its normal position then slide it rearward and remove it from the vehicle.

Installation is the reverse of the removal procedure. Check linkage adjustments and fluid level.

TURBO HYDRA-MATIC 400 TRANSMISSION
Description
The Turbo Hydra-Matic 400 transmission is a fully automatic unit consisting primarily of a 3-element hydraulic torque converter and a compound planetary gear set. Three multiple-disc clutches, one sprag unit, one roller clutch and two bands provide the friction elements required to place and keep the transmission in the proper gear.

External control connections to transmission are:

MANUAL LINKAGE — To select the desired operating range.

ENGINE VACUUM — To operate a vacuum modulator unit.

ELECTRICAL SIGNAL — To operate an electrical detent solenoid.

A vacuum modulator is used to automatically sense any change in the torque input to the transmission. The vacuum modulator transmits this signal to the pressure regulator for line pressure control, to the 1-2 accumulator valve, and to the shift valves so that all torque requirements of the transmission are met and smooth shifts are obtained at all throttle openings.

The detent solenoid is activated by an electric switch on the carburetor or accelerator linkage. When the throttle is fully opened, the switch on the carburetor is closed, activating the detent solenoid and causing the transmission to downshift at speeds below approximately 70 MPH.

N, D, L_2, L_1, (P, R, N, 3, 2, 1).

P. — Park position locks the output shaft to the transmission case by means of a locking pawl. The engine may be started in Park position.

R. — Reverse enables the car to be operated in a reverse direction.

N. — Neutral position enables the engine to be started and run without driving the car.

3-D. — Drive Range is used for all normal driving conditions and maximum economy. Drive Range has three gear ratios, from the starting ratio to direct drive. Downshifts are available by depressing the accelerator to the floor.

2-L_2.-L_2 Range has the same starting ratio as Drive Range, but prevents the transmission from shifting to third speed, to retain second speed acceleration when extra performance is desired. L_2 Range can also be used for engine breaking. L_2 Range can be selected at any car speed, and the transmission will shift to second gear and remain in second until the car speed or the throttle are changed to obtain first gear operation in the same manner as in D. Range.

1-L_1.-L_1 Range can be selected at any vehicle speed, and the transmission will shift to second gear and remain in second until vehicle speed is reduced to approximately 40 MPH, depending on axle ratio. L_1 Range position prevents the transmission from shifting out of first gear.

DETENT SWITCH
Adjustment procedure
Pull detent switch driver rearward until the hole in the switch body aligns with hole in driver. Insert a .092 inch diameter pin through the aligned holes to a depth of .10 inches. This will hold driver in position. Loosen mounting bolt. Depress accelerator to W.O.T. position. With accelerator pedal in this position move switch forward until driver contacts accelerator lever. Tighten mounting bolt. Remove pin.

Transmission fluid
Transmission fluid level should be checked with transmission warm and selector lever in "P" Park position, every time engine oil level is checked.

Fluid checking procedure
To determine proper fluid level, proceed as follows: with

the transmission hot (after vehicle has been driven at least 15 miles), the fluid level should be between the "FULL" mark and 1/4 inch below FULL. The vehicle should be level with the engine idling and the transmission in PARK.

If the vehicle has not been driven sufficiently to bring the transmission up to operating temperature, the fluid level should be checked as follows: Apply the parking brake, put the selector lever in PARK, and start the engine. Do Not Race the Engine. Move the selector lever through each range. Immediately check the fluid level with the selector lever in the PARK position. The engine should be running at a slow idle and the vehicle should be level. The fluid level on the dipstick should be between the "ADD" mark and 1/4 inch below. If additional fluid is required, add sufficient fluid to bring the level to 1/4 inch below the "ADD" mark on the dipstick. If the transmission fluid level can be correctly established at room temperature (80°F) as described, when the transmission reaches normal operating temperature, the fluid level will appear at the "FULL" mark. The fluid level is set at 1/4 inch below the "ADD" mark on the dipstick to allow for expansion of the fluid which occurs as the transmission temperature rises to its normal operating point of 180°F. DO NOT OVERFILL, as foaming and loss of fluid through the vent pipe might occur as fluid heats up. If fluid is too low, especially when Cold, complete loss of drive may result which can cause transmission failure.

Fluid level indicator
The fluid level indicator is located in the filler pipe at the right rear corner of the engine. To bring the fluid level from the add mark to the full mark add 1 pint. Fluid level should be to the full mark with transmission at normal operating temperature. With cold fluid the level should be approximately 1/4" below the add mark.

Neutral safety switch adjustment
The neutral safety switch must be adjusted so that the car will start in the park or neutral position, but will not start in the other positions.

Draining and refilling transmission
Drain oil immediately after operation before it has had an opportunity to cool. To drain oil raise car on hoist or place on jack stands, and provide container to collect draining fluid. Remove oil pan and gasket. Discard gasket. Drain fluid from oil pan. Clean pan with solvent and dry thoroughly with clean compressed air. Remove filter. Remove and discard oil filter to case o-ring. Install new oil filter to case o-ring. Install filter assembly. Install new gasket on oil pan and install pan. Tighten attaching bolts to 12 lb. ft. Lower car and add 5 pints of transmission fluid through filler tube. With manual control lever in Park position, start engine. DO NOT RACE ENGINE. Move manual control lever through each range. Immediately check fluid level with selector lever in Park, engine running, and vehicle on LEVEL surface. Add additional fluid to bring level to 1/4" below the "ADD" mark on the dipstick.

PRESSURE REGULATOR VALVE
Removal
Remove bottom pan and filter. Compress regulator boost valve bushing against pressure regulator spring and remove snap ring, using J-5403 pliers. Remove regulator boost valve bushing and valve. Remove pressure regulator spring.

Remove regulator valve, spring retainer, and spacer(s) if present.

Installation
Installation of the pressure regulator valve is the reverse of the removal. Adjust oil level.

CONTROL VALVE BODY
Removal
Remove bottom pan and filter. Disconnect lead wire from pressure switch assembly. Remove control valve body attaching screws and detent roller spring assembly. Do not remove solenoid attaching screws. Remove control valve body assembly and governor pipes. If care is taken in removing control valve body the six (6) check balls will stay in place above the spacer plate. Do not drop manual valve. Remove the governor pipes and manual valve from control valve body.

Installation
Installation of the control valve body is the reverse of the removal. Adjust oil level.

GOVERNOR
If difficulty is encountered when removing the governor assembly, it may be necessary to remove the oil pan and withdraw the governor pipes (approximately 1/8") to free governor assembly.

Removal
Remove governor cover attaching screws, cover, and gasket. Discard gasket. Withdraw governor assembly from case.

Installation
Installation of the governor assembly is the reverse of the removal. Use a new gasket under the governor cover. Adjust oil level.

MODULATOR AND MODULATOR VALVE
Removal
Remove modulator assembly attaching screw and retainer. Remove modulator assembly from case. Discard "o" ring seal. Remove Modulator valve from case.

Installation
Installation of the modulator assembly and modulator valve is the reverse of the removal. Use a new "O" ring seal on the modulator assembly. Adjust oil level.

PARKING LINKAGE
Removal
Remove bottom pan and oil filter. Unthread jam nut holding detent lever to manual shaft. Remove manual shaft retaining pin from case. Remove manual shaft and jam nut from case. DO NOT remove manual shaft seal unless replacement is required. Remove parking actuator rod and detent lever assembly. Remove parking pawl bracket attaching screws and bracket. Remove parking pawl return spring.

The following should not be completed unless part replacement is required. Remove parking pawl shaft retainer. Remove parking pawl shaft, cup plug, parking pawl shaft, and parking pawl.

Installation

Installation of the parking linkage is the reverse of the removal. Use new seal and cup plug, if removed, and new bottom pan gasket. Adjust oil level.

REAR SEAL
Removal

Remove propeller shaft. Pry seal out with screw driver or small chisel.

Installation

All Models except "CP", use a non-hardening sealer on outside of seal body; and using Tool J-21359, drive seal in place. Model "CP", use a non-hardening sealer on outside of seal body; and using Tool J-21464, drive seal in place. Re-install propeller shaft.

TRANSMISSION REPLACEMENT

Before raising the car, disconnect the negative battery cable and release the parking brake. Place car on hoist. Remove propeller shaft. Disconnect speedometer cable, electrical lead to case connector, vacuum line modulator, and oil cooler pipes. Disconnect shift control linkage.

Support transmission with suitable transmission jack. Disconnect rear mount from frame crossmember. Remove two bolts at each end of frame crossmember plus through bolt at inside of frame and parking brake pulley. Remove crossmember. Remove converter under pan. Remove converter to flywheel bolts. Lower transmission until jack is barely supporting it. Remove transmission to engine mounting bolts and remove oil filler tube at transmission. Raise transmission to its normal position, support engine with jack and slide transmission rearward from engine and lower it away from vehicle. Use converter holding Tool J-5384 when lowering transmission or keep rear of transmission lower than front so as not to lose converter.

The installation of the transmission is the reverse of the removal with the following added step. Before installing the flex plate to converter bolts, make certain that the weld nuts on the converter are flush with the flex plate and the converter rotates freely by hand in this position. Then, hand start all three bolts and tighten finger tight before torqueing to specification. This will insure proper converter alignment. After installation of transmission, remove car from hoist. Check linkage for proper adjustment.

TURBO HYDRA-MATIC 400 DIAGNOSIS PROCEDURE

In the event of a major transmission failure, replace filter assembly, and flush oil cooler and cooler lines.

Accurate diagnosis of transmission problems begins with a thorough understanding of normal transmission operation. In particular, knowing which units are involved in the various speeds or shifts so that the specific units or circuits involved in the problem can be isolated and investigated further.

The following sequence, based on field experience, provides the desired information quickly and in most cases actually corrects the malfunction without requiring the removal of the transmission.

Check oil level and condition. Check and correct detent switch. Check and correct vacuum line and fittings. Check and correct manual linkage.

Oil level and condition check

Always check the oil level before road testing. Oil must be visible on dip stick prior to operating the vehicle. Erratic shifting, pump noise, or other malfunctions can in some cases be traced to improper oil level. If oil level is low, refer to Oil Leaks. The condition of the oil is often an indication of whether the transmission should be removed from the vehicle, or to make further tests. When checking oil level, a burned smell and discoloration indicate burned clutches or bands and the transmission will have to be removed.

Manual linkage

Manual linkage adjustment and the associated neutral safety switch are important from a safety standpoint. The neutral safety switch should be adjusted so that the engine will start in the Park and Neutral positions only. With the selector lever in the Park position, the parking pawl should freely engage and prevent the vehicle from rolling. The pointer on the indicator quadrant should line up properly with the range indicators in all ranges.

Oil leaks

Before attempting to correct an oil leak, the actual source of the leak must be determined. In many cases, the source of the leak can be deceiving due to "wind flow" around the engine and transmission. The suspected area should be wiped clean of all oil before inspecting for the source of the leak. Red dye is used in the transmission oil at the assembly plant and will indicate if the oil leak is from the transmission. The use of a "black light" to identify the oil at the source of the leak is also helpful. Comparing the oil from the leak to that on the engine or transmission dip stick (when viewed by black light) will determine the source of the leak.

Oil leaks around the engine and transmission are generally carried toward the rear of the car by the air stream. For example, a transmission "oil filler tube to case leak" will sometimes appear as a leak at the rear of the transmission. In determining the source of an oil leak it is most helpful to keep the engine running.

Possible points of oil leaks

Transmission Oil Pan Leak: Attaching bolts not correctly torqued. Improperly installed or damaged pan gasket. Oil pan gasket mounting face not flat.

Rear Extension Leak: Attaching bolts not correctly torqued. Rear seal assembly — damaged or improperly installed. Gasket seal — (extension to case) damaged or improperly installed. Porous casting.

Case Leak: Filler pipe "O" ring seal damaged or missing; misposition of filler pipe bracket to engine — "loading" one side of "O" ring. Modulator assembly "O" ring seal — damaged or improperly installed. Governor cover, gasket and bolts — damaged, loose; case face leak. Speedo gear — "O" ring damaged. Manual shaft seal — damaged, improperly installed. Line pressure tap plug — stripped, shy sealer compound. Parking pawl shaft cup plug — damaged, improperly installed. Vent pipe. Porous case.

Front End Leak: Front seal — damaged (check converter neck for nicks, etc., also for pump bushing moved forward); garter spring missing from pump to converter seal. Pump attaching bolts and seals — damaged, missing, bolts loose. Converter — leak in weld. Pump "O" seal — damaged. (Also check pump groove and case bore.) Porous casting (pump or case).

Oil Comes Out Vent Pipe' Transmission over-filled. Water in oil. Pump to case gasket mispositioned. Foreign material between pump and case, or between pump cover and body. Case — porous, pump face improperly machined. Pump — shy of stock on mounting faces, porous casting.

Case porosity — repair

Transmission leaks caused by aluminum case porosity have been successfully repaired with the transmission in the vehicle. Road test and bring the transmission to operating temperature. Raise the car and, with the engine running, locate the source of the oil leak. Check for leaks in all operating positions. The use of a mirror will be helpful in finding leaks. Shut off engine and thoroughly clean area with a solvent and air dry. Using the instruction of the manufacturer, mix a sufficient amount of epoxy cement, part #1360016, to make the repair. While the transmission is still hot, apply the epoxy to the area, making certain that the area is fully covered. Allow epoxy cement to dry for three hours and retest for leaks.

OIL PRESSURE CHECK
With car stationary

Oil Pressure Check — Road or Normal: Operating Conditions — While road testing with the transmission oil pressure gauge attached and the vacuum modulator tube connected, the transmission pressures should check approximately as shown on chart. Transmission oil pressure gauge and engine tachometer should be connected and the oil pressures should check as follows: Pressures indicated below are 0 output speed with the vacuum modulator tube disconnected and with engine at 1200 rpm. Pressures indicated below are with the vacuum tube connected for normal modulator operation, and with the engine at approx. 1000 rpm. Pressures are not significantly affected by altitude or barometric pressure when the vacuum tube is connected.

Drive, Neutral, Park	(1 or 2) L_1 or L_2	Reverse
60	150	107

Approximate altitude of check. (ff above sea level)	Drive Neutral Park	L1 or L2	Reverse
0	150	150	244
2,000	150	150	233
4,000	145	150	222
6,000	138	150	212
8,000	132	150	203
10,000	126	150	194
12,000	121	150	186
14,000	116	150	178

VACUUM MODULATOR
assembly

After thorough investigation of field return modulator assemblies, it has been found that over 50% of the parts returned as defective were good. For this reason, the following procedure is recommended for checking Turbo Hydra-Matic 400 modulator assemblies in the field before replacement is accomplished.

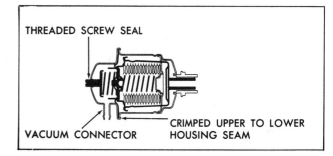

Vacuum modulator assembly

Vacuum diaphragm leak check

Insert a pipe cleaner into the vacuum connector pipe as far as possible and check for the presence of transmission oil. If oil is found, replace the modulator. Gasoline or water vapor may settle in the vacuum side of the modulator. If this is found without the presence of oil, the modulator should not be changed.

Atmospheric leak check

Apply a liberal coating of soap bubble solution to the vacuum connector pipe seam, the crimped upper to lower

	Minimum	Maximum
L₂-2nd Gear - Steady road load at approximately 25 mph	145 psi	155 psi

Gear	Selector Lever Position	Minimum	Maximum
1st	Drive		
2nd	("Zero" throttle to full throttle)	60	150
3rd			
3rd	Drive Range, Zero Throttle at 30 mph	60	
Reverse	Rev. (Zero to full throttle)	95	260

housing seam, and the threaded screw seal. Using a short piece of rubber tubing, apply air pressure to the vacuum pipe by blowing into the tube and observe for leak bubbles. If bubbles appear, replace the modulator. Do not use any method other than human lung power for applying air pressure, as pressures over 6 psi may damage the modulator.

Bellows comparison check
Using a comparison gauge, compare the load of a known good Hydra-Matic modulator with the assembly in question. Install the modulator that is known to be acceptable on either end of the gauge. Install the modulator in question on the opposite end of the gauge. Holding the modulators in a horizontal position, bring them together under pressure until either modulator sleeve end just touches the line in the center of the gauge. The gap between the opposite modulator sleeve end and the gauge line should then be 1/16" or less. If the distance is greater than this amount, the modulator in question should be replaced.

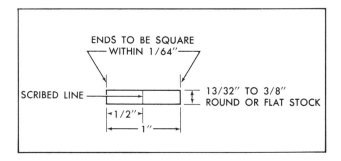

Bellows comparison gauge

Sleeve alignment check
Roll the main body of the modulator on a flat surface and observe the sleeve for concentricity to the cam. If the sleeve is concentric and the plunger is free, the modulator is acceptable. Once the modulator assembly passes all of the above tests, it is an acceptable part and should be re-used.

Pressure switch check
The pressure switch in the Turbo Hydra-Matic 400 transmission may be easily checked with a continuity tester which uses either the vehicle's battery or a separate source to power the test lamp or meter. To prevent damaging the pressure switch, the tester used must have sufficient resistance that is does not supply the switch with more than .8 amp of current at 12 volts. Test lamps which use size 1893 or smaller bulb, will not damage the switch.

Disconnect electrical wire harness from transmission electrical connector. If using a continuity tester which has a self contained power source, connect one tester lead to pressure switch terminal of transmission connector, and connect other lead to ground. Tester should show current flowing through the switch (switch closed). If the vehicle's battery is to be used to power a simple light bulb type tester, connect one lamp lead to pressure switch terminal of transmission connector and connect other lead to a

"Hot" terminal of the vehicle's wiring harness. Check that vehicle's battery cables are connected.

Set parking brake, apply service brakes, start engine and let it idle. Move transmission selector to "Drive" and tester should show current flowing through switch to ground (switch closed). With brakes still applied, move transmission selector to "Reverse" which supplies transmission oil pressure to pressure switch and tester should show current flowing through switch to ground (switch open). If above test indicates a defective pressure switch circuit, the transmission bottom pan must be removed to service the pressure switch or lead wire assembly.

FLOOR SHIFT CABLE
Check operation
If adjustment is required, loosen nut so that pin moves in slot of transmission lever and remove floor console cover. Move transmission lever counterclockwise to its maximum position and then clockwise five (5) detent positions (park). Position shift lever in park (P) position and insert a .040 inch spacer forward of pawl and tighten nut to 20 pound feet. With shift lever in park (P) position, position ignition switch to "Lock" position. Remove backdrive cable cotter pin and washer at column lever and disconnect cable from lever retaining pin.

From inside vehicle, loosen the two nuts at the steering column-to-dash bracket. Rotate the lock tube lever counterclockwise (viewed from front of column) to remove any free play from the column. Reposition the bracket until the cable eye passes freely over the retaining pin on the bracket. Holding the bracket in position, have an assistant tighten the bracket retaining nuts inside the vehicle. Reinstall the cotter pin and washer retaining the cable to the lever retaining pin.

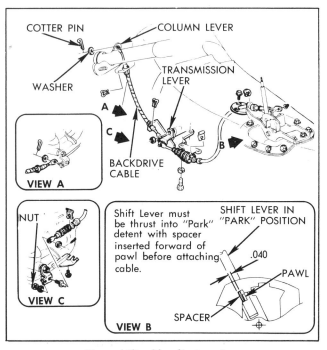

Floor shift cable adjustment

DRIVE AXLE & DRIVELINE 9

INDEX

9

DESCRIPTION

The independent three-link type rear suspension consists of a fixed differential carrier which is rubber mounted to the frame at three points, with the strut rods, drive shafts and torque control arms forming the three links at each wheel, and a transversely mounted multi-leaf spring.

The box section trailing torque control arms are mounted at the forward end into frame side member openings through pivot bolts and rubber bushings, and extend rearward to connect to the leaf spring. The wheel spindles and spindle supports are attached to the torque control arms through four bolts pressed into the arm. Rear wheel toe-in angle is adjusted through the use of variable thickness shims inserted between the torque arm and the frame side member web at the forward pivoting joint.

The rear wheel spindles are driven through double"U" jointed, tubular drive shafts which are flange mounted to a splined spindle flange at their outboard end and bolted to the differential side gear yokes at their inboard end. The wheel spindle support houses the inner and outer tapered roller bearings, two to each wheel. Bearing adjustment is made through the use of a spacer and variable thickness shims between the bearings.

The spindle supports also incorporate integrally forged, fork-shaped mounting brackets to accept the outer ends of the rubber-bushed strut rods. The strut rods are mounted laterally from the spindle support to a bracket bolted to the lower surface of the axle carrier. The strut rod connection at this point is with an eccentric cam arrangement and provides for rear wheel camber adjustment.

The direct double-acting shock absorbers are attached at the upper eye to a frame bracket and at the lower eye to the strut rod mounting shaft which incorporates a threaded stud for the shock absorber lower eye. The transversely mounted multi-leaf spring is clamp bolted at the center section to a lower mounting surface on the differential carrier cover. The outer ends of the main leaf are provided with a hole through which the spring is link bolted to the rear of the torque control arms. The spring assembly is provided with full length liners.

Models equipped with a Mark IV V8 engine incorporate a stabilizer shaft which attaches to the upper rear section of the torque control arms, and stretches rearward where it is connected to the frame by two rubber-bushed mounting brackets.

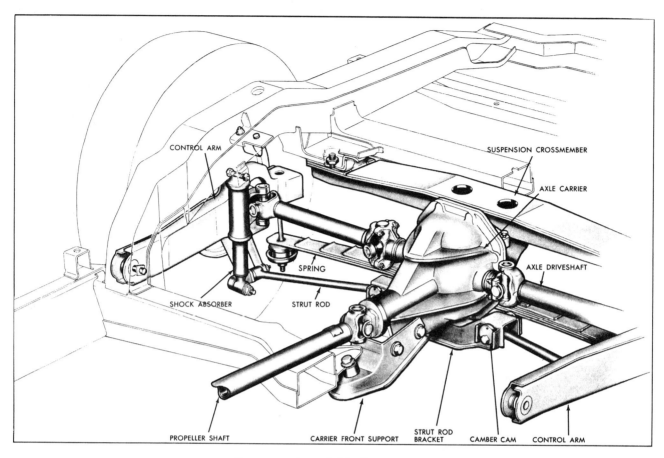

Rear suspension and driveline components

CAMBER
Adjusting

Wheel camber angle is obtained by adjusting the eccentric cam and bolt assembly located at the inboard mounting of the strut rod. Place rear wheels on alignment machine and determine camber angle. To adjust, loosen cam bolt nut and rotate cam and bolt assembly until specified camber is reached. Tighten nut securely and torque to specifications.

TOE-IN
Adjustment

Wheel toe-in is adjusted by inserting shims of varying thickness inside the frame side member on both sides of the torque control arm pivot bushing. Shims are available in thicknesses of 1/64", 1/32", 1/8" and 1/4". To adjust toe-in, remove torque control arm pivot bolt; then position torque control arm to obtain specified toe-in. Shim gap toward vehicle centerline between torque control arm bushing and frame side inner wall.

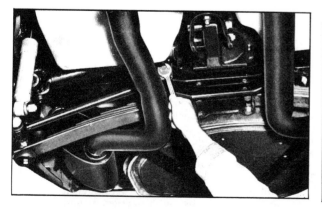

Adjusting rear wheel camber

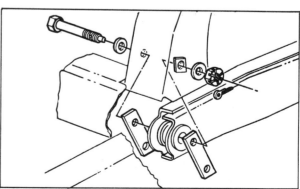

Toe-in adjusting shim location

Do not use thicker shim than necessary, and do not use undue force when shimming inner side of torque control arm — to do so may cause toe setting to change.

Shim outboard gap as necessary to obtain solid stack-up between torque control arm bushing and inner wall of frame side member. After correct shim stack has been selected, install pivot bolt making sure that all shims are retained — torque nut to specifications and install cotter pin. If specified torque does not permit cotter pin insertion, tighten nut to next flat.

WHEEL BEARING

Inspection

The tapered-roller spindle bearings should have end play of .001" to .008". During inspection, check end play and, when necessary, adjust. Raise vehicle on hoist. Disengage bolt lock tabs and disconnect outboard end of axle drive shaft from wheel spindle flange. Mark camber cam in relation to bracket. Loosen and turn camber bolt until strut rod forces torque control arm outward. Position loose end of axle drive shaft to one side for access to spindle.

Checking wheel bearing adjustment

Remove wheel and tire assembly. Mount dial indicator on torque control arm adjacent surface and rest pointer on flange or spindle end. Grasp brake disc and move axially (in and out) while reading movement on dial indicator. If end movement is within the .001" to .008" limit, bearings do not require adjustment. If not within .001" to .008" limit, record reading for future reference and adjust bearings.

Adjustment

149 Apply parking brake to prevent spindle from turning and remove cotter pin and nut from spindle. Release parking brake and remove drive spindle flange from splined end of spindle. Remove brake caliper and brake disc. Install

Thread Protector J-21859-1 over spindle threads; then remove drive spindle from spindle support, using Tool J-22602. *When using Tool J-22602 to remove drive spindle, make sure puller plate is positioned vertically in the torque control arm before applying pressure to the puller screw.* Remove shim and bearing spacer from spindle support.

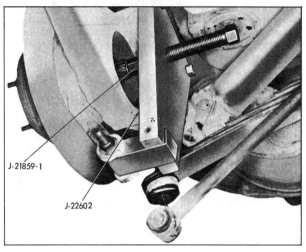

Removing drive spindle from support

Note size of shim used. If dial indicator reading was more than .008", select a shim thinner by the amount needed to bring end play within limits. If dial indicator reading was less than .001", select a shim thicker by the amount needed to bring end play within limits. Shims are available in thicknesses from .097" to .048" in increments of .003".

EXAMPLE: Bearing end play reading obtained on dial indicator was .011", .003" over limit. Bearing shim removed from spindle measures .145". New shim installed measures .139", .006" smaller. End play is now decreased by .006" and is .005", which is within the .001" to .008" limit.

After determining shim thickness, install bearing spacer and shim on spindle. Position spindle in spindle support. Place large end of spacer facing outer bearing with shim located between spacer and inner bearing. Press inner bearing race and roller assembly on spindle. Position Tool J-4731 over spindle and against bearing inner race. Position washer and spindle nut on spindle and proceed to tighten nut until bearing is forced on spindle sufficiently to allow spindle drive flange to be installed. Remove spindle nut, washer and Tool J-4731. Discard nut and use a new one for final assembly.

This drive flange to spindle fastener is an important attaching part in that it could affect the performance of vital components and systems, and/or could result in major repair expense. It must be replaced with one of the same part number or with an equivalent part if replacement becomes necessary. Do not use a replacement part of lesser quality or substitute design. Torque values must be used as specified during reassembly to assure proper retention of this part.

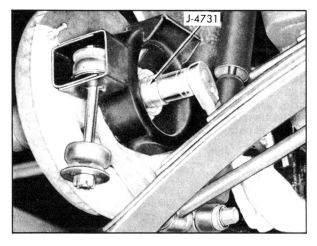

Installing drive spindle to support

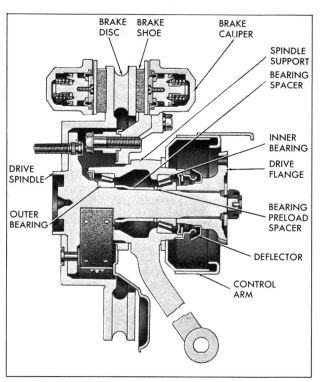

Wheel spindle and support cross section

Position drive flange over spindle, making sure flange is aligned with spindle splines. Install washer and nut on spindle then tighten nut to specifications and install cotter pin. If specified torque does not permit cotter pin insertion, tighten nut to next flat. Seat spindle support outer seal in bore by using screw driver, or other suitable tool, to press against metal portion of seal. Install brake disc and caliper. Install axle drive shaft, wheel and tire assembly, adjust camber cam to original position and torque all components to specifications.

This wheel to spindle fastener is an important attaching part in that it could affect the performance of vital components and systems, and/or could result in major repair expense. It must be replaced with one of the same part number or with an equivalent part if replacement becomes necessary. Do not use a replacement part of lesser quality or substitute design. Torque values must be used as specified during reassembly to assure proper retention of this part.

Lower vehicle and remove from hoist.

WHEEL SPINDLE AND SUPPORT
Removal
Remove wheel drive spindle as outlined under "Wheel Bearing Adjustment." Make bearing removers out of 3/8" square steel bar stock. After removing deflector, spindle inner grease seal and inner bearing race, remove bearing cups while spindle support is still mounted to the torque arms, by inserting remover tool and tapping cup out. New bearing cups are installed using Tool J-7817 cup installer and handle J-8092. To remove spindle support from torque arm, disconnect parking brake cable from actu-

ating lever. Remove four nuts securing support to torque arm and withdraw brake backing plate and position it out of the way. Disconnect shock absorber lower eye from strut rod mounting shaft. It may be necessary to support spring outer end before disconnecting shock absorber, as shock absorber has internal rebound control. Remove cotter pin and nut from strut rod mounting shaft, then pull shaft from support and strut rod. Spindle support may then be removed and bearing cups serviced.

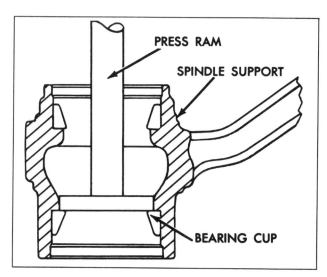

Bearing cup remover

150

To remove drive spindle outer bearing and seal, position Tool J-8331 between chambered edge of bearing seat and inner race of bearing. Clamp Tool J-8331 in a vise and apply pressure to upseat bearing. *Use extreme caution when positioning Tool J-8331 against machined surface of spindle. Make sure all tool imperfections such as nicks and burrs are removed from spindle contact area before applying pressure to unseat bearing.*

Unseating drive spindle outer bearing

Position Tool J-8331 and spindle in an arbor press and press bearing from spindle. Remove outer seal and inspect for damage — replace if necessary. Other seal must be placed on spindle before outer bearing inner race and roller assembly are installed. Pack outer bearing with a high-melting point wheel bearing lubricant and position on spindle. Large end of bearing should be toward shoulder on spindle. Press bearing on spindle using a bearing spacer and Tool J-9436 as installers.

Pack spindle inner bearing with a high-melting point wheel bearing lubricant and position bearing in spindle support — small end of bearing should be positioned inward. Tap new spindle inner grease seal into seal bore and install deflector over support inner end.

Removing and installation drive spindle outer bearing

Installation
Position support over torque arm bolts with strut rod fork toward center of vehicle and downward. Place backing plate over studs and torque nuts to specifications. Connect parking brake cable to actuating lever. Install drive spindle assembly as outlined under "Wheel Bearing Adjustment". If new spindle support or associated parts are installed, determine correct shim size. Assemble spindle to support, using a .145" shim. Check bearing adjustment and correct as necessary.

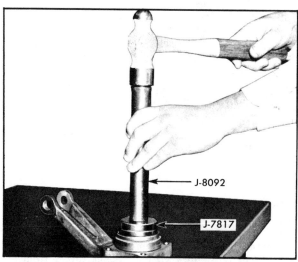

Installing spindle bearing cup

DIFFERENTIAL CARRIER SUPPORT BRACKET
Replacement
Raise vehicle on hoist. Place a 1/2" thick block of wood or steel between nose of differential carrier and floor pan. This will prevent carrier from twisting upward when support bracket is disconnected. Remove carrier support-to-crossmember attaching bolt. Remove nuts from both bracket-to-carrier through bolts and remove bolts. Assemble carrier support bracket by reversing above procedure. Torque all bolts to specifications. Lower vehicle and remove from hoist.

SHOCK ABSORBER
Removal
Raise vehicle on hoist. Disconnect shock absorber upper mounting bolt. Remove lower mounting nut and lock washer. Slide shock upper eye out of frame bracket and pull lower eye and rubber grommets off strut and mounting shaft. Inspect grommets and shock absorber upper eye from excessive wear.

Installation
Slide upper mounting eye into frame mounting bracket and install bolt, lock washer and nut. Place rubber grommet, shock lower eye, inboard grommet, washers and nut over strut rod shaft. Install washer with curve pointing inboard. Torque nuts to specifications. Lower vehicle and remove from hoist.

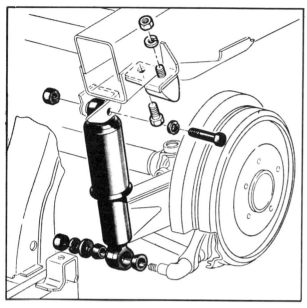

Shock absorber installation

cam bolt nut and cam and bolt assembly. Pull strut down out of bracket and remove bushing caps. Inspect strut rod bushings for wear and replace where necessary. Replace strut rod if it is bent or damaged in any way.

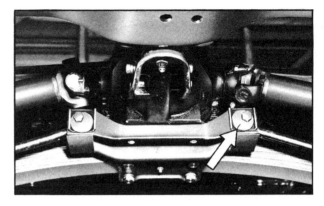

Marking camber cam and bracket

STRUT ROD AND BRACKET
Removal
Raise vehicle on hoist. Disconnect shock absorber lower eye from strut rod shaft. Remove strut rod shaft cotter pin and nut. Withdraw shaft by pulling toward front of vehicle.

Repairs
With strut rod bushing centered over Tool J-7877-2 and with strut rod supported horizontally, press or drive bushing from rod, using Tools J-7877-1 and J-7079-2. With strut rod end centered over Tool J-7877-2 and rod supported horizontally, press or drive bushing into arm using Tools J-7877-3 and J-7079-2. Tool J-7877-3 should bottom on strut rod when bushing is fully installed.

Removing strut rod shaft

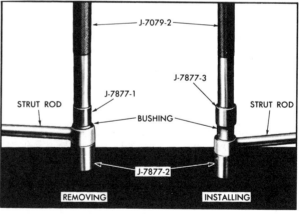

Strut rod bushing replacement

Installation
Place bushing caps over inboard bushing and slide rod into bracket. Install cam and bolt assembly and adjust cam to line up with mark on bracket. Tighten nut but do not torque at this point.

Mark relative position of camber adjusting cam and bracket, so they may be reassembled in same location. Loosen camber bolt and nut. Remove four bolts securing strut rod bracket to carrier and lower bracket. Remove

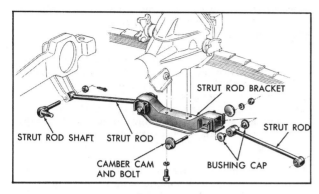

Strut rods

block is used to protect the "C" clamp threads from being distorted due to contact with the jack pad. Raise jack until all load is off link. Remove link cotter key and link nut. Remove cushion. Carefully lower jack until spring tension is released. Repeat for other side. Remove four bolts and washers securing spring center clamp plate. Slide spring out from under vehicle.

Raise bracket and assemble to carrier lower mounting surface. Torque bolts to specifications. Raise outboard end of strut rod into spindle support fork and insert strut rod shaft into fork so that flat on shaft lines up with corresponding flat in spindle fork. Install retaining nut, but do not torque. Place shock absorber lower eye and bushing over strut shaft, install washer and nut and torque to specifications. With weight on wheels torque camber cam nut and strut rod shaft nut to specifications. Then install cotter pin through rod bolt. Check rear wheel camber and adjust where necessary. Lower vehicle and remove from hoist.

SPRING
Removal
Raise vehicle on hoist allowing axle to hang. Remove wheels and tires. Install a "C" clamp on spring approximately 9" from one end. Tighten securely. Place adjustable lifting device under spring with lifting pad of jack inboard of link bolt near the "C" clamp. Place a suitable piece of wood between jack pad and "C" clamp screw. The "C" clamp is merely acting as a "stop" so that the jack won't slip when the spring is released. The wood

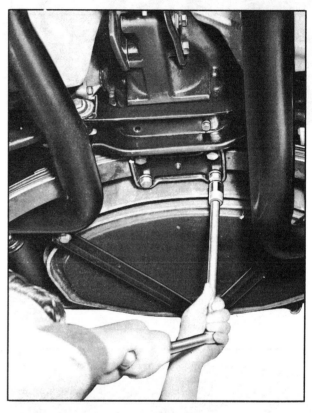

Removing clamp plate

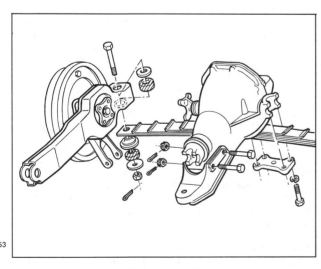

Spring mounting

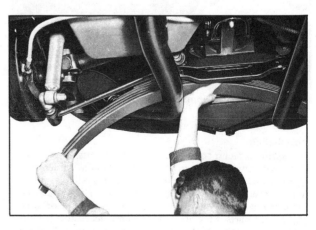

Lowering spring

Removing rear spring

Place spring on carrier cover mounting surface, indexing center bolt head with hole in cover. Place center clamp plate in position and install four bolts and washers. Snug bolts to position spring and torque to specifications. Install "C" clamp as in removal procedure. Place adjustable lifting device inboard of link bolt near "C" clamp. Add wooden block as in removal procedure. Raise spring outer end until spring is nearly flat, aligning torque arm with spring end. Install new attaching parts. Whenever servicing spring or removing spring attaching parts, always install new link bolts, rubber cushions, retainers, nuts and cotter pins.

Lower jack making sure cushions remain indexed in retainers. Remove "C" clamp. Remove jack and repeat for other side. Place vehicle weight on wheels and torque center clamp bolts to specifications. Lower vehicle and remove from hoist.

Disassembly

Clamp spring center section in vise and remove center bolt. Release vise, remove spring and separate leaves. Replace worn or damaged liners as necessary and replace any broken leaves. Replace main leaf spring cushion retainers by chiseling over flared portion until retainer may be knocked out of leaf. Place new retainers into position and flare over with a ball peen hammer or other suitable tool. Insert drift into center bolt holes in leaves to align spring leaves. Install center bolt and tighten securely.

Aligning spring leaves

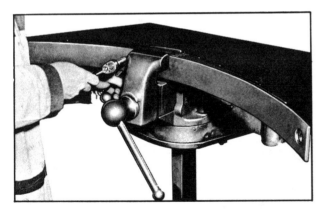

Removing center bolt

Installation

All spring attachments, including center bolts, are important attaching parts in that they could affect the performance of vital components and system, and/or could result in major repair expense. They must be replaced with parts of the same part numbers or with equivalent parts if replacement becomes necessary. Do not use replacement parts of lesser quality or substitute design. Torque values must be used as specified during reassembly to assure proper retention of these parts.

TORQUE CONTROL ARM
Removal

Disconnect spring on side torque arm is to be removed. If vehicle is so equipped, disconnect stabilizer shaft from torque arm – refer to "Stabilizer Shaft Replacement". Remove shock absorber lower eye from strut rod shaft. Disconnect and remove strut rod shaft and swing strut rod down. Remove four bolts securing axle drive shaft to spindle flange and disconnect drive shaft. It may be necessary to force torque arm outboard to provide clearance to drop drive shaft.

Disconnect brake line at caliper and from torque arm. Disconnect parking brake cable. Remove torque arm pivot bolt and toe-in shims and pull torque arm out of frame. Tape shims together and identify for correct reinstallation. For service operations pertaining to the spindle support assembly, refer to service operations under Wheel Spindle and Support.

Disassembly

Using 11/16" drill, drill out flared end of bushing retainer. Remove special retainer plate and tap retainer out of

154

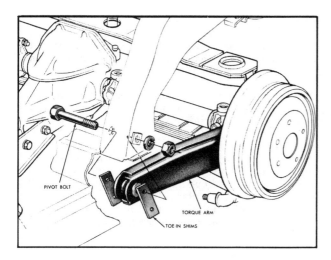

Torque control arm

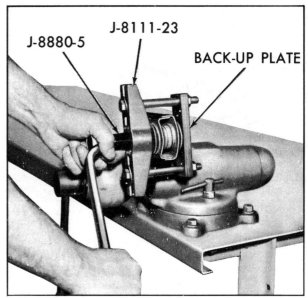

Flaring retainer

bushing. Remove bushings by spreading them apart with a chisel and tap out of arm. If bushing diameters are severaly rusted in torque arm and arm tends to spread during removal, clamp arm in a C-clamp to prevent spreading. Lightly oil new bushing diameters (not rubber portion) and start bushings squarely in arm. Install Tool J-7055-1 and press bushings into place. When bushings are fully installed, place special plate over flared portion of new bushing retainer and insert retainer into bushing. Out of 5/8" thick steel plate 1-1/2" wide, make flaring tool support and drill clearance holes for 2-1/2" bolts. Place fabricated back-up plate on flared end of bushing retainer and assemble Tool J-8111-23 to plate with two 1/2" x 5" bolts. Make sure threaded hole in J-8111-23 is centered over unflared end of bushing retainer and that chamfered retainer plate is centered over retainer tube. Lightly oil pointed end of J-8880-5 screw and thread into J-8111-23 until pointed end contacts bushing retainer. Continue threading J-8111-23 until retainer is flared.

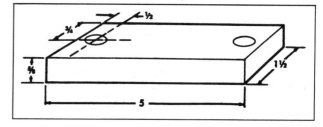

Flaring tool back-up plate

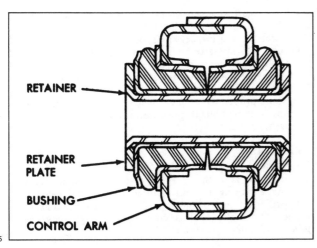

Torque control arm bushing cross section

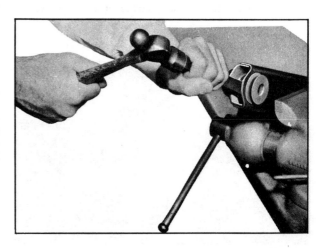

Removing bushing

155

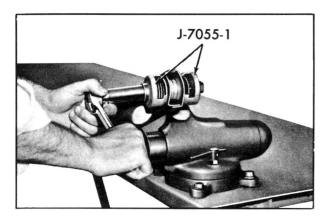

Installing bushing

Installation

Place torque arm in frame opening. Place toe-in shims in original position on both sides of torque arm, install pivot bolt and tighten but do not torque at this point. Raise axle drive shaft into position and install to drive flange. Torque bolts to specifications. Raise strut rod into position and insert strut rod shaft so that flat lines up with flat in spindle support fork. Install nut and torque to specifications. Install shock absorber lower eye and tighten nut to specifications. Connect spring end. If vehicle is so equipped, connect stabilizer shaft to torque arm — refer to "Stabilizer Shaft Replacement." Install brake line at caliper and torque arm. Bleed brakes. Install wheel and tire. Torque, torque arm pivot bolt to specifications and install cotter pin with weight on wheels. Lower vehicle and remove from hoist.

SUSPENSION CROSSMEMBER
Removal

Disconnect and remove spring as outlined under Spring Removal. Remove differential carrier assembly as outlined under Differential Carrier-Removal. Support crossmember, remove bolts securing crossmember isolation mounts to frame and lower crossmember. Remove bolts securing carrier cover to crossmember. Inspect rubber isolation mounts for aging and replace where necessary.

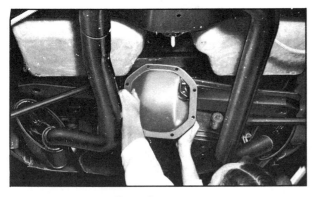

Removing crossmember

Repairs

Bend back isolation mount tabs to allow mount removal. Place crossmember on a suitable support and press mount out of arm using a piece of suitable size pipe or tubing on outer shell or inner insert.

Place new mount into position on crossmember compress outer sleeve and press mount into place until it is full and squarely seated. After installation, bend over locking tabs.

Installation

Install carrier cover to crossmember and torque bolts to specifications. Raise crossmember into position and install mounting bolts. Torque bolts to specifications. Install differential carrier assembly as outlined under Differential Carrier-Installation in this section. Install spring, wheels and tires. Lower vehicle and remove from hoist.

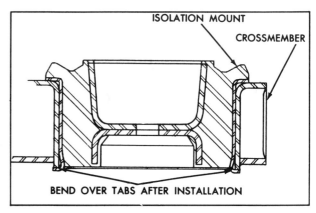

Crossmember mount cross-section

STABLILIZER SHAFT
Replacement

Raise vehicle on hoist. Support at frame side rail. Disconnect stabilizer shaft at both torque arms. Remove stabilizer shaft brackets from the frame and withdraw assembly from vehicle. Inspect bushings for signs of deterioration, and inspect shaft for bends, breaks or other defects — do not attempt to straighten shaft — replace parts as deemed necessary. Position bushings on shaft and loosely install shaft to torque arms and at frame brackets. Install flange of bushing on wheel side of link. Align shaft to assume proper replacement when bolts are torque, and torque attaching bolts to specifications. Lower vehicle and remove from hoist.

REAR AXLE
Description

The rear axle is of the type where the differential carrier housing enclosing the differential and hypoid gears is supported on a crossmember mounted to the chassis frame. The differential is connected through universal joints to the drive shafts and wheels.

The internal components of the carrier are of conventional design, incorporating a hypoid gear set with an

156

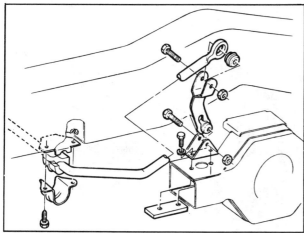

Stabilizer shaft installation

Scored hypoid ring gear

overhung pinion supported on two pre-loaded, tapered roller bearing assemblies, and a two-pinion differential assembly supported on tapered roller bearings. Pinion mounting distance adjustments are made through the use of shims, as are the differential bearing pre-load and backlash adjustments. The differential side gears drive two splined yokes which are retained laterally by snap rings located on the yoke splined end. The yokes are supported on caged needle bearings pressed into the carrier, adjacent to the differential bearings. A lip seal, pressed in outboard of the bearings, prevents oil leakage and dirt entry. The carrier cover is bolted to the carrier and provides accessibility to the internal parts. The cover incorporates two integral, reinforced mounting pads which serve as the carrier attaching point to the suspension crossmember, and an attaching point for the spring center section. The filler plug is located on the right side of the cover near the bolting flange.

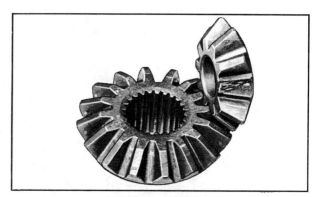

Differential gear failure

All service operations allow carrier removal without removing the carrier cover. Cover removal is not necessary in any of the service procedures except in the case of complete carrier housing replacement, as the carrier and cover are serviced as an assembly.

AXLE DRIVE SHAFT
Removal
Raise vehicle on hoist. Disconnect inboard driveshaft trunnion from side gear yoke. Bend bolt lock tabs down. Wire brush bolt heads to remove grit and scale. Remove bolts securing shaft flange to spindle drive flange. Pry driveshaft out of outboard drive flange pilot and remove by withdrawing outboard end first.

Disassembly and assembly
Remove bearing lock ring from trunnion yoke. Support trunnion yoke on a piece of 1-1/4" pipe on arbor press bed. Using suitable socket or rod, press trunnion down far enough to drive opposite bearing cup from yoke. Remove

157

Cracked hypoid ring gear

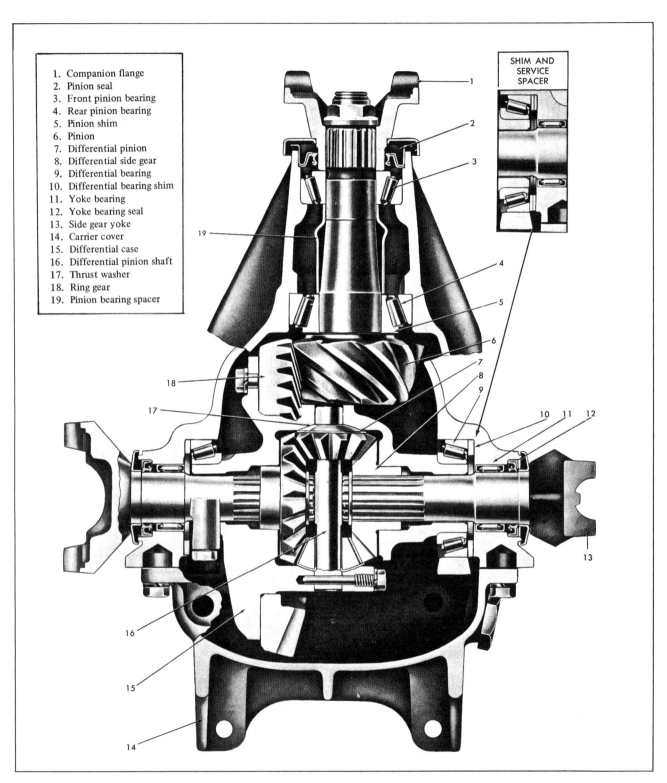

1. Companion flange
2. Pinion seal
3. Front pinion bearing
4. Rear pinion bearing
5. Pinion shim
6. Pinion
7. Differential pinion
8. Differential side gear
9. Differential bearing
10. Differential bearing shim
11. Yoke bearing
12. Yoke bearing seal
13. Side gear yoke
14. Carrier cover
15. Differential case
16. Differential pinion shaft
17. Thrust washer
18. Ring gear
19. Pinion bearing spacer

SHIM AND
SERVICE
SPACER

Rear axle cross-section

Removing driveshaft

trunnion and press other bearing cup from yoke, being careful not to drop cup or lose bearing rollers. Remove trunnion and yoke from other joint in a similar manner. Remove dust seals from trunnion, clean and inspect bearing rollers and trunnion. Relubricate bearings with a high-melting point wheel bearing type lubricant. In addition to packing the bearings, make sure that the lubricant reservoir at the end of each trunnion is completely filled with lubricant. In filling these reservoirs, pack lubricant into the hole so as to fill from the bottom. This will prevent air pockets and ensure an adequate supply of lubricant.

Place new dust seals on trunnion, cavity of seal toward end of trunnion — then position Tool J-21556 over end of trunnion and into cavity portion of seal. Press seal onto trunnion until tool bottoms against trunnion. Installation of seal is critical to proper sealing — use specified tool during installation to prevent seal distortion, and to assure proper seating of seal on trunnion.

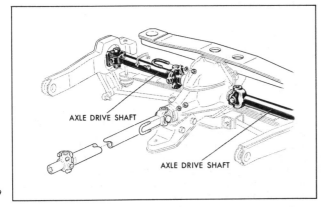

159

Axle driveshaft

Partially install one bearing cup into yoke. Place trunnion in yoke and into bearing cup. Install other bearing cup and press both bearing cups into yoke, being careful to keep trunnion aligned in bearing cups. Press bearing cups far enough to install lock rings, and install lock rings. It may be more convenient to use a bench vise for removal and installation, instead of an arbor press. In this case, proceed with disassembly and assembly procedure as with an arbor press.

Installation

Place driveshaft inboard trunnion into side gear yokes and assemble "U" bolts. Rotate yokes so that trunnion seats are phased 90° apart. Install outboard drive flange into spindle drive flange pilot, position bolt lock over bolt holes and install four bolts. Torque bolts to specification and bend lock tabs flat against bolt heads. Lower vehicle and remove from hoist.

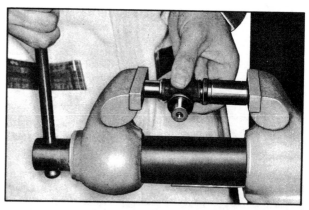

"U" joint trunnion seal installation

PINION FLANGE, DUST DEFLECTOR AND/OR OIL SEAL
Removal

Raise vehicle on hoist. Allow wheels to hang free. Place 1/2" thick block of wood or steel between carrier upper surface to rear of companion flange, and body floor. This will prevent carrier assembly from twisting upward when front support bracket is disconnected. Disconnect carrier front mounting bracket bolt from frame crossmember. Remove nuts from both bracket-to-carrier through bolts and remove bolts and mounting bracket. Disconnect propeller shaft at companion flange. Slide transmission yoke forward and lower propeller shaft down and out. Disconnect axle drive shafts from axle.

Using Tool J-5833 with adapter J-5810 and a suitable socket on the pinion flange nut, rotate the pinion through several complete revolutions and record the torque required to keep the pinion turning. If flange is to be reused, mark pinion and flange for assembly in the same position. Attach J-8614-1 companion flange holder and remove flange nut and washer. Remove companion flange by driving off with brass drift and hammer. Using screwdriver, pry oil seal out of carrier.

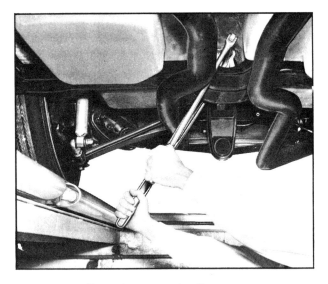

Removing companion flange nut

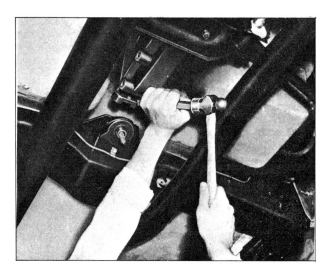

Removing companion flange

flange splines and tap into place. Install a new nut on the pinion shaft. Tighten nut to remove end play and continue alternately tightening in small increments and checking preload with torque wrench until it is the same as previously recorded. Raise propeller shaft into position and connect to companion flange and transmission yoke. Reinstall axle drive shafts. This propeller shaft to pinion flange fastener is an important attaching part in that it could affect the performance of vital components and systems, and/or could result in major repair expense. It must be replaced with one of the same part number or with an equivalent part if replacement becomes necessary. Do not use a replacement part of lesser quality or substitute design. Torque values must be used as specified during reassembly to assure proper retention of this part.

Installing pinion oil seal

Place rubber cushion on carrier front mount, and place bracket into position and loosely install rear carrier-to-bracket bolt. Install front mounting nut and front carrier-to-bracket bolt. Torque all affected parts to specifications and install cotter pin to carrier bracket front bolt. Lower vehicle and remove from hoist.

DIFFERENTIAL CARRIER
Removal

Raise vehicle on hoist. Disconnect spring and link bolts as outlined in Spring-Removal. Disconnect axle drive shafts at carrier by removing "U" bolts securing trunnion to side gear yoke. Disconnect carrier front support bracket at frame crossmember. Disconnect propeller companion flange. Slide transmission yoke forward into transmission. Drop propeller shaft down and out toward the rear. Mark camber cam and bolt relative location on strut rod bracket and loosen cam bolts. Remove four bolts securing bracket to carrier lower surface and drop bracket. Remove camber cam bolts and swing strut rods up and out of the way. Remove eight carrier-to-cover bolts. Loosen bolts gradually to allow lubricant to drain out. With mounting bolts removed, pull carrier partially out of cover, drop nose to clear crossmember and gradually work carrier down and out.

Inspect companion flange splines for excessive wear or twisting and check deflector for looseness. If deflector is loose or damaged, break stake marks and remove. Install new deflector and stake in place.

Installation

Pack cavity between the seal lips of the pinion flange oil seal with a Lithium base extreme pressure lubricant. Position seal in bore and place gauge plate J-22804-1 over seal and against seal flange. Gauge plate insures proper seating of seal in carrier bore. Using Tool J-21057, press seal into carrier bore until gauge is flush with the carrier shoulder and the seal flange. Turn gauge plate 180° from installed position; seal must be square in carrier bore to seal properly against pinion flange. Lubricate companion

Installation

Clean inside of carrier cover and liberally grease gasket surface. Place new gasket on cover. Cut heads of two 1/2"-13 x 1-1/4" bolts and slot unthreaded end. Install these aligning studs into two below-center carrier bolt holes, one on each side. Raise carrier into position aligning studs into cover. Install carrier-to-cover bolts and tighten securely. Connect propeller shaft to companion flange and transmission yoke. This propeller shaft to pinion flange fastener is an important attaching part in that it could affect the performance of vital components and systems, and/or could result in major repair expense. It must be replaced with one of the same part number or with an equivalent part if replacement becomes necessary. Do not use a replacement part of lesser quality or substitute design. Torque values must be used as specified during reassembly to assure proper retention of this part.

Install rubber cushion on bracket and position to frame crossmember. Install nut and torque to specifications. Raise axle drive shafts into position and assemble inboard trunnion to side gear yokes with "U" bolts. Assemble strut rods to bracket and raise bracket into position under carrier. Install four bolts and torque to specifications. Move camber cams to marked location and tighten cam nuts.

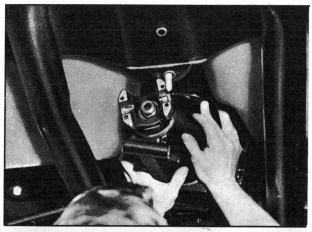

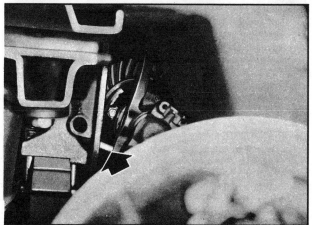

161

Installing carrier assembly

Connect spring end link bolts as outlined under Spring-Installation, in this section. Remove filler plug, located on right side of cover, and fill with hypoid lubricant to level of filler hole. Lower vehicle and remove from hoist.

POSITRACTION DIFFERENTIAL UNIT

The optionally available Positraction differential unit is installed in the conventional carrier to replace the standard differential unit. Service procedures for the Positraction equipped axle are the same as on a conventional axle except for the following operations.

On the car check

If vehicle is equipped with a manual transmission, shift transmission into neutral. Raise rear of vehicle until wheels are off the ground, remove one wheel and tire assembly. Attach Adapter J-5748 to axle shaft flange and install a 1/2-13 bolt into adapter. With wheel and tire assembly still on vehicle held firmly to prevent turning, measure torque required to rotate opposite axle shaft with a 0-150# torque wrench attached to J-5748. Torque should be 70 ft. lbs. minimum new, and no less than 40 ft. lbs. if used.

Reinstall wheel and tire assembly. This wheel to drive spindle fastener is an important attaching part in that it could affect the performance of vital components and systems, and/or could result in major repair expense. It must be replaced with one of the same part number or with an equivalent part if replacement becomes necessary. Do not use a replacement part of lesser quality or substitute design. Torque values must be used as specified during reassembly to assure proper retention of this part.

PROPELLER SHAFT AND UNIVERSAL JOINTS
Description

There are two types of propeller shafts and before working on the universal joints, the type of propeller shaft must be determined because disassembly and assembly procedures are different. The two shafts differ in that one uses snap rings (Cleveland type) for trunnion retention and the other does not use snap rings (Saginaw design). Saginaw design propeller shaft trunnions are retained by a nylon material which is injected into a groove in the yoke.

Universal joint repair kits for the Saginaw design use snap rings which install inboard of the yoke. Repair kits for the Cleveland type are identical to parts being replaced. Some propeller shafts incorporate a damper. This damper is not replaceable. If replacement becomes necessary, the damper and sleeve are to be replaced as an assembly.

Removal

Two methods are used to retain the propeller shaft to the differential pinion flange. One method utilizes "U" bolts and the other is a strap attachment.

Raise vehicle on hoist. Mark relationship of shaft to companion flange and disconnect the rear universal joint by removing trunnion bearing "U" bolts or straps. Tape bearing cups to trunnion to prevent dropping and loss of bearing rollers. Withdraw propeller shaft front yoke from transmission by moving shaft rearward, passing it under the axle housing. Watch for oil leakage from transmission output shaft housing.

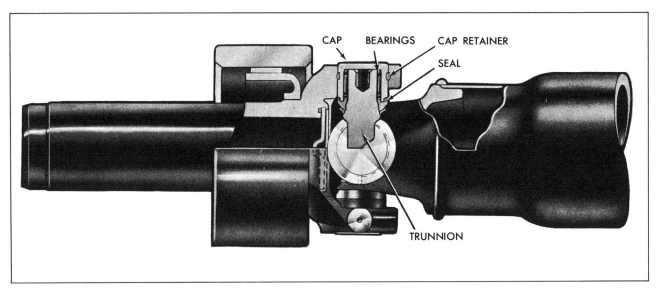

Propeller shaft cross section (saginaw type)

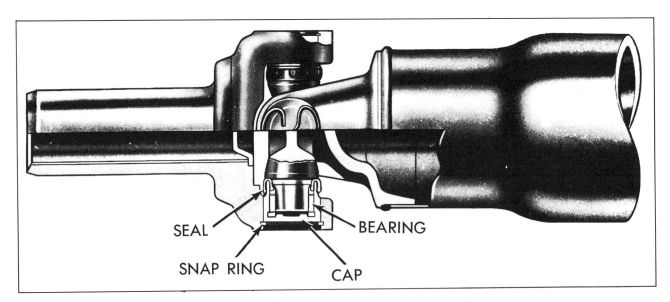

Propeller shaft cross section (cleveland type)

Repairs (except Saginaw)

The universal joints are of the extended-life design and do not require periodic inspection or lubrication; however, when these joints are disassembled, repack bearings and lubricate reservoir at end of trunnions with high-melting point wheel bearing lubricant and replace the dust seals.

Remove bearing lock rings from trunnion yoke. Support trunnion yoke on a piece of 1-1/4" I.D. pipe on an arbor bed. Due to length of the propeller shaft it may be more convenient to use a bench vise, for removal and installation, instead of an arbor press. In this case, proceed with disassembly and assembly procedure as with an arbor press.

Using a suitable socket or rod, press on trunnion until bearing cup is almost out. Grasp cup in vise and work cup out of yoke. The bearing cup cannot be fully pressed out.

Press trunnion in opposite direction and remove other cup. Clean and inspect dust seals, bearing rollers, and trunnion. Relubricate bearings. In addition to packing the bearings, make sure that the lubricant reservoir at the end of each trunnion is completely filled with lubricant. In filling these reservoirs, pack lubricant into the hole so as to fill from bottom (use of squeeze bottle is recommended). This will prevent air pockets and ensure an adequate supply of lubricant.

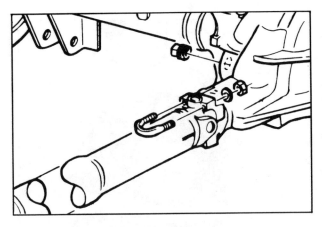

"U" bolt attachment

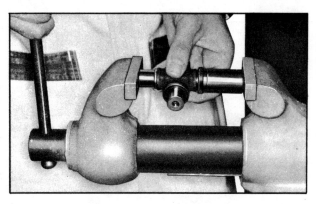

"U" joint trunnion seal installer

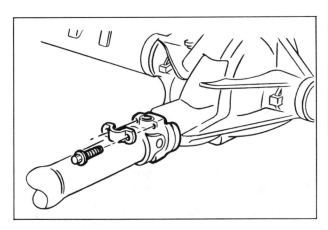

Removing propeller shaft from companion flange

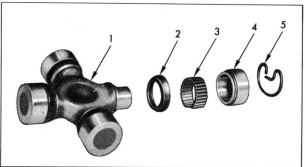

Universal joint repair kit (except saginaw)

If not installing a "U" joint service kit, place dust seals on trunnions — cavity of seal toward end of trunnion. Press seal onto trunnion exercising caution during installation to prevent seal distortion and to assure proper seating of seal on trunnion. If installing seal on small size trunnion, seal installer J-21548 should be used. Position trunnion into yoke. Partially install one bearing cup into yoke. Start trunnion into bearing cup. Partially install other cup. Align trunnion into cup, and press cups into yoke. Install lock rings.

Repairs (Saginaw type)

Because of the elastic properties of the nylon retainers, the trunnions must be pressed from the yokes. Pressing the trunnions from the yokes will shear the retainers which renders the bearing caps unsuitable for reuse. A repair kit, which employs a snap ring to retain the trunnion, must be used when reassembling the propeller shaft.

Remove trunnion at differential end of propeller shaft. Support trunnion on a press bed so that the propeller shaft yoke can be moved downward. Support front of propeller shaft so that shaft is in a horizontal position.

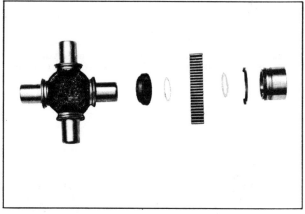

Trunnion repair kit (saginaw)

Using a piece of pipe or similar tool, with an inside diameter slightly larger than 1-1/8", press bearing from yoke. Apply force on yoke around bearing until nylon retainer breaks. Continue to apply force until the downward movement of the yoke forces the bearing as far as possible from the yoke. Complete removal of bearing by tapping around circumference of exposed portion with a small hammer. Rotate propeller shaft so that opposite bearing may be removed in the same manner. Remove trunnion from yoke.

163

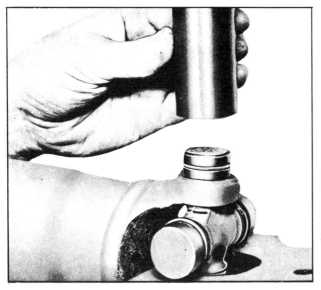

Bearing partially pressed from propeller shaft

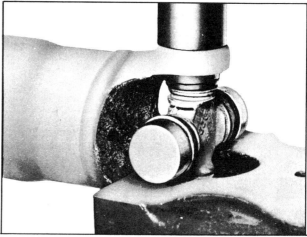

Pressing trunnion bearing from propeller shaft

Remove trunnion at transmission end of propeller shaft. Support splined yoke on a press bed and the rear of the propeller shaft on a stand so that shaft is horizontal. Be sure that weight is evenly distributed on each side of the splined yoke and that the fixed yoke half of the "U" joint is free to move downward. Using a piece of pipe or similar tool, with an inside diameter slightly larger than 1-1/8", press bearing from yoke. Apply force on yoke around bearing until nylon retainer breaks. Continue to apply force until the downward movement of the yoke forces the bearing as far as possible from the yoke. Complete removal of bearing by tapping around circumference of exposed portion with a small hammer. Rotate propeller shaft so that opposite bearing may be removed in the manner described above. Remove splined yoke and trunnion from propeller shaft. Support exposed trunnion fingers on press bed and press splined yoke from bearing caps. Remove trunnion from splined yoke.

Install trunnion and bearing assembly to the propeller shaft yokes. The trunnion and bearing assembly used at the differential end of the propeller shaft incorporates two different size bearing caps. The larger bearing caps (with the annular grooves) must be mated with the propeller shaft yoke.

Installing bearing cup and trunnion

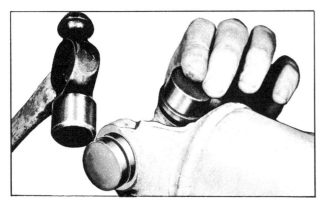

Removing bearing from propeller shaft

Install bearing cap and seal about one-fourth way in on one side of yoke, using a soft-face hammer to tap the bearing into position. Check bearing cap during installation to ensure that it does not become cocked in yoke. Insert trunnion into yoke. Firmly seat trunnion into bearing cup and press bearing cup into yoke until it is flush with yoke. Install opposite bearing cap and seal making sure that rollers do not become jammed on trunnion. Check for free movement of trunnion in yoke. Install bearing retainer snap rings making sure that gap in ring is toward yoke.

164

Installation

Inspect yoke seal in the transmission extension, replace if necessary as described in the transmission section. Insert propeller shaft front yoke into transmission extension, making sure that output shaft splines mate with propeller shaft yoke splines. Align propeller shaft with companion flange using reference marks established in "Removal" procedure; remove tape used to retain trunnion bearing caps; connect exposed bearing caps to companion flange by installing retaining strap and screws — torque screws to specifications.

This propeller shaft to pinion flange fastener is an important attaching part in that it could affect the performance of vital components and systems, and/or could result in major repair expense. It must be replaced with one of the same part number or with an equivalent part if replacement becomes necessary. Do not use a replacement part of lesser quality or substitute design. Torque values must be used as specified during reassembly to assure proper retention of this part.

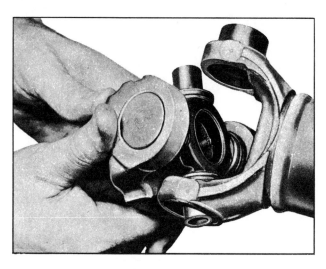
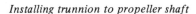

Installing trunnion to propeller shaft

Installing snap ring to retain trunnion

BRAKES 10

INDEX

10

DESCRIPTION

The 1966 was last year of single system brake system. Servicing is same as dual system except as noted in Text 6. The system is designed with separate hydraulic systems for the front and rear brakes using a single dual master cylinder. If a wheel cylinder or brake line should fail at either the front or rear brake system of the vehicle, the operator can still bring the vehicle to a controlled stop.

The master cylinder has two entirely separate reservoirs and outlets in a common body casting. The front reservoir and outlet is connected to the front wheel brakes, and the rear reservoir and outlet is connected to the rear wheel

brakes. Two pistons within the main cylinder receive mechanical pressure from the brake pedal push rod and transmit it through the brake lines as hydraulic pressure to the wheel cylinders. The filler cap is accessible from inside the engine compartment.

A brake pipe distribution and switch assembly is mounted below the main cylinder. The front and rear hydraulic brake lines are routed from the main cylinder, through the brake pipe distribution and switch assembly, to the front and rear brakes. The switch is wired electrically to the brake alarm indicator light on the instrument panel. If a leak in either front or rear system should occur, the

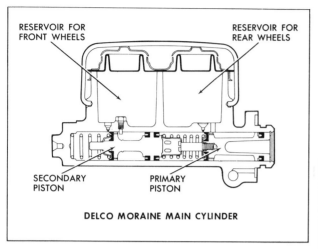

Main cylinder

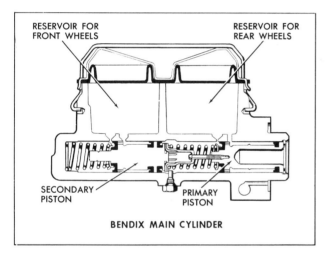

Main cylinder

member, then back to an equalizer near the rear frame crossmember. A single piece rear cable passes through the equalizer and back to the rear service brakes.

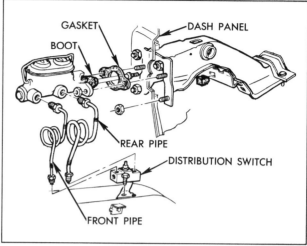

Brake pipe distribution and switch assembly (3)

Maintenance and adjustments

In any service operation it is extremely important that absolute cleanliness be observed. Any foreign matter in the hydraulic system will tend to clog the lines, ruin the rubber cups of the main and wheel cylinders and cause inefficient operation or even failure of the braking system. Dirt or grease on a brake lining may cause that brake to grab first on brake application and fade out on heavy brake application. The split system consists basically of two separate brake systems. When a failure is encountered on either, the other is adequate to stop the vehicle. If one system is not functioning, it is normal for the brake pedal lash and pedal effort to substantially increase. This occurs because of the design of the master cylinder which incorporates an actuating piston for each system. When the rear system loses fluid, its piston will bottom against the front piston. When the front system loses fluid, its piston will bottom on the end of the main cylinder body. The pressure differential in one of the systems causes an uneven hydraulic pressure balance between the front and rear systems. The brake pipe distribution and switch assembly, near the main cylinder, detects the loss of pressure and illuminates the brake alarm indicator light on the instrument panel. The pressure loss is felt at the brake pedal by an apparent lack of brakes for most of the brake travel and then, when failed chamber is bottomed, the pedal will harden.

If a vehicle displays these symptoms, it is a good indication that one of the systems contains air or has failed and it is necessary to bleed or repair the brakes.

Bleeding system
Clean all dirt from the top of the main cylinder and

pressure differential, during brake application, will cause the piston to compress the springs and move in the bore until it touches the electrical contact which causes the parking brake lamp on the instrument panel to light. This lamp will also be illuminated when the parking brake is applied.

The front brake hydraulic line is routed through a pressure metering valve. The valve controls the hydraulic pressure to the front brakes resulting in the correct pressure balance between the front and rear hydraulic systems.

A ratchet-type parking brake lever is located behind the console. The parking brake cable runs forward from the lever to a pulley assembly mounted on a frame cross-

remove the cylinder cover and rubber diaphragm. Fill main cylinder (if necessary) and reinstall the cover. Install brake bleeder wrench on a bleeder valve at a wheel cylinder and install a bleeder hose on the bleeder valve. If the master cylinder is equipped with bleeder valves, bleed these valves first, then proceed to the wheel cylinder nearest the master cylinder then, the next nearest and so on until all cylinders have been bled and there is no evidence of air.

Bleeding brake using tool J-21472

Pour a sufficient amount of brake fluid into a transparent container to ensure that the end of the bleeder hose will remain submerged during bleeding. Place the loose end of the bleeder hose into the container. Carefully monitor the fluid level at the main cylinder during bleeding. Do not bleed enough fluid at one time to drain the reservoir. Replenish as needed to ensure a sufficient amount of fluid is in the main cylinder at all times.

Open wheel cylinder bleeder valve by turning tool counter-clockwise approximately 1/3 of a turn. Have helper depress the brake pedal. Just before the brake pedal reaches the end of its travel, close the bleeder valve and allow the brake pedal to return slowly to the released position. Repeat until expelled brake fluid flows in a solid stream without the presence of air bubbles, then close the bleeder valve tightly. Remove brake bleeder wrench and hose from the bleeder valve and repeat Steps 2 through 6 on the remaining wheel cylinders. Fill the main cylinder. Install the main cylinder diaphragm and cover. Note: Rear calipers have two bleed valves, one inside and one outside. The wheel must be removed to get at outside valve.

HYDRAULIC BRAKE LINES
Description
The flexible hoses which carry the hydraulic pressure from the steel lines to the wheel cylinders are carefully designed and constructed to withstand all conditions of stress and twist which they encounter during normal vehicle usage. The hoses require no service other than periodic inspection for damage from road hazards or other like sources.

Removal
Should damage occur and replacement become necessary, separate hose from steel line by turning double flare

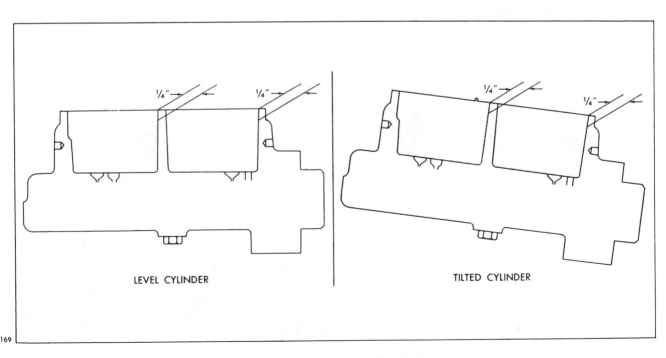

LEVEL CYLINDER   TILTED CYLINDER

Correct main cylinder fluid level

connector out of hose fitting. Remove "U" shaped retainer from hose fitting and withdraw hose from support bracket. Turn hose fitting out of wheel cylinder inlet.

Installation

Install new copper gasket on cylinder end of hose (male end). Moisten threads with brake fluid and install hose in wheel cylinder. Tighten hose fitting (22 ft.lb.). With weight of car on wheels and suspension in normal position (front wheels straight ahead) pass female end of hose through support bracket, allowing hose to seek its own position. Insert hex of hose fitting into the 12 point hole

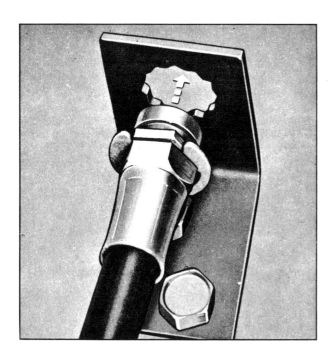

Brake line support bracket

in support bracket in position which induces least twist to hose. Do not twist hose unduly during this operation as its natural curvature is absolutely necessary to maintain proper hose-to-suspension clearance through full movement of the suspension and steering parts. Install "U" shaped retainer to secure hose in support bracket. Inspect by removing weight completely from wheel, turn wheels from lock to lock while observing hose position. Be sure that hose does not touch other parts at any time during suspension or wheel travel. If contact does occur, remove hose retainer and rotate the female hose end in the support bracket one or two points in appropriate direction, replace retainer, and re-inspect. Place steel tube connector in hose fitting and tighten securely (100 in. lb.). Bleed all brakes. Do not tighten male end (Wheel cylinder end) once the other end is fixed. If necessary to tighten male end, disconnect hose at opposite end, then reconnect following above procedure.

PARKING BRAKE
Lever assembly removal

Place parking brake lever in the fully released position. Raise vehicle on hoist, unhook and remove return spring. Remove rear nut from cable stud at equalizer and allow front cable to hang down. Lower vehicle to floor.

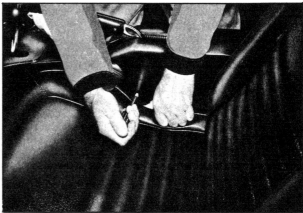

Removing or installing cover lower mounting screw

Inside vehicle, remove screws securing cover to underbody. It is necessary to push the seat cushion down to gain access to the lower mounting screw on each side. Remove the retainer. Lift the cover upward and to the rear to allow seal to slide out of cover slot and remove cover. Remove trim seal from lever. Remove screw and washer securing parking brake alarm switch to side of lever assembly. Remove four bolts securing lever assembly to underbody and lift lever assembly upward. Remove lever forward mounting bracket. Remove front cable from lever assembly using long nose pliers. Remove lever from vehicle.

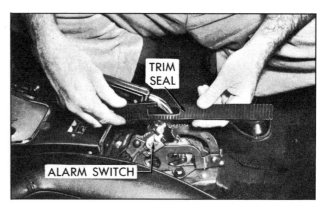

Removing or installing trim seal

170

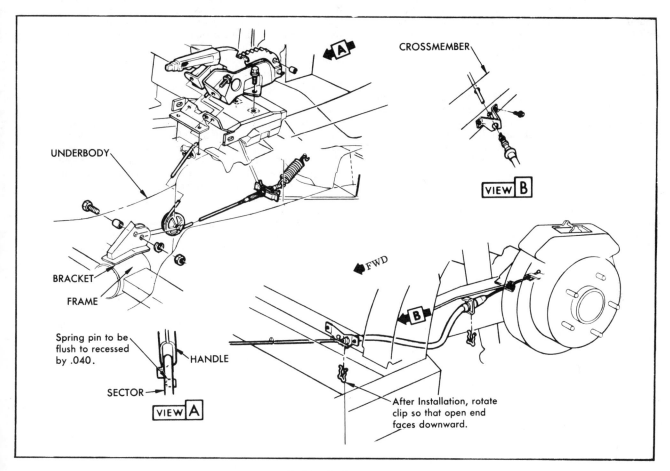

Parking brake system

Installation

Install front cable in lever assembly using long nose pliers. Slide slack cable through hole and place lever assembly in position. Place forward mounting bracket in position and secure plate and lever assembly to underbody with four mounting bolts. Secure parking brake alarm switch to side of lever assembly with washer and screw. Install trim seal on lever. Insert the end of the trim seal into the slot in the cover and slide the cover forward and down into position. Place the retainer in position. Secure retainer and cover to underbody with six screws. It is necessary to push the seat cushion down, to gain access to the lower mounting screw hole on each side. Raise vehicle on hoist. Under vehicle, insert front cable stud through equalizer assembly and secure with nut. Install return spring. Adjust parking brake as outlined under Maintenance and Adjustments. Lower vehicle to floor.

FRONT CABLE
Removal

Perform the complete parking brake lever removal procedure. Under vehicle, remove nut, washer, and bolt securing pulley to pulley bracket. Remove seal grommet from underbody cable hole. Pull front cable out of vehicle.

Installation

Install seal grommet around front cable. Slide front cable

Removing or installing front cable

into position through cable hole and work seal grommet into installed position. Under vehicle, loop front cable around pulley wheel and secure pulley wheel to bracket with bolt, washer, and nut. Perform the complete parking brake lever installation procedure.

REAR CABLE
Removal

Raise vehicle on hoist. Remove the cable clip retainers on the back side of the frame rail from each of the rear brake cable ends. Disconnect the cable at the rear wheel flange plate. Remove the cable ball out of the recess in the brake lever assembly clevis. Disconnect the cables at the equalizer connector and remove the cables.

Installation

Attach the rear cable balls on the rear cable ends to the rear wheel recess in the brake lever assembly clevis. Attach the cable assembly to the rear wheel flange plate. Connect the cable assemblies along the frame rail area. Insert the cable end through the equalizer and attach to the equalizer connector. Adjust as outlined under Maintenance and Adjustments. Lubricate the cables and all moving parts for trouble-free operation. Lower vehicle to floor.

BRAKE PEDAL
Removal

Remove air conditioning components if necessary. Disconnect clutch pedal return spring (manual transmission models only). Disconnect clutch push rod at pedal. Disconnect brake pedal return spring on manual brake cars. Disconnect brake pedal from main cylinder push rod by removing retainer and clevis pin. Remove steering column from vehicle. Support main cylinder from inside

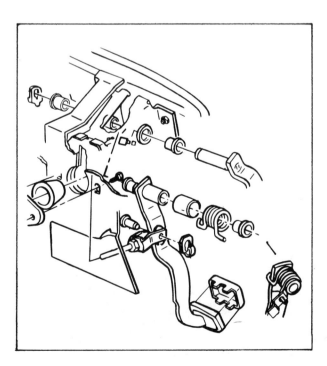

Brake pedal installation

engine compartment and remove four support brace nuts. Remove four nuts and bolts securing support plate to bracket and remove support plate. Remove two screws securing bracket to underside of instrument panel and lower bracket and pedals to floor. Remove retainer from right side of pedal pivot shaft. Slide clutch pedal assembly to the left and remove from support brace. Withdraw brake pedal and all nylon bushings.

Inspection

Clean all metal parts with a good non-toxic cleaning solvent. Wipe nylon bushings clean with a clean cloth. Nylon bushings should not be treated with cleaning agent of any nature. Inspect all nylon bushings for wear and damage. Inspect mating surface of bushings for wear and damage — replace parts as required.

Installation

Lubricate and install nylon bushings on pedal pivot shaft, right side of support brace cutout, and through both ends of brake pedal bore. Position brake pedal return spring on pedal arm and place pedal assembly in support brace. Index return spring in support brace cutout. Slide pedal pivot shaft through support brace and brake pedal bore. Install retainer to right side of pedal pivot shaft. Install bracket with pedal assemblies to underside of instrument panel with two screws. Install support plate on bracket with four bolts and nuts. Place main cylinder in position and install four bracket and cylinder mounting bolts; secure entire assembly with four nuts. Install steering column in vehicle.

Install brake pedal return spring on manual brake cars. On manual transmission models, connect clutch pedal push rod to pedal bracket and install retainer. Install clutch pedal return spring. Adjust brake pedal free travel. Adjust stoplight switch. Adjust clutch pedal travel.

MASTER CYLINDER
Cleaning precautions

Always use denatured alcohol or brake fluid to clean any main cylinder or wheel cylinder parts or when flushing a brake system. Never use mineral-base cleaning solvent such as gasoline, kerosene, carbon-tetrachloride, acetone, paint thinner or units of like nature as these solvents deteriorate rubber parts, causing them to become soft and swollen in an extremely short time.

DUALMASTER CYLINDER
Description

The master cylinder is designed and built to satisfy brake system displacement requirements. Therefore, it is necessary that correct parts be used when replacing either complete main cylinder assemblies or the component pistons of these assemblies. There are two sources for dual main cylinders: Delco Moraine and Bendix. The Bendix unit can be readily identified by a "Secondary Piston Stop" bolt on the bottom of the casting. Delco Moraine main cylinders have a stop bolt on the inside. The length of the secondary pistons in these main cylinders is a critical factor in the displacement capabilities of a particular main cylinder. These secondary pistons are coded, using rings or grooves in the shank or center section of the piston. The primary pistons are of two types. One has a deep socket for the push rod and the other a very shallow socket.

172

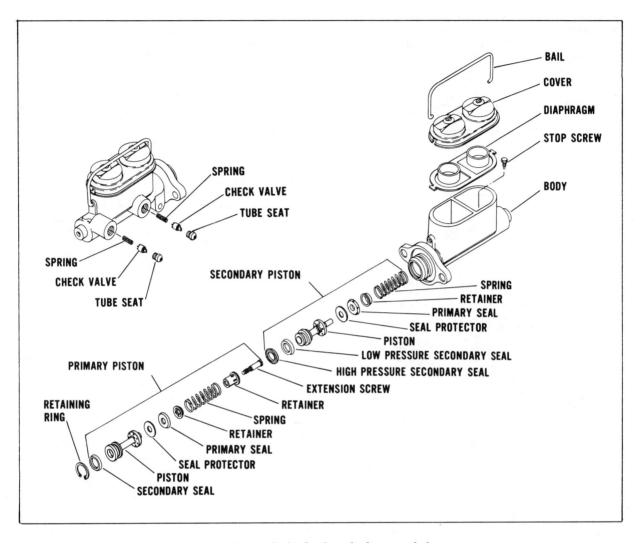

Main cylinder for drum brakes – explode

Replacement pistons must be identical to the original piston. This can be verified by checking the identification marks on the secondary piston and the socket contour of the push rod end of the primary piston.

Removal
Wipe master cylinder and lines clean with a cloth. Place dry cloths below cylinder area to absorb any fluid spillage. Disconnect hydraulic lines at cylinder. Cover line ends with clean lint-free cloth to prevent foreign matter from entering the system. Disconnect the push rod from the brake pedal. Unbolt and remove the cylinder from the firewall. Remove the cylinder mounting gasket and boot. Remove the cylinder cover and dump out the fluid. Pump the remaining fluid from the cylinder by depressing the push rod.

Disassembly
173
Clamp master cylinder in a bench vise. Remove push rod retainer. Remove secondary piston stop bolt from bottom of front fluid reservoir. Remove the snap ring retainer and primary piston assembly. Remove the secondary piston, piston spring, and retainer by blowing air through the stop bolt hole. (If no air is available, a piece of wire may be used. Bend approximately ¼ inch of one end into a right angle, hook the secondary piston and pull it out.) Position main cylinder in vise with outlet holes facing up. Remove the primary seal, primary seal protector, and secondary seals from the secondary piston. Remove the piston extension screw securing the primary piston spring to the primary piston. Remove the spring retainer, primary seal, primary seal protector and secondary seal from the primary piston.

Cleaning and inspection
Inspect the bore for corrosion, pits, and foreign matter. Be sure that the outlet ports are clean and free. Inspect the fluid reservoirs for foreign matter. Check the bypass and compensating ports to the cylinder bore to ensure that they are not restricted. Do not use wire to check

ports. Before washing parts, hands must be clean. Do not wash hands in gasoline or oil before cleaning parts. Use soap and water only. Use denatured alcohol to clean all metal parts thoroughly. Immerse parts in the cleaning fluid and brush with hair brush to remove foreign matter. Blow out all passages, orifices, and valve holes. Air dry the parts and place on clean paper or lint-free clean cloth. Be sure to keep parts clean until reassembly. Rewash parts, if there is any occasion to doubt cleanliness. Check pistons for scratches or other visual damage; replace if necessary.

Main cylinder (disc brakes) used with angle mountings

Assembly

Use care when reassembling the main cylinder check valves. Improper assembly of the check valve seats will result in distortion of the seats. If this occurs, there will be no check valve action and a loss of brake pedal travel will result; the pedal will have to be pumped one or more times before actual car braking occurs.

Place the main cylinder in a vise with the outlet hole facing up. (If check valves were removed) place the check valve springs in the outlet holes. Be sure the springs are seated in the bottom of the holes. Place new rubber check valves over the springs, being careful not to displace the springs from the spring seats.

Place new brass tube seats in the outlet holes. Be sure seats are not cocked as this would cause burrs to be turned up as the tube seats are pressed in. Thread a spare brake line tube nut into the outlet hole and turn the nut down until the tube seat bottoms. Remove the tube nut and check the outlet hole for loose burrs, which might have been turned up when the tube seat was pressed down. Repeat this process to bottom the second seat.

Put new secondary seals in the two grooves in the end of the secondary piston assembly. The seal which is nearest the end will have its lips facing toward that end. The seal

in the second groove should have its lips facing toward the portion of the secondary piston which contains the small compensating holes.

Assemble a new primary seal protector and primary seal over the end of the secondary piston with the flat side of the seal seats against the seal protector, and the protector against the flange of the piston which contains the small compensating holes. Assemble the new secondary seal into the groove on the push rod end of the primary piston. The lips of this seal should face toward the small compensating holes in the opposite end of the primary piston.

Assemble the new primary seal protector and primary seal on the end of the primary piston with the flat side of the seal seated against the seal protector, and the protector against the flange on the piston which contains the compensating holes. Assemble the spring retainer in one end of the primary piston spring and the secondary piston stop in the other end. Place the end of the spring over the end of the primary piston with the spring retainer seats inside of the lips of the primary seal.

Remove all cleaning liquid from the threaded hole in the primary piston. Place the piston extension screw down through the secondary piston stop and the primary spring retainer and screw it into the primary piston until it bottoms out. Coat the bore of the master cylinder with clean brake fluid. Coat the primary and secondary seals on the secondary piston with clean brake fluid. Insert the secondary piston spring retainer into the secondary piston spring. Place the retainer and spring down over the end of the secondary piston until the retainer locates inside of the lips of the primary cup. Hold the master cylinder with the open end of the bore down. Push the secondary piston into the bore until the spring seats against the closed end of the bore. Position the master cylinder in a vise with the open end of the bore up. Coat the primary and secondary seal on the primary piston with clean brake fluid. Push the primary piston assembly, spring end first, into the bore of the master cylinder. Hold the piston down and snap the lock ring into position in the small grooves in the I.D. of the bore. Push the primary piston down to move the secondary piston forward far enough to clear the stop screw hole in the bottom of the front fluid reservoir. Install the stop screw. Install reservoir diaphragm in the reservoir cover and install the cover on the main cylinder. Push bale wire into position to secure the reservoir cover.

Installation

Assemble the push rod through the push rod retainer, if it has been disassembled. Push the retainer over the end of the main cylinder. Assemble new boot over push rod and press it down over the push rod retainer. Slide new mounting gasket into position. Secure the main cylinder to the firewall with mounting bolts.

Connect the push rod clevis to the brake pedal with pin and retainer. Connect the brake lines to the main cylinder. Fill the main cylinder reservoirs. Bleed the brake system. If necessary, adjust the brake pedal free play.

MASTER CYLINDER (SINGLE SYSTEM — 1966)
Removal

Disconnect hydraulic line at master cylinder. Remove the two nuts and lock washers holding cylinder. Remove the cylinder, gasket and rubber boot.

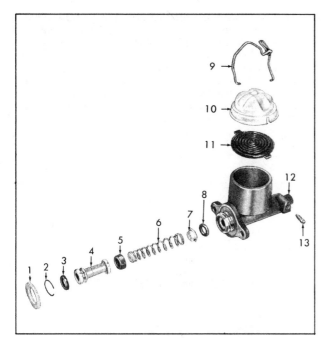

Main cylinder

1. Seal
2. Lock ring
3. Secondary cup
4. Piston
5. Primary cup
6. Spring
7. Valve assembly
8. Valve seat
9. Bail wire
10. Reservoir cover
11. Seal
12. Body
13. Bleeder

Disassembly

Remove boot from cylinder. Place cylinder in vise so that the lock ring can be removed from the small groove in the I.D. of bore. Remove lock ring, piston assembly, primary cup, spring and valve assembly and valve seat from cylinder bore. Pry wire off cover with screw driver and remove cover and seal.

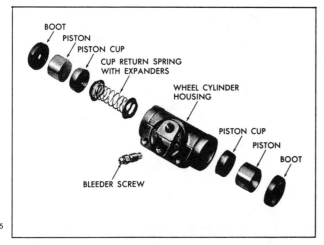

Wheel cylinder

Inspection

Wash all parts in clean alcohol. Make sure that compensating and bypass ports in cylinder body and bypass holes in piston are clean and open.

NOTE: Use soap and water to clean hands.

Inspect cylinder bore for pits and foreign matter. Inspect primary and secondary cups, check valve and valve seat for damage and swelling. Check piston fit in cylinder bore. The clearance between piston and wall of cylinder should be 0.001-0.005 inch.

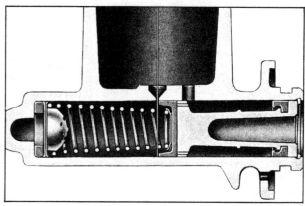

Checking compensating port clearance

Assembly

Care must be taken to reassemble the check valve correctly. Improper assembly of the check valve seat will result in its distortion. When the check valve seat is distorted, there will be no check valve action and there will be a loss of brake pedal travel, also, the pedal will have to be depressed or pumped one or more times before actual car braking occurs.

Install valve seat in cylinder bore so that flat portion of seat rests against end of cylinder bore. Position valve and spring assembly into bore. Dip primary cup into clean brake fluid and install into main cylinder with the flat side toward push rod end. Make sure cup seats over end of spring. Assemble the secondary seal in the groove on the piston so that the lip faces toward the end of the piston that contains the bypass holes. Dip piston in clean brake fluid and place piston in cylinder bore. Install piston stop ring. Check clearance between the edge of the primary cup and the center of compensating port.

NOTE: This check is made easily by using a wire and inserting it through the reservoir and into piston chamber.

Install a new seal in cover and place cover on cylinder. Secure by snapping bail wire in place. Install rubber boot, making certain boot seals tightly on cylinder body. This seal must be maintained to keep water and other foreign matter from entering the main cylinder. Install mounting gasket to main cylinder.

Installation

Position master cylinder on mounting studs and secure to dash wall. Make sure push rod goes through rubber boot and into piston. Connect brake line to cylinder. Check, and if necessary, adjust brake pedal free play. Bleed brakes.

BRAKE PIPE DISTRIBUTION AND SWITCH ASSEMBLY
Removal

Disconnect battery cable. Disconnect electrical lead from switch assembly. Place dry rags below the switch to absorb any fluid spilled during removal of switch. Disconnect four hydraulic lines from connections at switch. If necessary, loosen line connections at main cylinder to loosen lines. Cover open line ends with clean, lint-free material to prevent foreign matter from entering the system. Remove mounting screw and remove switch from vehicle.

Installation

Make sure new switch is clean and free of dust and lint. If any doubt exists, wash switch in denatured alcohol and dry with air. Place switch in position and secure to bracket with mounting screw. Remove protective material from open hydraulic brake lines and connect lines to switch. If necessary, tighten brake line connections at main cylinder. Connect switch electrical lead. Connect battery cable. Bleed the brake systems.

BRAKE WARNING LIGHT
Checking

Disconnect wire from switch terminal and use a jumper to connect wire to a good ground. Turn ignition key to "ON". Warning lamp should light. If lamp does not light, bulb is burned out or electrical circuit is defective. Replace bulb or repair electrical circuit as necessary. When warning lamp lights, turn ignition switch off. Disconnect jumper and reconnect wire to switch terminal.

WARNING LIGHT SWITCH
Testing

Raise vehicle on hoist and attach a bleeder hose to a rear brake bleed screw and immerse other end of hose in a container partially filled with clean brake fluid. Be sure master cylinder reservoirs are full. Turn ignition switch to "ON". Open bleed screw while a helper applies heavy pressure to brake pedal. Warning lamp should light. Close bleed screw before helper releases pedal. Attach bleeder hose to front brake bleed screw and repeat above test. Again warning lamp should light. Turn ignition switch off. Lower vehicle to floor.

If warning lamp does not light, but does light when a jumper is connected to ground, warning light switch is defective. Do not attempt to disassemble switch. A defective switch must be replaced with a new switch assembly. Caution should be taken to prevent air from entering hydraulic system during checks on switch.

The recommended checking interval should be 24 months or 24,000 miles, any time major brake work is done or when there is excessive pedal travel.

4 WHEEL DISC BRAKES
Description

The disc brake consists of a fixed caliper, splash shield, mounting bracket and rotating disc. The caliper assembly contains four pistons and two shoe and lining assemblies. A seal and dust boot are installed on each piston, with a piston spring in the caliper cylinder bore beneath each piston. A retaining pin extends through each caliper half and both shoes to hold the shoes and linings in position in the caliper. Machined surfaces within the caliper prevent the shoe and lining assembly from rotating with the brake disc when pressure is applied.

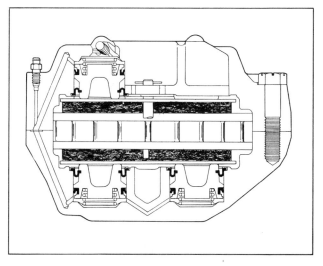

Disc brake – cutaway view

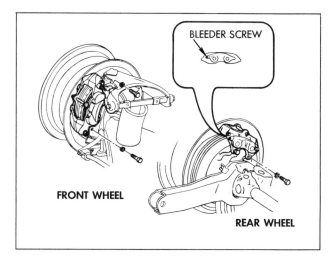

Disc brake

The disc, which has a series of air vent louvers to provide cooling, is mounted on the front wheel-hub. The caliper straddles the disc and mounts on a mounting bracket attached to the steering knuckle or rear axle flange.

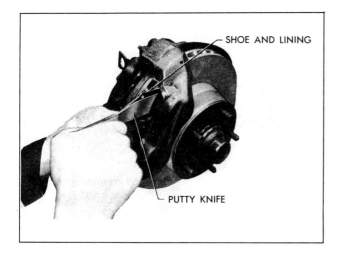

Replacing shoe and lining assembly

BRAKE CALIPER
Cleaning precautions
Always use denatured alcohol or brake fluid to clean any main cylinder or wheel cylinder parts. Never use mineral-base cleaning solvent such as gasoline, kerosene, carbon-tetrachloride, acetone, paint thinner or units of like nature as these solvents deteriorate rubber parts, causing them to become soft and swollen in an extremely short time.

Removal
Place car on hoist. Remove wheel. On front caliper, disconnect the brake hose at brake line support bracket. On rear caliper, disconnect the brake tubing from the inboard caliper. Tape the open tube or line end to prevent foreign matter from entering the system. Pull cotter pin from end of shoe assembly retaining pin. Remove the pin and shoe assembly from the caliper. Identify the inboard and outboard shoe if they are to be reused. Remove the end of brake hose at bracket by removing U-shaped retainer from the hose fitting and withdrawing the hose from bracket. Remove the caliper assembly from the mounting bracket by removing two hex head bolts.

Disassembly
Separate the caliper halves by removing the two large bolts. Remove the two small "O" rings from the cavities around the fluid transfer holes in the two ends of the caliper halves. To free the piston boots so that the pistons may be removed, push the piston down into the caliper as far as it will go. Insert a screw driver blade under the inner edge of the steel ring in the boot, and using the piston as a fulcrum, pry the boot from its seat in the caliper half. Use care not to puncture seal when removing pistons from caliper. Remove the pistons and piston springs from the caliper half. Remove the boot and seal from their grooves in the piston.

Cleaning and inspection
Clean all metal parts using denatured alcohol. Remove all traces of dirt and grease. Do not use mineral base solvents to clean brake parts. Using an air hose, blow out all fluid passages in the caliper halves, making sure that there is no dirt or foreign material blocking any of these passages. Discard all rubber parts. Boots, seals, and "O" rings should be replaced with new service kit parts. Carefully inspect the piston bores in the caliper halves. They must be free of scores and pits. A scored or otherwise damaged bore will cause leaks and unsatisfactory brake operation. Replace the caliper half if either bore is damaged to the extent that polishing with very fine crocus cloth will not restore it. Check the fit of the piston in the bore using a feeler gauge. Clearance should be as follows:

2-1/16 inch Bore	.0045 to .010
1-7/8 inch Bore	.0045 to .010
1-3/8 inch Bore	.0035 to .009

If the bore is not damaged, and the clearance exceeds either of the upper limits, a new piston will be required.

Assembly
Assemble the seal in the groove in the piston which is closest to the flat end of the piston. The lip on the seal must face toward the large end of the piston. Be sure lips are in the piston groove and do not extend over the step in the end of the groove. Place the spring in the bottom of the piston bore. Lubricate the seal with clean brake fluid. Install the piston assembly in the bore using applicable piston ring compressor Tool J-22639, 22629, or 22591. Use care not to damage the seal lip as piston is pressed past the edge of the bore.

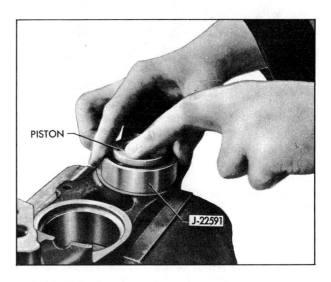

Installing piston in caliper bore using tool J-22591

Assemble the boot in the groove of the piston closest to the concave end of the piston. The fold in the boot must face toward the end of the piston with the seal on it. Depress the pistons and check that they slide smoothly into the bore until the end of the piston is flush with the end of the bore. If not, recheck piston assembly and location of the piston spring and the seal.

Position applicable boot seal installer Tool J-22592, J-22628, or J-22638 over the piston and seat the steel boot retaining ring evenly in the counterbore. The boot retaining ring must be flush or below the machined face of the caliper. Any distortion or uneven seating could allow contaminating and corrosive elements to enter the bore. Position the "O" rings in the small cavities around the brake fluid transfer holes in both ends of the outboard caliper halves. Lubricate the hex head bolts with Delco Brake Lube or equivalent or dip in clean brake fluid. Fit caliper halves together and secure with bolts.

Installing caliper on disc

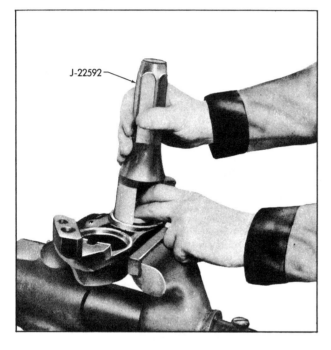

Installing boot seal in caliper using tool J-22592

Installation
Carefully mount the assembled caliper over the edge of the disc. Use a putty knife to depress pistons so that the caliper can be lowered into position on the disc. Use care to prevent damage to boots on the edge of the disc as the caliper is mounted. Secure the caliper to the mounting bracket with two hex head bolts. Refer to torque specifications in rear of manual for correct torque values. All brake attachments are important attaching parts in that they could affect the performance of vital components and systems, and/or could result in major repair expense. They must be replaced with parts of the same part numbers or with equivalent parts if replacement becomes necessary. Do not use replacement parts of lesser quality or substitute design. Torque values must be used as specified during reassembly to assure proper retention of these parts. If replacing old shoe assemblies, be sure to get the shoes in the same position from which they were removed. Install the shoe and lining assemblies. Place a new copper gasket on the male end of the front wheel

brake hose. Install brake hose in the calipers. With the wheels straight ahead, pass the female end of the brake hose through the support bracket. Make sure the tube seat is clean and connect the brake line tube nut to the caliper. Tighten securely. Allowing the hose to seek a normal position, without twist, insert hex of the hose fitting into the 12-point hole in the support bracket and secure it in place with the "U"-shaped retainer. Turn the steering geometry from lock to lock while observing the hose. Check that the hose does not touch other parts at any time during suspension or geometry travel. If contact does occur, remove the "U"-shaped retainer and rotate the end of the hose in the support bracket one or two points in a direction which will eliminate hose contact. Reinstall the retainer and recheck for hose contact. If it is satisfactory, place the steel tube connector in the hose fitting and tighten securely.

If rear brake caliper is being serviced, connect brake line to caliper. Bleed brakes. Install wheels and lower vehicle. Do not move car until a firm pedal is obtained.

BRAKE DISC SERVICING
Minimum requirements
Both frictional surfaces must be flat within .001 inch total indicator reading when measured across the surface (radially). Both frictional surfaces must be flat within .0005 inch when measured around the disc (circumferentially) at any radius. Both frictional surfaces must be parallel within .001 inch total indicator reading. When mounted on bearing cups, lateral runout must not exceed .002 inch total indicator reading. Both frictional surfaces must be free from scratch marks and pits (porosity). Finish of the frictional surface must be 30–50 micro-inches measured around the disc (circumferentially) and 50 micro-inches measured across the disc (radially) at any radius. The minimum disc thickness allowable after refinishing is 1.215 inch for the 1¼ inch thick disc. Refinishing of the disc surfaces can be performed if precision equipment is available and the minimum specifications can be maintained.

Light scoring of the disc surfaces not exceeding .015 inch in depth, which may result from normal use, is not detrimental to brake operation.

FRONT HUB
Removal
Raise vehicle on hoist. Remove wheel and tire assembly. Remove brake caliper. Remove hub and disc as an assembly.

Disassembly and assembly
Drill out the rivets attaching the disc to the hub. When reassembling the disc to the hub it is necessary to rivet the two assemblies. The lug nuts will supply ample retention. When checking runout after the rivets have been removed it will be necessary to install the lug nuts.

Checking
Place car on hoist and raise. Tighten the adjusting nut of the wheel bearing until all play has been removed. It should be just loose enough to allow the wheel to turn. Clamp a dial indicator to the caliper so that its button contacts the disc at a point about 1 inch from the outer edge. When the disc is turned, the indicator reading should not exceed .002 inches. If runout exceeds this amount the hub and disc assembly should be repaired. Due to the close tolerances involved it is not recommended that the front hub and discs be machined or serviced separately. After checking the runout, readjust the wheel bearings.

Installation
Follow removal procedure in reverse. Bleed the system. Install wheel and tire assemblies and lower vehicle.

REAR HUB
Removal
Raise vehicle on hoist. Remove wheel and tire. Remove brake caliper. Drill out rivets attaching the disc to the axle spindle. Follow the same procedure as in front hub for reassembling the disc. Be sure the emergency brake adjusting holes of the spindle and disc are in alignment.

Checking
Check the rear wheel bearing end play. Then dial indicate the disc face. If lateral runout of the disc exceeds the bearing end play by .002 inches, the disc should be repaired.

Installation
Follow removal procedure in reverse. Bleed the system. Install wheel and tire assemblies and lower vehicle.

PARKING BRAKE SHOES
Removal
Remove tire and wheel assemblies and brake disc. Remove retractor spring at the top of the shoes. Remove hold down springs on primary and secondary shoes. Remove shoes by pulling them away from the anchor pin. Remove the adjusting screw spring and adjusting screw from the shoes.

Installation
Put light coat of lubriplate or equivalent on pads, backing plate, and the threads of the adjusting screw. Attach adjusting screw spring to the bottom hole in each shoe. Insert the star wheel between the shoes. (On left hand brakes, the star wheel goes next to the rear shoe; on right

hand brakes, the star wheel goes next to the forward shoe). Install the shoes on the backing plate by spreading them and placing them around the anchor pin. Install the hold down springs on the hold down nails. Install retractor spring on one shoe and stretch to other shoe. Make sure that the lever assembly which spreads the shoe is located so that the notches on the lever fit against the shoes. Install disc and caliper as outlined in this section. Bleed brakes. Adjust the parking brake. Install wheels and lower vehicle. When replacing the parking brake shoes, it is necessary to "break in" the new shoes in the following manner. (Brakes should be adjusted before beginning.)

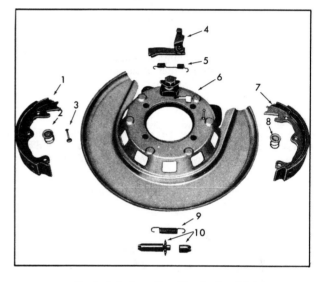

Exploded view of parking brake shoes

1. Parking brake shoe	6. Backing plate
2. Hold down spring and cap	7. Parking brake shoe
3. Hold down pin	8. Hold down spring and cap
4. Actuating lever	9. Adjusting screw spring
5. Retractor spring	10. Adjusting screw assembly

Burnishing new linings (parking brake)
Count the number of notches required to pull 40 lbs. on the parking brake handle. (Pull scale on the second finger notch). Release the brake and drive the vehicle at 30 m.p.h. Apply the parking brake the same number of notches gained above. Maintain 30 m.p.h. for ¼ mile and then release the brake.

POWER BRAKES
Description
The Power Brake Unit is a self-contained hydraulic and vacuum unit, utilizing manifold vacuum and atmosphere pressure for its power. This unit permits the use of a low brake pedal as well as less pedal effort than is required with the conventional (nonpower) hydraulic brake system. Only two external line connections are necessary — one a vacuum connection from manifold to check valve located on front shell; the other, a hydraulic connection from the main cylinder outlet directly into the hydraulic

system. The unit is mounted on the engine side of the fire wall and directly connected to the brake pedal.

MAINTENANCE AND ADJUSTMENTS
Inspections
Check vacuum line and vacuum line connections as well as vacuum check valve in front shell of power unit for possible vacuum loss. Inspect all hydraulic lines and connections at the wheel cylinders and main cylinder for possible hydraulic leaks. Check brake assemblies for scored drums, grease or brake fluid on linings, worn or glazed linings, and make necessary adjustments. Check brake fluid level in the hydraulic reservoirs. The reservoirs should be filled to the proper level. Check for loose mounting bolts at main cylinder and at power section. Check air cleaner filter in power piston extension and replace filter if necessary. Check brake pedal for binding and misalignment between pedal and push rod.

Lubrication
The power brake unit is lubricated at assembly and needs no further lubrication other than maintaining normal reservoir fluid level. The reservoir should be filled as described in this section.

Bleeding instructions
The power system may be bled manually or with a pressure bleeder. Use only GM Supreme 11 Brake Fluid or equivalent. Do not use the power assist while bleeding.

The engine should not be running and the vacuum reserve should be reduced to zero by applying the brake several times before starting the bleeding procedure.

Air cleaner service
Servicing of the air cleaner is recommended and the element replaced when restriction becomes severe enough to affect power brake response. At any other time, if cleaning of the filter is felt necessary, it should be shaken free of dirt or washed in soap and water and thoroughly dried.

POWER BRAKE CYLINDER
Removal
Remove vacuum hose from vacuum check valve. Disconnect hydraulic lines at main cylinder. Disconnect push rod at brake pedal assembly. Remove nuts and lock washers securing power unit to fire wall, and remove power unit from engine compartment.

POWER BRAKE CYLINDER
Removal and installation
Remove vacuum hose from vacuum check valve. Disconnect hydraulic lines at main cylinder. Disconnect push rod at brake pedal assembly. Remove nuts and lock washers securing power unit to fire wall, and remove power unit from engine compartment.

Install in reverse sequence. Check brake lights and bleed system.

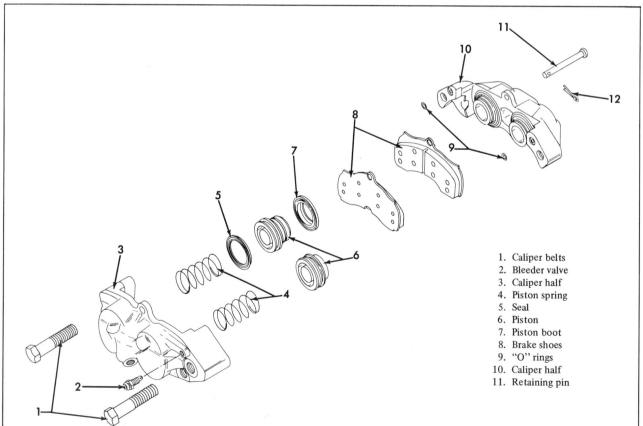

1. Caliper belts
2. Bleeder valve
3. Caliper half
4. Piston spring
5. Seal
6. Piston
7. Piston boot
8. Brake shoes
9. "O" rings
10. Caliper half
11. Retaining pin

180

Caliper assembly — exploded view

FRONT SUSPENSION 11

INDEX

DESCRIPTION

The front suspension is of the short-long arm type with independent coil springs. The springs ride on the lower control arms. Ball joints connect the upper and lower control arms to the steering knuckle. Tapered roller wheel bearings are used. Camber angle is adjusted by means of upper control arm inner support shaft shims.

Caster angle is adjusted by means of upper control arm inner support shaft shims. A stabilizer bar is used on all models.

FRONT WHEEL BEARING
Adjustment

Proper front wheel bearing adjustment has a definite bearing on the operation of a vehicle. Improperly adjusted front wheel bearings will result in a lack of steering stability causing wheel wander, shimmy and excessive tire wear. Very accurate adjustment is possible because the spindles are drilled both vertically and horizontally and the adjusting nuts are slotted in all six sides.

With wheel raised, remove hub cap and dust cap and then remove the cotter pin from the end of the spindle. While rotating wheel, tighten spindle nut to 12 lbs. ft. torque. Back off adjusting nut one flat and insert cotter pin. If slot and pin hole do not line up, back off the adjusting nut an additional ½ flat or less as required to insert cotter pin. Spin the wheel to check that it rolls freely and then lock the cotter pin by spreading the end and bending inboard to avoid possibility of damaging the static collector in the dust cap. Bearings should have zero preload and .001" to .008" end movement when properly adjusted. Install dust cap, hub cap or wheel disc and lower wheel. Perform the same operation on each front wheel.

FRONT END ALIGNMENT
Preliminary steps

Check and correct any conditions which could affect front end alignment; steering gear loose or improperly adjusted, steering gear housing loose at frame, excessive wear or play in ball joints or steering shaft coupling, tie rod or steering connections loose, improper front spring heights, unbalanced or underinflated tires, improperly adjusted wheel bearings, shock absorbers not operating properly.

Wheel alignment should always be made with the car rolled forward taking out any slack in the same manner as when the car is traveling forward.

Caster and camber adjustment

Before adjusting caster and camber angles, the front bumper should be raised and quickly released to allow car to settle to its normal height. Caster and camber adjustments are made by means of shims inserted between the upper control arm inner support shaft and the support bracket attached to the frame. Shims may be added, subtracted or transferred to change the readings as follows:

Caster — transfer shims, front to rear or rear to front. The transfer of one shim to the front bolt from the rear bolt will decrease positive caster. One shim (1/32") transferred from the rear attaching bolt to the front attaching bolt will change caster (approx.) 1/2°.

Camber — change shims at both the front and rear of the shaft. Adding an equal number of shims at both front and rear of the support shaft will decrease positive camber. One shim (1/32") at each location will move camber

11

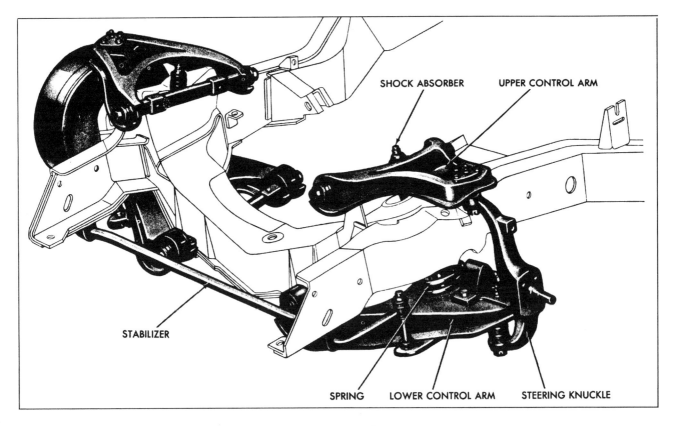

Front suspension

(approx.) $1/6°$. To adjust for caster and camber, loosen the upper support shaft to crossmember nuts, add or subtract shims as required and retighten nuts. Caster and camber can be adjusted in one operation.

Toe-in adjustment

Toe-in is checked with the wheels in the straight ahead position. It is the difference of the distance measured between the extreme front and the extreme rear of both front wheels. Toe-in must be adjusted after caster and camber adjustment.

If the equipment being used measures the toe-in of each wheel individually, set the steering gear on the high point, mark 12 o'clock position on the steering shaft and position the steering wheel for straight ahead driving. Loosen the clamp bolt at each end of each tie rod and adjust to the total toe-in as given in the specifications at the end of this book.

If a tram gauge is being used, set the front wheels in the straight ahead position. Loosen the clamp bolts on one tie rod and adjust for the proper toe-in. Loosen the clamp bolts on the other tie rod. Turn both rods the same amount and in the same direction to place the steering gear on its high point and position the steering wheel in its straight ahead position.

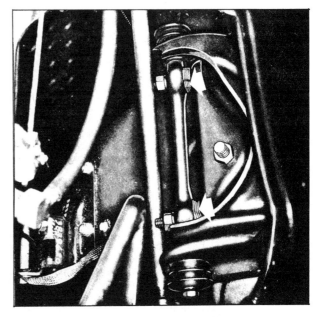

182

Caster and camber adjustment

After the adjustment has been made, position inner tie rod clamps with open end down and the bolt horizontal. Position outer clamps with open end toward the rear and bolt vertical.

FRONT WHEEL HUB
Replacement
Raise vehicle on hoist, remove hub caps, break loose the wheel stud nuts. Remove wheel nuts, wheel and tire and brake caliper. Remove bolts holding brake caliper to its mounting and insert a fabricated block (1-1/16 x 1-1/16 x 2 inches in length), between the brake pads as the caliper is being removed. Once removed the caliper assembly can be wired or secured in some manner away from the concerned area. Pry out hub grease cap, cotter pin, spindle nut and washer, and remove hub. Do not drop wheel bearings. Reverse the procedure to install and lower vehicle on hoist.

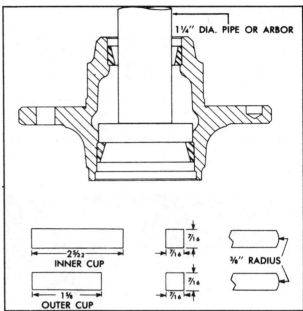

Front wheel bearing cup removers

Installation
Pack both inner and outer bearings using a high melting point wheel bearing lubricant. Place inner bearing in hub, then install a new inner bearing lip seal assembly. Seal flange should face bearing cup. Carefully install wheel hub over steering spindle. Install outer bearing, pressing it firmly into the hub by hand. Install spindle washer and adjusting nut. Draw up tight and adjust wheel bearings.

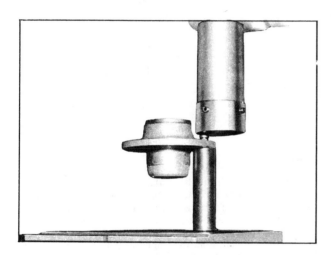

Pressing front hub bolts

FRONT WHEEL BEARINGS
Removal
Remove wheel hub. Discard cotter pin. Install new one when assembling. Remove outer roller bearing assembly from hub with fingers. The inner bearing assembly will remain in the hub and may be removed after prying out the inner bearing lip seal assembly. Discard seal. Wash all parts thoroughly in cleaning solvent and blow dry.

Check bearings for cracked separators or worn or pitted rollers and races. Check brake disc for cracks, or scoring. Check fit of bearing outer cups in hub.

Replacement of bearing cups
Using steel bar stock, make press-out tools. Insert removers through hub, indexing ends into slots in hub shoulder behind bearing cup. Using a suitable extension pipe or rod, press bearing cups from hub. Install new bearing cups in hub. Make sure that the bearing cups are not cocked and are fully seated against shoulder in hub.

183

Installing front hub inner bearing

STEERING KNUCKLE
Removal

Raise car on hoist and support lower control arm. Remove hub cap, wheel hub dust cover, cotter pin, adjusting nut and washer. Remove wheel, caliper, wheel hub, and bearing assembly. Remove steering arm and splash shield from steering knuckle. Wire splash shield to frame. Do not disconnect brake line. Remove upper and lower ball studs cotter pins and loosen ball stud nuts. Free steering knuckle from ball studs by rapping on steering knuckle bosses. Remove ball stud nuts and remove steering knuckle.

Backing plate removed

Installation

Place steering knuckle in position and insert upper and lower ball studs into knuckle bosses. Install ball stud nuts and tighten nut to specifications. If necessary, tighten one more notch to insert cotter pins.

Place splash shield in position on steering knuckle. Place steering arm in position on back of steering knuckle and insert two bolts through splash shield, steering knuckle and steering arm. Install locknuts and tighten. Install wheel hub, brake caliper and disc and hub assembly, wheel and tire assembly over spindle. Insert outer wheel bearing race and roller assembly, washer and nut. Adjust front wheel bearing. Install new cotter pin, dust cap and hub cap. Lower vehicle on hoist, recheck and readjust wheel alignment where necessary.

SHOCK ABSORBER
Removal

Raise vehicle on hoist and with an open end wrench hold the shock absorber upper stem from turning, and then remove the upper stem, retaining nut, retainer and rubber grommet. Remove the two bolts retaining the lower shock absorber pivot to the lower control arm and pull the shock absorber assembly out from the bottom.

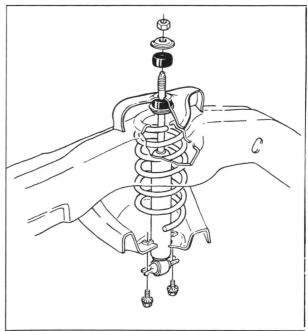

Shock absorber

Installation

With the retainer and rubber grommet in place over the upper stem, install the shock absorber (fully extended) up through the lower control arm and spring so that the upper stem passes through the mounting hole in the upper support arm. Install the rubber grommet, retainer and attaching nut over the shock absorber upper stem. With an open end wrench, hold the upper stem from turning and tighten the retaining nut. Install the two bolts attaching the shock absorber lower pivot to the lower control arm, tighten and lower vehicle on hoist.

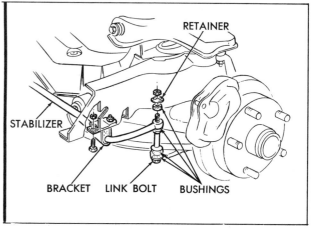

184

Stabilizer bar

STABILIZER BAR
Removal

Raise car on hoist and support both front wheels. Disconnect stabilizer bar from lower control arm. Remove stabilizer bar brackets from the frame and remove stabilizer. Remove stabilizer link bolts, spacers and rubber bushings from lower control arms. Inspect rubber stabilizer link bushings and stabilizer insulator bushings for aging and wear. Replace if necessary.

Installation

If new insulators are necessary, coat stabilizers with recommended rubber lubricant and slide frame bushings into position. Insert stabilizer brackets over bushings and connect to frame. Do not torque at this point. Connect stabilizer ends to link bolts on lower control arms. Never get lubricant on outside of frame stabilizer bar bushings or they may slip out of brackets. Lower car on hoist.

FRONT SPRING
Removal

Raise car and remove nut, retainer and grommet from top of shock absorber. Support front of car so control arms swing free. Disconnect stabilizer bar from lower control arm and remove bolts retaining shock absorber to lower control arm, remove shock absorber. Bolt Tool J-22944 to a suitable jack and place tool under lower control arm bushings so that the control arm bushings seat in the grooves of the tool. As a safety precaution install a chain through the spring and lower control arm. Remove cross shaft rear retaining nut and the two front retaining bolts. Lower control arm by slowly releasing jack until spring can be removed. Remove chain. Swing control arm forward and remove spring.

Installation

Properly position spring on the control arm and lift control arm. Using a large drift position control arm to cross-member and install cross shaft retaining bolts. Torque to specifications. Connect the stabilizer bar link and install bolts retaining shock absorber to the lower control arm. Lower car on hoist and install the shock absorber upper grommet, retainer and nut.

LOWER CONTROL ARM BALL JOINT
Inspection

The lower control arm ball joint should be replaced whenever wear is indicated in the upper joint inspection. The lower control arm ball joint is a loose fit in the assembly when not connected to the steering knuckle.

Support car weight on wheels or wheel hubs. With outside micrometer or caliper, measure distance from top of lubrication fitting to bottom of ball stud, and record the dimensions for each side. Then support car weight at outer end of each lower control arm, so that wheels or wheel hubs are free, then repeat above. If the difference in dimensions on either side is greater than 1/16" (.0625"), the joint is excessively worn and both lower joints should be replaced.

If inspection of lower joints does not indicate excessive wear, inspect further. Examine lubrication hole in each joint assembly after cleaning out hole. Look for evidence of the liner partially or fully blocking lubrication opening. Such evidence indicates that wear is sufficiently advanced that both lower joints should be replaced.

Another indication of lower joint excessive wear is

185

Removing front coil spring

indicated when difficulty is experienced when lubricating the joint. If the liner has worn to the point where the lubrication grooves in the liner have worn away, then abnormal pressure is required to force lubricant through the joint. This is another reason to replace both lower joints.

If the above inspections do not indicate any reason for joint replacements, test the torque tightness of the lower ball stud in the knuckle on each side. Wirebrush off nut and cotter pin attaching sperical joint ball stud to steering knuckle and examine for evidence of looseness of stud in knuckle. If no evidence of looseness, remove cotter pin and with prick punch or equivalent, mark nut stud and knuckle to identify relative location. Tighten nut as installed and observe torque reading. If less than 45 lbs. ft., stud may have been loose in steering knuckle and replacement of both lower sperical joints may be recommended. Check to see if torque of 60-94 lbs. ft. can be obtained without bottoming stud or ball joint against knuckle. If bottoming occurs, replace ball joint or steering knuckle.

Removal
Raise car and support lower control arm at outer end on floor jack, with hoist or jack pad clear of lower ball stud nut and seal. Remove caliper assembly. Remove upper and lower ball stud nuts, free ball studs from steering knuckle and wire knuckle and disc assembly up to fender skirt to preclude interference while performing next step. Be careful not to enlarge the holes in control arm; cut off rivets.

Installation
Install new joint against underside of control arm and hold in place with special bolts and nuts supplied with new joint. Use only alloy bolts supplied for this operation. The special thick headed bolt must be installed in the forward side of the control arm. Tighten bolts and nut on ball stud to specification. Lubricate the joint. Lower car on hoist.

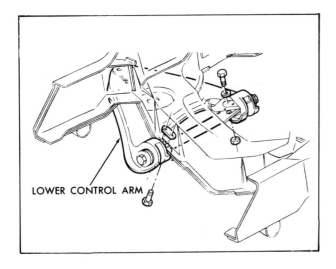

Lower control arm attachment

LOWER CONTROL ARM
Removal
Remove the front coil spring. Remove lower control arm ball stud from the steering knuckle boss. Remove the lower control arm assembly.

Installation
Install lower control arm ball stud into steering knuckle boss. Torque to specifications. Install coil spring. Tighten cross shaft bolts to specifications and lower vehicle on hoist. With unit on floor, tighten cross shaft bushing bolts to torque shown in specifications.

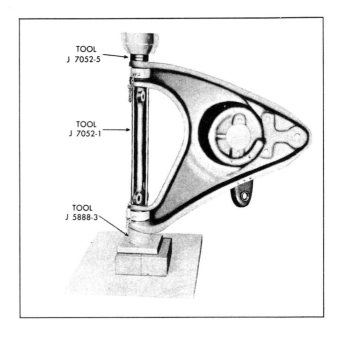

Installing lower control arm cross shaft bushing

UPPER CONTROL ARM BALL JOINT (OFF CAR)
Inspection
The upper ball joint is checked for wear by checking the torque required to rotate the ball stud in the assembly. Install a stud nut on the stud and measure the torque required to turn the stud in the assembly with a torque wrench. Specified torque for a new joint is 3 to 10 lbs. ft. If torque readings are excessively high or low, replace the ball joint. If excessive wear is indicated in upper joint, both upper and lower joints should be replaced. This inspection does not necessitate upper control arm removal.

Replacement
Raise car on hoist and support its weight at the outer end of the lower control arm. Remove the wheel and tire assembly. Remove cotter pin and nut from upper control arm ball stud. Remove the stud from knuckle.

Cut off the ball joint rivets with a chisel. Enlarge ball stud attaching holes in control arm to 21/64" to accept 5/16" bolts included in unit. Install new joint and retain in place with the special nuts and bolts supplied. Reassemble ball stud to the steering knuckle and lower vehicle on hoist.

UPPER CONTROL ARM
Removal
Raise car and support its weight at outer end of lower control arm. Remove wheel and tire assembly. Remove cotter pin and nut from upper control arm ball stud. Remove the stud from knuckle. Remove two nuts retaining upper control arm shaft to front crossmember. (Note the number of shims at each bolt.) If necessary, remove the bolts attaching the control arm to the frame to allow proper clearance for control arm removal. Remove upper control arm from vehicle.

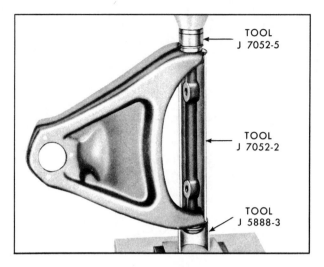

Installing upper control arm cross shaft bushings

Installation
Install upper control arm. Install locknuts, bolts retaining upper control arm shaft to frame. (Install same number of shims as removed at each bolt.) Torque nuts as shown in the specifications. Install ball stud through knuckle, install nut, tighten and install cotter pin.

Install wheel and tire assembly. Lower car to floor. Bounce front end of car to centralize bushings and tighten bushing collar bolts as shown in the specifications and lower car on hoist.

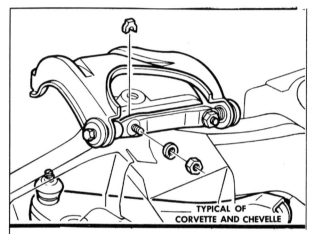

Upper control arm attachment

STEERING 12

INDEX

12

STANDARD STEERING

Description

The steering gear is the recirculating ball type, transmitting forces from worm to sector gear through ball bearings. The steering linkage is of the relay type with the pitman arm connected to one end of the relay rod. The outer end of the relay rod is connected to an idler arm which is connected to the frame side rail opposite the steering gear. Two adjustable tie rods connect the relay rod to the steering arms.

Lubrication

The steering gear is factory-filled with steering gear lubricant. Seasonal change of this lubricant should not be performed and the housing should not be drained — no lubrication is required for the life of the steering gear.

Every 36,000 miles, the gear should be inspected for seal leakage (actual solid grease — not just oily film). If a seal is replaced or the gear is overhauled, the gear housing should be refilled with #1051052 (13 oz. container) Steering Gear Lubricant which meets GM Specification GM 4673M, or its equivalent. Do not use EP Chassis Lube, which meets GM Specification GM 6031M, to lubricate the gear. DO NOT OVERFILL the gear housing.

The steering linkage should be lubricated with water resistant EP chassis lubricant every 6,000 miles or four months, whichever occurs first.

Adjustments

Before any adjustments are made to the steering gear a careful check should be made of front end alignment, shock absorbers, wheel balance and tire pressure for possible cause. Remove pitman arm nut and mark relation

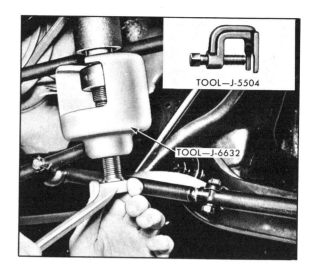

Removing pitman arm

of pitman arm position to sector shaft. Remove pitman arm. Loosen the pitman shaft lash adjuster screw locknut and turn the adjuster screw a few turns in a counter-clockwise direction. This removes the load imposed on the worm bearings by the close meshing of rack and sector teeth. Turn steering wheel gently in one direction until stopped by gear, then back away about one turn. Do not turn steering wheel hard against stops when steering relay rod is disconnected as damage to ball guides may result.

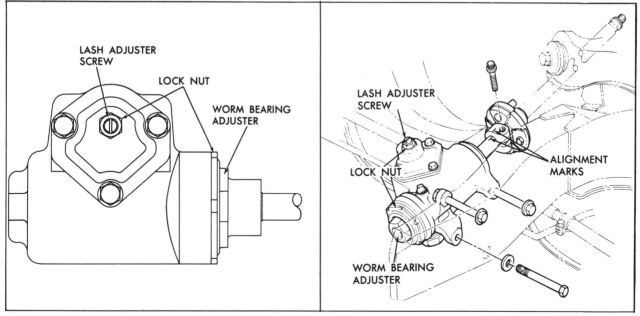

190

Steering gear adjustment points

Disconnect battery ground cable. Check preload by removing horn button or shroud and applying a torque wrench with a 3/4 inch socket on the steering wheel nut. Do not use torque wrench having maximum torque reading of more than 100 inch pounds.

On cars with tilt steering columns, it will be necessary to disconnect the steering coupling to obtain a torque reading of the steering column. This reading should then be subtracted from any reading taken on the gear. On vehicles with cushioned rim steering wheels, it will be necessary to remove the upper horn contact to gain access to the steering wheel nut. On cars with telescopic steering columns, check preload by removing the horn button and applying torque wrench, with a phillips-head adapter socket, on the star-headed screw in center of steering wheel. Apply a light vertical pressure to torque wrench end to keep phillips-head engaged in screw. If torque does not lie within limits, adjustment is necessary.

Checking wheel pull (typical)

To adjust worm bearings, loosen worm bearing adjuster locknut and turn worm bearing adjuster until there is no perceptible end play in worm. Check pull at steering wheel, readjusting if necessary to obtain proper torque. Tighten locknut and recheck torque. If the gear feels "lumpy" after adjustment of worm bearings, there is probably damage in the bearings due to severe impact or to improper adjustment and the gear must be disassembled for replacement of damaged parts. After proper adjustment of worm is obtained and all mounting bolts securely tightened, adjust lash adjuster screw. First turn the steering wheel gently from one stop all the way to the other, carefully counting the total number of turns. Then turn wheel back exactly half way, to center position. Turn lash adjuster screw clockwise to take out all lash in gear teeth, and tighten locknut. Check torque at steering wheel taking highest reading as wheel is turned through center position. Correct torque should be 1/2 to 1-1/4 lbs.

191

Readjust if necessary to obtain proper torque. If maximum specification is exceeded, turn lash adjuster screw counter-clockwise, then come up on adjustment by turning the adjuster in a clockwise motion.

Tighten locknut then recheck toruqe, as it must lie between specified readings. Reassemble pitman arm to sector shaft, lining up marks made during disassembly. Install horn cap or shroud and connect battery ground cable.

Steering wheel alignment and high point centering
Set front wheels in straight ahead position. This can be checked by driving vehicle a short distance on a flat surface to determine steering wheel position at which vehicle follows a straight path. With front wheels set straight ahead, check position of mark on wormshaft designating steering gear high point. This mark should be at the top side of the shaft at 12 o'clock position. If gear has been moved off high point when setting wheels in straight ahead position, loosen adjusting sleeve clamps on both left and right hand tie rods, then turn both sleeves an equal number of turns in the same direction to bring gear back on high point. Turning the sleeves an unequal number of turns or in different directions will disturb the toe-in setting of the wheels. Readjust (if necessary). With wheels in a straight ahead position and the steering gear on highpoint, check the steering wheel alignment. If the spokes are not within the limits specified, the wheel should be removed and centered.

STEERING RATIO
Changing
The steering ratio may be changed as follows: Do not use the rearward hole in the steering arm with power steering equipment or interference may result.

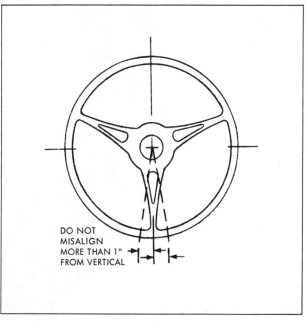

DO NOT MISALIGN MORE THAN 1" FROM VERTICAL

Steering wheel alignment

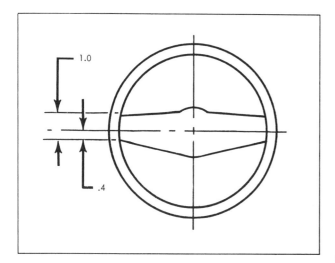

Steering wheel alignment

Place vehicle on hoist. Remove tie rod ball stud nut at steering arm and disconnect tie rod from steering arm. Move tie rod end to forward hole for 17.6:1 ratio (fast

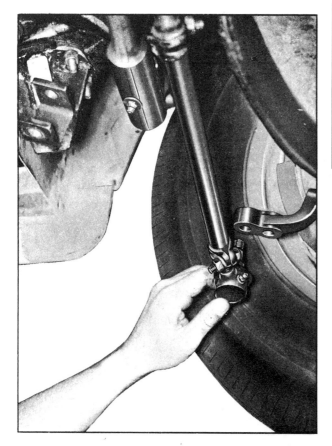

Adjusting steering ratio

ratio) or rear hole for 20.2:1 ratio (standard ratio). Install tie rod stud nut and tighten securely. Repeat operation on opposite steering arm. Remove vehicle from hoist.

STANDARD STEERING WHEEL
Removal
Disconnect battery ground cable. Pry off horn button. Remove contact assembly screws and assembly. Remove remaining screws from steering wheel and remove wheel from hub assembly.

To remove steering wheel hub, remove steering wheel nut. Use gear puller, thread puller anchor screws into threaded provided in hub assembly. Turn center bolt of tool against steering shaft and remove hub.

Removing steering wheel with Tool J-2927

Installation
Place turn signal in neutral position; install hub assembly on steering shaft and secure with nut. Torque to 30 ft/lbs. Attach steering wheel to hub assembly using the attaching screws and tighten securely. Replace horn contact on steering wheel. Snap horn button in place and reconnect battery ground cable.

TELESCOPING STEERING WHEEL
Removal
Pry off horn button ornament and retaining with small screwdriver, in one of notches. Remove horn button. Turn the lock knob approx. 90° counterclockwise and remove the screws holding the lock knob to the lock bolt. Turn the lock bolt counterclockwise and remove with knob. Remove six attaching screws and lift off wheel.

Installation

Position steering wheel on hub, align marks and secure with the screws. Position locking rod into upper end of steering shaft and place lock handle into position. Coat lock bolt threads with Lubriplate and screw into upper end of steering shaft. Torque the lock bolt to 40 in. lbs., and position handle fully clockwise. Align the nearest holes in the handle with the lock bolt by backing off the handle slightly counterclockwise, and secure.

Handle must lock the telescoping mechanism in the lock of full clockwise position and release when fully counterclockwise.

Install horn button onto horn cap. Be sure lug on horn button is in line with the double hole in the cap. Then secure it. Install the horn cap to the horn button; be sure ornament is in line with the marked screw hole of the horn button.

TILT TELESCOPING WHEEL
Removal

Disconnect battery ground cable. Pry off horn button cap. Remove three screws securing the upper horn contact and remove the contact. Remove shim. Remove two screws securing the center star screw. Remove the star screw, lock lever and spacer.

If the steering wheel is being removed to perform work on the internal parts of the column only, omit removing wheel from hub. Remove six screws securing steering wheel to hub and remove wheel. Remove nut from shaft and using steering wheel puller Tool J-2927, remove steering wheel, extension and hub from vehicle. It is not necessary to install a centering adapter onto the shaft — simply butt the tool center screw against the steering shaft. Remove lower horn contact.

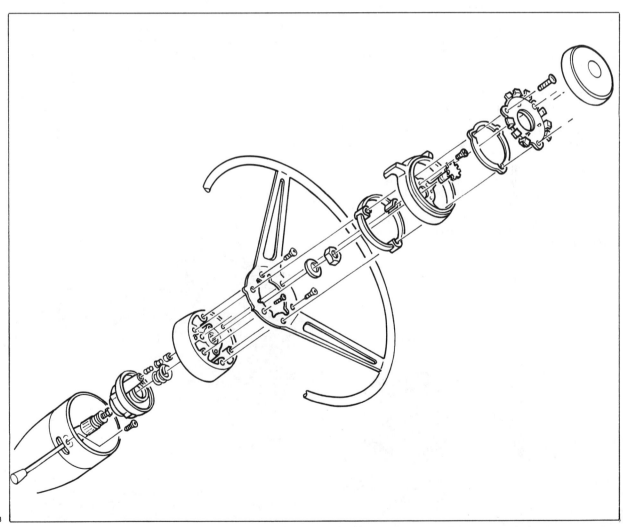

193

telescoping wheel option

Installation

Place lower horn contact on end of shaft. If hub was removed from steering wheel, attach hub and extension to steering wheel with screws removed during disassembly. Position the horn contact tower so that it will engage the hole in the steering wheel before installing the wheel. Place steering wheel and hub assembly in position and secure to column with nut.

Position spacer on steering wheel. Position lock lever on steering wheel. Install star screw through lock lever, turn into shaft, and adjust to lock position. Attach spacer to wheel with three screws. Place lock lever in lock position and attach star screw with two screws. Remove three screws holding spacer.

Attach upper horn contact with three screws and tighten securely. Add shims before installing contact. Install horn button cap. Connect battery ground cable.

STEERING COUPLING
Removal

Before removing the flexible or pot joint coupling, the mast jacket must be lowered and pulled rearward far enough to permit coupling removal. Do not allow the steering column to hang down from the dash panel or distortion to the column will result. When the instrument panel mounting bracket is removed, be sure the column is supported.

Remove the bolt from the coupling clamp. This is a special bolt and will require a 12-point socket or box end wrench. Tap the coupling with a soft hammer to remove it from the wormshaft. The coupling is splined to the wormshaft.

Installation

Install the coupling assembly onto the wormshaft, aligning the flat on the coupling with the flat on the shaft. Install the coupling clamp bolt and torque to specifications. Be sure the coupling reinforcement is bottomed on the wormshaft so that the clamp bolt passes through the undercut on the wormshaft.

STEERING GEAR
Removal

Remove retaining nuts, lock washers, and bolts at steering coupling. Remove pitman arm nut and washer from pitman shaft and mark relation of arm position to shaft. Remove pitman arm. Remove screws securing steering gear to frame and remove gear from vehicle.

Installation

Place gear into position so that steering coupling mounts properly to flanged end of steering shaft. Secure gear to frame with bolts, washers, and nuts. Be sure the coupling reinforcement is bottomed on the worm shaft so that the coupling bolt passes through the undercut on the wormshaft.

Secure steering coupling to flanged end of steering column with lock washers, and nuts. Maintain coupling adjustments. Install pitman arm, aligning marks made during removal, and secure with washer and retaining nut.

PITMAN SHAFT SEAL
Replacement

A faulty seal may be replaced without removal of steering gear from car by removing pitman arm. Loosen lash adjuster lock nut and turn lash adjuster screw several turns counter-clockwise. Remove three cap screws holding side cover to gear housing. Pull side cover and pitman shaft from gear housing as a unit. Do not separate side cover from pitman shaft. Pry the pitman shaft seal from gear housing using a screw driver being careful not to damage housing bore. Inspect the lubricant in the gear for contamination. If the lubricant is contaminated in any way, the gear must be removed from the vehicle and completely overhauled.

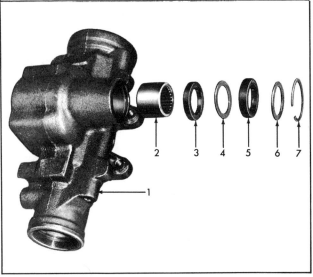

Pitman shaft seals

1. Housing
2. Bearing
3. Oil seal
4. Steel washer
5. Oil seal (double lip)
6. Steel wahser
7. Retaining ring

194

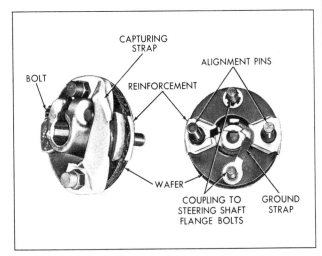

Flexible type steering coupling (typical)

Coat new seal with steering gear lube and position in pitman shaft bore. Place a socket or piece of pipe of suitable diameter on top of seal and drive seal into bore by tapping pipe or socket with soft hammer. Install pitman shaft side cover assembly, being careful not to damage new seal with splines on end of shaft; splines may be wrapped with a few turns of tape to prevent this. Before lowering the side cover all the way into the housing, add Steering Gear Lubricant meeting GM Specification GM4673M. Do not over-fill the gear housing. Install new side cover gasket and align side cover on gear housing and install cap screws. Perform steering gear adjustment and install pitman arm.

Removing lock plate retaining ring using Tool J-23131

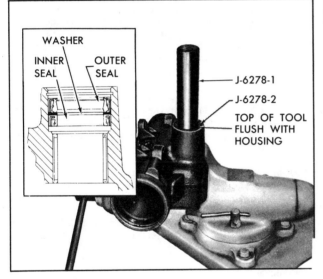

Installing pitman shaft seals

Slide the directional signal cancelling cam, upper bearing preload spring and thrust washer off the end of the shaft. Remove the directional signal lever screw and remove the lever. Push the hazard warning knob in and unscrew the knob. All Colums — Pull the switch connector out of the bracket on the jacket and wrap the upper part of the connector with tape to prevent snagging the wires during switch removal.

DIRECTIONAL SIGNAL SWITCH
Removal, all except tilt-telescoping column

The directional signal switch can be removed with the steering column in place and without disturbing any of the column mountings.

Remove the steering wheel. Remove the column to instrument panel trim cover. Remove the three cover screws and lift the cover off the shaft. The cover screws have plastic retainers on the back of the cover so it is not necessary to completely remove these screws.

Place Lock Plate Compressing Tool J-23131 on the end of the steering shaft and compress the lock plate as far as possible using the shaft nut. Pry the round wire snap ring out of the shaft groove and discard the ring. Remove Tool J-23131 and lift the lock plate off the end of the shaft. If the column is being disassembled on the bench, with the snap ring removed the shaft could slide out of the lower end of the mast jacket, damaging the shaft assembly.

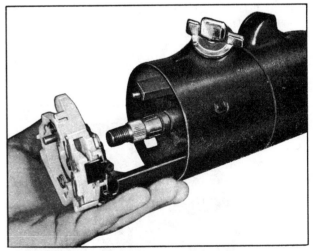

Removing directional signal switch assembly

Installation

It is extermely important that only the specified screws, bolts and nuts be used at assembly. Use of overlength screws could prevent a portion of the assembly from compressing under impact.

All except Tilt-Telescoping — Be sure that the wiring harness is in the protector. Feed the connector and cover down through the housing and under the mounting bracket (column in car).

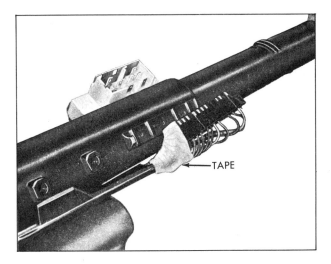

Taping directional signal connector and wires

TILT-TELESCOPING COLUMN DIRECTIONAL SWITCH
Removal
Remove the steering wheel. Remove the column to instrument panel trim cover. Remove the "C" ring plastic retainer. Place Tool J-23131 over the end of the steering shaft. Place a 5/16" nut under each leg of the tool. Reinstall star screw to prevent shaft from moving out of columns. Using the shaft nut, compress the lock plate just enough to remove the "C" ring. Remove the "C" ring. Remove the shaft nut, Tool J-23131, the two 5/16" nuts and the star screw.

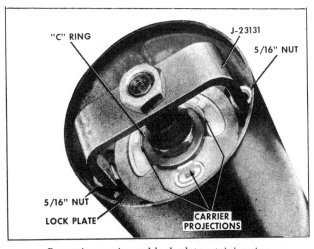

Removing carrier and lock plate retaining ring

using Tool J-23231

Lift the lock plate, horn contact carrier and upper bearing preload spring off the end of the shaft. Pull the switch connector out of the bracket on the jacket, remove the harness cover and wrap the upper part of the connector with tape to prevent snagging the wires during switch removal. Push the hazard warning knob in and remove the knob. Remove the directional signal lever. Place the directional signal and shifter housing in "low" position. Remove the three directional switch screws and pull the switch straight up, guiding the wiring harness out of the housing.

Installation
It is extremely important that only the specified screws, bolts and nuts be used at assembly. Use of over-length screws could prevent a portion of the column from compressing under impact. Feed the harness connector down through the housing and under the mounting bracket, reinstall the cover on the harness and clip the connector to the jacket. Install the three switch mounting screws, the switch lever and the hazard warning knob. Make certain that the switch is in "Neutral" and that the hazard knob is out. Place the upper bearing preload spring, horn contact carrier and lock plate onto the upper end of the shaft.

Place Tool J-23131 over the end of the steering shaft. Place a 5/16" nut under each leg of the tool. Reinstall the star screw to prevent the shaft from moving out of the column.

Using the shaft nut, compress the lock plate just enough to install the "C" ring. Install the "C" ring. Remove the shaft nut, Tool J-23131, the two 5/16" nuts and the star screw. Install the "C" ring plastic retainer. Install the column to instrument panel trim cover. Install the steering wheel as outlined under "Steering Wheel — Installation".

LOCK CYLINDER
Removal
The lock cylinder is located on the upper right hand side of the column. The lock cylinder may be removed in any position from "Accessory" to "On"; however, the "Lock" position is recommended because of its positive location. The lock cylinder cannot be disassembled. If replacement is required, a new cylinder, coded to the old key, must be installed.

Remove the steering wheel as outlined under "Steering Wheel — Removal". Remove the directional signal switch as outlined under "Directional Signal Switch — Removal". Insert a small screw driver or similar tool into the turn signal housing slot. Keeping the tool to the right side of the slot, break the housing flash loose and at the same time depress the spring latch at the lower end of the lock cylinder. With the latch depressed, the lock cylinder can be removed from the housing.

Assembly
Place the key part way into the lock cylinder assembly. Place the wave washer and anti-theft ring onto the lower end of the lock cylinder. If the key is installed all the way into the lock cylinder, the plastic keeper in the lock cylinder protrudes and prevents installation of the sleeve assembly. Make sure that the plastic keeper in the sleeve assembly protrudes from the sleeve. Align the lock bolt on the lock cylinder and the tab on the anti-theft washer and

Removing flash and depressing lock cylinder retainer

the slot in the sleeve assembly. Push the sleeve all the way onto the lock cylinder assembly, push the ignition key the rest of the way in and rotate the lock cylinder clockwise. Rotate the lock counter-clockwise into "LOCK" position. Place the lock in a brass jawed vise or between two pieces of wood. If a vise is used, place cloth around the knob to prevent marring the knob surface. Place the adapter ring onto the lower end of the cylinder so that the finger of the adapter is located at the step in the sleeve and the serrated edge of the adapter is visible after assembly to the cylinder and before "staking". The key must be free to rotate at least 1/3 of a circle (120°). Tap the adapter onto the cylinder until it is stopped at the bottom of the cylinder flats (cylinder will extend above adapter approximately 1/16"). Using a small flat punch, at least 1/8" in diameter, stake the lock cylinder over the adapter ring in four places just outboard of the four dimples. Check lock operation before reinstalling in car.

Installation

Hold the lock cylinder sleeve and rotate the knob clockwise against the stop. Insert the cylinder into the housing bore with the key on the cylinder sleeve aligned with the keyway in the housing. Push the cylinder into abutment of cylinder and sector. Hold an .070" drill between the lock bezel and housing. Rotate the cylinder counter-clockwise, maintaining a light pressure until the drive section of the cylinder mates with the sector. Push in until the snap ring pops into the grooves and the lock cylinder is secured in the housing. Remove the .070" drill. Check lock cylinder for freedom of rotation. Install the direction signal switch and steering wheel.

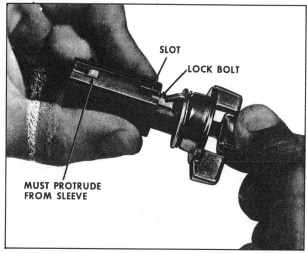

Ignition lock-assembly

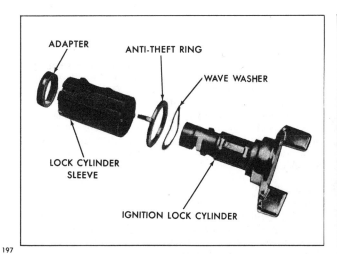

Ignition lock assembly — exploded view

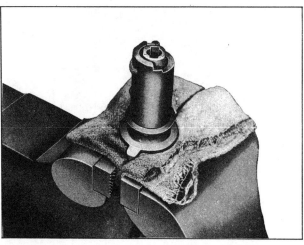

Ignition lock installed in a vise

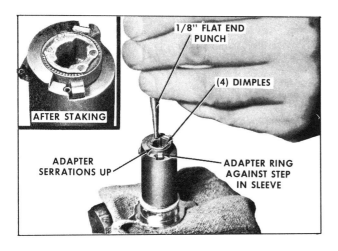

Installing adapter ring

IGNITION KEY WARNING SWITCH
Removal
The ignition key warning switch (buzzer switch) is located within the column housing and cannot be removed until the steering wheel and directional signal switch are removed.

Remove the steering wheel and directional signal switch. With a pair of needle nose pliers, remove the buzzer switch assembly. If the lock cylinder is in the housing, it must be in the "On" position. Do not drop the spring clip into the housing.

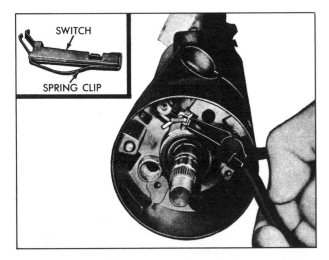

Removing ignition key warning switch

Installation
If the lock cylinder is in the column, the buzzer switch

actuating button on the lock cylinder must be depressed before the buzzer switch can be installed. Install the buzzer switch with the contacts toward the upper end of the steering column and with the formed end of the spring clip around the lower end of the switch. Reinstall the directional signal switch and steering wheel as previously outlined in this section.

IGNITION SWITCH ASSEMBLY
Description
The ignition switch is mounted on top of the mast jacket near the front of the dash. For anti-theft reasons, the switch is located inside the channel section of the brake pedal support and is completely inaccessible without first lowering the steering column. The switch is actuated by a rod and rack assembly. A portion of the rack is toothed and engages a gear on the end of the lock cylinder, thus enabling the rod and rack to be moved axially (with respect to the column) to actuate the switch when the lock cylinder is rotated.

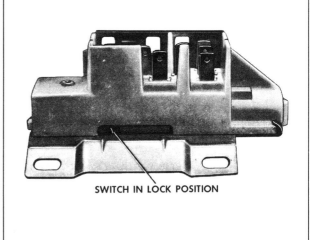

Ignition switch assembly

Removal
Remove or lower the steering column. If the steering column is not removed from the car, be sure that it is properly supported, before proceeding. The switch should be positioned in "Lock" position before removing. If the lock cylinder has already been removed, the actuating rod to the switch should be pulled up until there is a definite stop, then moved down one detent, which is the "Lock" position. Remove the two switch screws and remove the switch assembly.

Installation
Before replacing the switch, be sure that the lock is in the "Lock" position. Make certain that the switch is in "Lock" position; if it is not, a screw driver (placed in the locking rod slot) can be used to move the switch to "Lock". Install the activating rod into the switch and assemble the switch on the column; tighten the mounting

screws. Use only the specified screws since over-length screws could prevent a portion of the assembly from compressing under impact. Reinstall the steering column assembly following the mandatory installation sequence.

STEERING COLUMN

To perform service procedures on the steering column upper end components, it is not necessary to remove the column from the car. The steering wheel, horn components, directional signal switch, ignition lock cylinder and ignition key warning switch may be removed with the column remaining in the vehicle as described earlier. It is absolutely necessary to handle the column with care when performing any service operation. Avoid hammering, jarring, dropping or leaning on any portion of the column. When reassembling the column components, use only the specified screws, nuts and bolts and tighten to specified torque.

ENERGY ABSORBING COLUMN
Inspection

To determine if the energy absorbing steering column components are functioning as designed, or if repairs are required, a close inspection should be made. An inspection is called for in all cases where damage is evident or whenever the vehicle is being repaired due to a front end collision. Whenever a force has been exerted on the steering wheel or steering column, or its components, inspection should also be made. If damage is evident, the affected parts must be replaced.

Column support bracket

Damage in this area will be indicated by separation of the mounting capsules from the bracket. The bracket will have moved forward toward the engine compartment and will usually result in collapsing of the jacket section of the steering column.

Column jacket

Inspect jacket section of column for looseness, and/or bends.

Steering shaft

If the steering shaft plastic pins have been sheared, the shaft will rattle when struck lightly from the side and some lash may be felt when rotating the steering wheel while holding the rag joint. It should be noted that if the steering shaft pins are sheared due to minor collision with no appreciable damage to other components, that the vehicle can be safely steered; however, steering shaft replacement is recommended.

Column removal

Disconnect the battery ground cable. Remove the steering wheel. Remove the nuts and washers securing the flanged end of the steering shaft to the flexible coupling. Disconnect the back-drive linkage. Disconnect the steering column harness at the connector. Disconnect the neutral-safety switch and back-up lamp switch connectors if so equipped. Remove the floor pan trim cover screws and remove the cover. Remove the two nuts securing the floor pan bracket to the mounting studs. Remove the instrument panel trim cover screws and remove the trim cover. Remove the transmission indicator cable, if so equipped. More the front seat as far back as possible to provide maximum clearance. Remove the two column bracket-to-

instrument panel nuts and carefully remove from vehicle. Additional help should be obtained to guide the lower shift levers through the firewall opening.

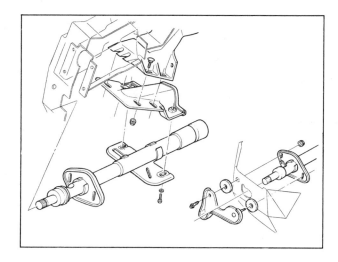

Steering column to dash panel

STANDARD
Disassembly

Place the column in a vise using both weld nuts of either Set A or B. The vise jaws must clamp onto the sides of the weld nuts indicated by arrows shown on Set B. Do not place the column in a vise by clamping onto only one weld nut, by clamping onto one weld nut of both Sets A and B or by clamping onto the sides not indicated by arrows since damage to the column could result. Remove the directional Signal Switch, Lock Cylinder, Ignition Key Warning Switch and Ignition Switch.

Remove the lower bearing retainer, bearing adapter assembly, shift tube thrust spring and washer. The lower bearing may be removed from the adapter by light pressure on the bearing outer race. Slide out the transmission control lock tube housing. Remove the spring plunger.

Assembly

Apply a thin coat of lithium soap grease to all friction surfaces. Install the sector shaft and sector into the turn signal and lock cylinder housing. Install the sector in the lock cylinder hole over the sector shaft with the tang end to the outside of the hole. Press the sector over the shaft with a blunt tool. Install the shift lever detent plate onto the housing. Insert the rack preload spring into the housing from the bottom side. The long section should be toward the handwheel and hook onto the edge of the housing.

Assemble the locking bolt onto the crossover arm on the rack and insert the rack and lock bolt assembly into the housing from the bottom with the teeth up (toward handwheel) and toward the centerline of the column. Align the 1st tooth on the sector with the 1st tooth on

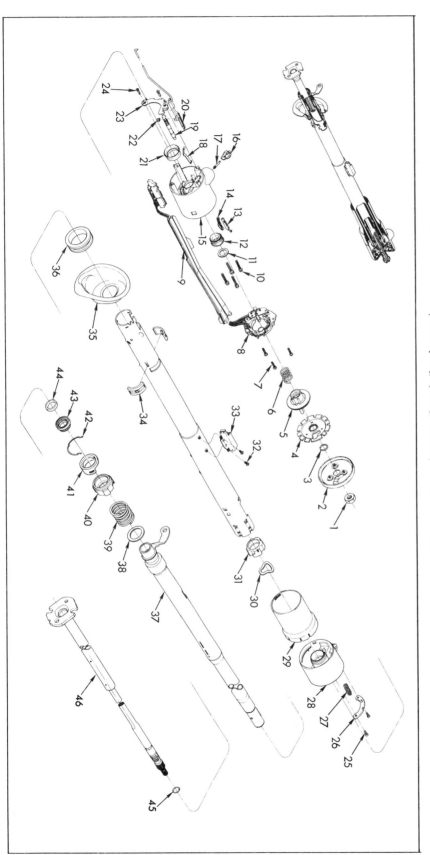

Steering column (except tilt and tilt-telescope)

1. Shaft nut
2. Cover
3. Lock plate retaining ring
4. lock plate
5. Cancelling cam
6. Bearing preload spring
7. Turn signal screws
8. Turn signal switch
9. Protector cover
10. Turn signal housing screws
11. Bearing thrust washer
12. Upper bearing

13. Key warning switch
14. Switch clip
15. Turn signal housing
16. Ignition switch sector
17. Sector shaft
18. Switch rack preload spring
19. Shaft lock bolt
20. Switch rod and rack assembly
21. Trust cup
22. Shaft lock bolt washer
23. Shift lever detent plate
24. Detent plate screws

25. Shift tube lock plate screws
26. Shift tube lock plate
27. Shift tube spring
28. Gearshift lever housing
29. Shift shroud
30. Wave washer
31. Gearshift housing bearing
32. Ignition switch screws
33. Ignition switch
34. Neutral safety or back-up switch retainers
35. Dash seal

36. Jacket collar
37. Shift tube
38. Thrust spring washer
39. Shift tube thrust spring
40. Lower bearing adapter
41. Lower bearing reinforcement
42. Retainer
43. Lower bearing
44. Spacer
45. Shaft stop ring
46. Steering shaft

200

the rack; if aligned properly, the block teeth will line up when the rack assembly is pushed all the way in.

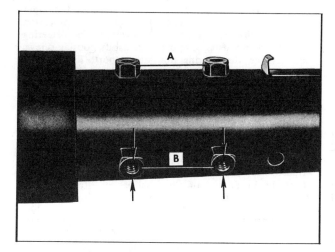

Installing steering column in vise

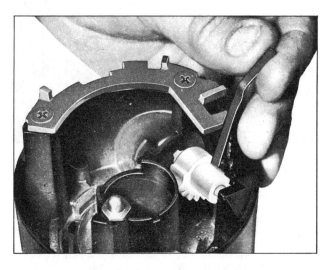

Installing rack preload spring

Assemble the spring and lower bearing and adapter assembly into the bottom of the jacket. Holding the adapter in place, install the lower bearing reinforcement and retainer clip. Be sure the clip snaps into the jacket and reinforcement slots.

TELESCOPING
Removal
Disconnect battery ground cable, and direction signal

wiring at connector. Remove 2 screws securing jacket escutcheon to dash panel, mast jacket to dash bracket bolts. Remove odometer reset control. Remove clutch return spring, mast jacket to firewall clamp and bolt, and 2 bolts retaining mast jacket cover plate and seal to firewall. Lower bolt has clutch spring retaining clip.

Remove steering coupling to steering shaft clamp bolt. Mark steering coupling to steering shaft alignment. Lock column in full down position then pull steering wheel and mast jacket assembly from opening in firewall. Remove spring and cancelling cam from steering shaft and turn signal lever and retaining screw. Remove the three directional signal switch retaining screws, wiring clamps and cover from directional wires, and wire terminals from the two plastic connectors using a small thin bladed screw driver. Note the color codes of the wires.

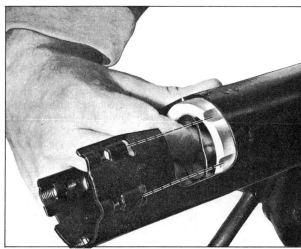

Installing gearshift housing lower bearing and wave washers

Carefully pull the directional signal switch out of the housing. Remove snap ring from lower end of mast jacket. Remove the retainer, felt seal and spring from lower end of mast jacket. Pull steering shaft out of the upper end of the mast jacket, remove the cap screw previously installed and allow the locking rod to slide out the upper end of the shaft. The two snap rings, stop and spring may be removed from the shaft at this time. Remove the mast jacket guide bolt and pull the inner mast jacket out of the outer mast jacket. The directional housing and dash escutcheon can now be removed from the outer mast jacket. Remove the lower bearing retaining screws and retainer from the lower end of the mast jacket and the lower bearing from the mast jacket.

Installation
Lock column in full down position then install mast

201

jacket and wheel assembly through firewall. Install seal, plate and loosely install bolts. Be sure clutch pull back spring clip is in place on lower bolt. Install clamp over steering shaft. With the help of an assistant position steering shaft to coupling using previously made alignment marks. Position shaft so notch lines up with bolt hole and install coupling clamp bolt; do not tighten at this time. Install upper mast jacket bracket bolts. Leave support bolts finger tight until mast jacket is correctly positioned. Position escutcheon and install attaching screws. Position steering column so that a minimum gap of 1/16" remains between the lower edge of the directional housing and the dash escutcheon with the column in the full down position. Then tighten all column bolts.

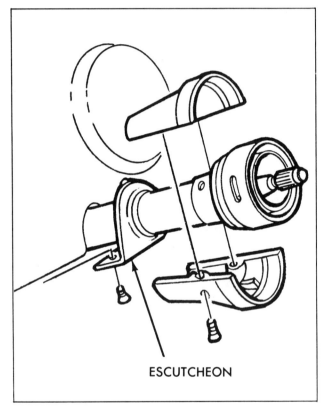

ESCUTCHEON

Steering column escutcheon

Install trip odometer cable to bracket and retaining nut. Connect directional signal harness to instrument panel harness assembly. Tighten steering shaft to coupling bolt. Tighten firewall bracket bolts, band type clamp bolt. Install clutch pull back spring.

Disassembly

Push the upper steering shaft in sufficiently to remove the steering shaft upper bearing inner race and seat. Remove the lower steering shaft flange. Pry off the lower bearing retainer clip and remove the bearing reinforcement, bearing and bearing adapter assembly from the lower end of the mast jacket. Remove the upper bearing housing pivot pins using Tool J-21854-1. Install the tilt release lever and disengage the lock shoes. Remove the bearing housing by pulling upward to extend the rack full down,

and then move the housing to the left to disengage the ignition switch rack from the actuator rod. Remove the steering shaft assembly from the upper end of the column.

Disassemble the steering shaft by removing the centering spheres and the anti-lash spring. Disassemble the upper steering shaft, locking wedge, locking rod and bumper stop from the upper yoke. Remove the transmission indicator wire, if equipped. Remove the four steering shaft bearing housing support to gearshift housing screws and remove the bearing housing support. Remove the ignition switch actuator rod. Remove the shift tube retaining ring with a screw driver and then remove the thrust washer.

Install Tool J-23072 into the lock plate, making sure that the tool screws have good thread engagement in the lock plate. Then, turning the center screw clockwise, force the shift tube from the housing. Remove the shift tube (transmission control lock tube on floor shift models) from the lower end of the mast jacket. Remove Tool J-23072. When removing the shift tube, be sure to guide

Removing bearing housing pivot pins using Tool J-21854

the lower end through the slotted opening in the mast jacket. If the tube is allowed to interfere with the jacket in any way, damage to the tube and jacket could result. Remove the bearing housing support lock plate by sliding it out of the jacket notches, tipping it down toward the housing hub at the 12 o'clock position and sliding it under the jacket opening. Remove the wave washer.

Pry the transmission control lock tube housing cover tabs out of the slots in the housing extension and remove the cover. Remove the three housing extension screws and remove the housing extension. Disassemble the bearing housing as follows: Remove the tilt lever opening shield. Remove the turn signal lever opening shield. Remove the lock bolt spring by removing the retaining screw and moving the spring clockwise to remove it from the bolt. Remove the snap ring from the sector drive shaft. With a small punch, lightly tap the drive shaft from the sector. Remove the drive shaft, sector and lock bolt. Remove the rack and rack spring. Remove the tilt release lever pin with a punch and hammer. Remove the lever and release

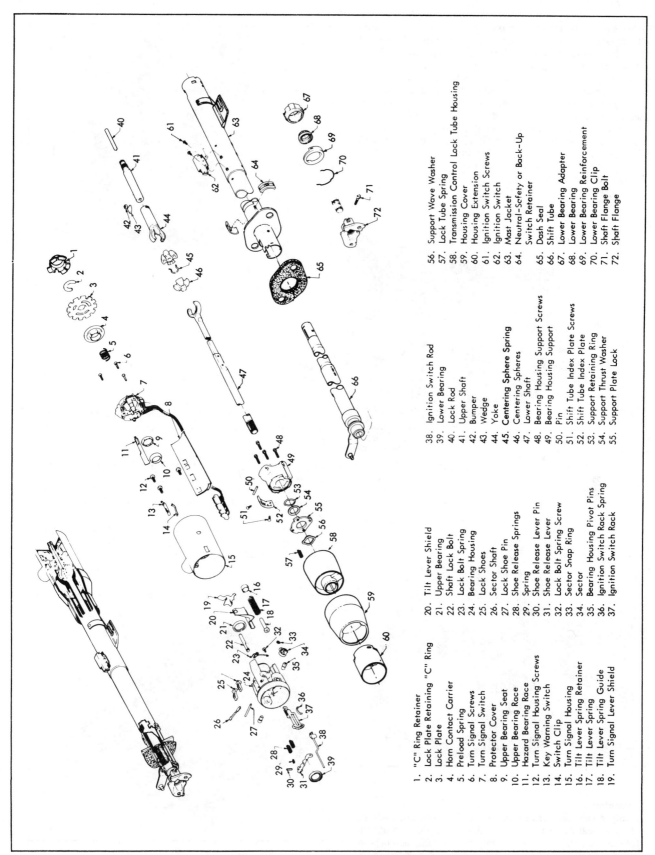

1. "C" Ring Retainer
2. Lock Plate Retaining "C" Ring
3. Lock Plate
4. Horn Contact Carrier
5. Preload Spring
6. Turn Signal Screws
7. Turn Signal Switch
8. Protector Cover
9. Upper Bearing Seat
10. Upper Bearing Race
11. Hazard Bearing Race
12. Turn Signal Housing Screws
13. Key Warning Switch
14. Switch Clip
15. Turn Signal Housing
16. Tilt Lever Spring Retainer
17. Tilt Lever Spring
18. Tilt Lever Spring Guide
19. Turn Signal Lever Shield
20. Tilt Lever Shield
21. Upper Bearing
22. Shaft Lock Bolt
23. Lock Bolt Spring
24. Bearing Housing
25. Lock Shoes
26. Sector Shaft
27. Lock Shoe Pin
28. Shoe Release Springs
29. Spring
30. Shoe Release Lever Pin
31. Shoe Release Lever
32. Lock Bolt Spring Screw
33. Sector Snap Ring
34. Sector
35. Bearing Housing Pivot Pins
36. Ignition Switch Rack Spring
37. Ignition Switch Rack
38. Ignition Switch Rod
39. Lower Bearing
40. Lock Rod
41. Upper Shaft
42. Bumper
43. Wedge
44. Yoke
45. **Centering Sphere Spring**
46. Centering Spheres
47. Lower Shaft
48. Bearing Housing Support Screws
49. Bearing Housing Support
50. Pin
51. Shift Tube Index Plate Screws
52. Shift Tube Index Plate
53. Support Retaining Ring
54. Support Thrust Washer
55. Support Plate Lock
56. Support Wave Washer
57. Lock Tube Spring
58. Transmission Control Lock Tube Housing
59. Housing Cover
60. Housing Extension
61. Ignition Switch Screws
62. Ignition Switch
63. Mast Jacket
64. Neutral-Safety or Back-Up Switch Retainer
65. Dash Seal
66. Shift Tube
67. Lower Bearing Adapter
68. Lower Bearing
69. Lower Bearing Reinforcement
70. Lower Bearing Clip
71. Shaft Flange Bolt
72. Shaft Flange

Tilt and telescope steering column assembly

lever spring. To relieve the load on the release lever, hold the shoes inward and wedge a block between the top of the shoes (over slots) and bearing housing. Remove the lock shoe retaining pin with a punch and hammer. Remove the lock shoes and lock shoe springs. With the tilt lever opening on the left side and shoes facing up, the four slot shoe is on the left. Remove the bearings from the bearing housing only if they are to be replaced. Remove the separator and balls from the bearings. Place the housing on work bench and with a pointed punch against the back surface of the race, carefully hammer the race out of the housing until a bearing puller can be used. Repeat for the other race.

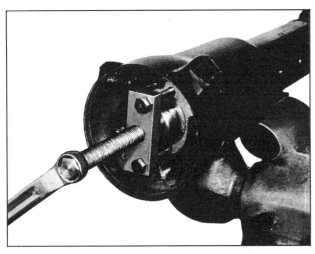

Removing shift tube using Tool J-23072

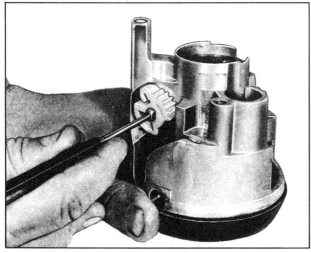

Removing sector drive shaft

Assembly
Apply a thin coat of lithium grease to all friction surfaces. If the bearing housing was disassembled, repeat the following: Press the bearings into the housing, if removed, using a suitable size socket. Be careful not to damage the

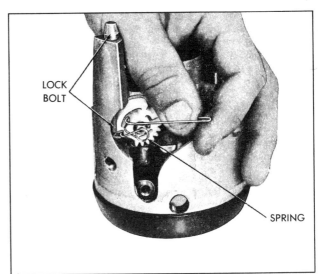

Replacing lock bolt spring

housing or bearing during installation. Install the lock shoe springs, lock shoes and shoe pin in the housing. Use an approximate .180" rod to line up the shoes for pin installation. Install the shoe release lever, spring and pin. To relieve the load on the release lever, hold the shoes inward and wedge a block between the top of the shoes (over slots) and bearing housing.

Install the sector drive shaft into the housing. Lightly tap the sector onto the shaft far enough to install the snap ring. Install the snap ring. Install the lock bolt and engage it with the sector cam surface. Then install the rack and spring. The block tooth on the rack should engage the block tooth on the sector. Install the external tilt release lever. Install the lock bolt spring and retaining screw. Tighten the screw to 35 in. lbs.

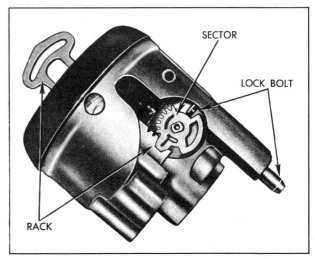

204

Installing lock bolt and rack assemblies

Install the shift lever spring into the housing by winding it up with pliers and pushing it into the housing. Install the plunger. Install the transmission control lock tube housing extension and torque the screws to 40 in. lbs. Install the housing cover so that the three cover tabs enter the slots in the housing extension. Install the housing onto the mast jacket assembly.

Install the bearing support lock plate wave washer. Install the bearing support lock plate. Work it into the notches in the jacket by tipping it toward the housing hub at the 12 o'clock position and sliding it under the jacket opening. Slide the lock plate into the notches in the jacket. Carefully install the shift tube into the lower end of the mast jacket. Align keyway in the tube with the key in the shift lever housing. Install the wobble plate end of Tool J-23073 into the upper end of the shift tube far enough to reach the enlarged portion of the tube. Then install the adapter over the end of the tool, seating it against the lock plate. Place the nut on the threaded end of the tool and pull the shift tube into the housing. Remove Tool J-23073. Do not push or tap on the end of the shift tube. Be sure that the shift tube lever is aligned with the slotted opening at the lower end of the mast jacket or damage to the shift tube and mast jacket could result.

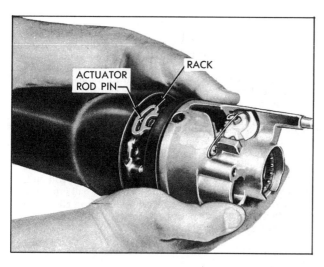

Installing shift tube using Tool J-23073

Install the bearing support thrust washer and retaining ring by pulling the shift lever housing up far enough to compress the wave washer. Install the bearing support by aligning the "V" in the support with the "V" in the jacket. Insert the screws through the support and into the lock plate and torque to 60 in. lbs. Install the ignition switch actuator rod before tightening the support screws. Install the rod through the transmission control lock tube housing and extension; insert it in the slot provided in the bearing support.

205 Align the lower bearing adapter with the notches in the jacket and push the adapter into the lower end of the mast jacket. Install lower bearing, bearing reinforcement

and retaining clip, being sure that the clip is aligned with the slots in the reinforcement, jacket and adapter. Install the centering spheres and anti-lash spring in the upper shaft. Install the lower shaft from the same side of the spheres that the spring ends protrude.

Install the upper steering shaft, locking wedge, locking rod and bumper in the upper yoke. Install the steering shaft assembly into the shift tube from the upper end. Carefully guide the shaft through the shift tube and bearing.

Install the upper bearing inner race and race seat. Install the tilt lever opening shield and on tilt-telescoping columns, install the turn signal lever opening shield. Remove the tilt release lever, install the turn signal housing and torque the three retaining screws to 45 in. lbs. Install the tilt release lever and shift lever. Drive the shift lever pin in. Install the ignition key warning switch, lock cylinder, turn signal switch and ignition switch as outlined previously in this section. Install the lower steering shaft flange and torque bolt to 30 ft. lbs. Install the neutral-safety switch or back-up switch as outlined in Section 12 of this manual.

TILT COLUMNS
Installation — mandatory sequence
Loosely assemble support (U) to bracket (T) with shoulder bolts (W) and nuts (Y). Assemble upper end of rag joint (Z) to steering gear, torquing screws. Position column assembly into vehicle, sliding splined end of steering shaft into upper end of rag joint (Z). Connect all electrical wiring.

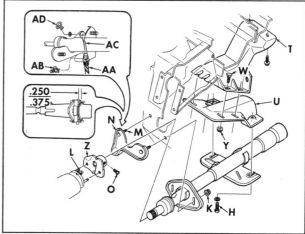

Steering column mounting

Start two steering column to support (U) attaching screws (H). Install transmission control cable bracket (N) with bolt (M) and hold in place. Maintaining a gap of .30 ± .05 between mast jacket and inside semi-circular surface of instrument panel, install nuts (K) and tighten to specified torque. Then pull support (U) rearward to accommodate height differential of instrument panel to bracket (T).

With column position now defined, tighten screws (H) and then nuts (Y) to specified torque. Loosen nuts (K). Do not tighten column to instrument panel until vehicle is on its wheel or suspension. Maintaining rag joint dimension shown, torque upper rag joint flange pinch bolt (O). Attach back drive cable sheath (AA) to transmission control cable bracket (N) with spring clip (AB), then attach inner cable (AC) to lever on steering column with retainer (AD). Secure steering column lower plate to front of dash by tightening nuts (K) to specified torque. Remove plastic spacers (L) from flexible coupling pins. Install the steering wheel. Connect the battery ground cable.

TIE RODS
Description

There are two tie rod assemblies used on all models. Each assembly is of three piece construction, consisting of a sleeve and two tie rod ends. The ends are threaded into the sleeve and locked with clamps. Right and left hand threads are provided to facilitate toe-in adjustment and steering gear centering. The tie rod ends are self-adjusting for wear and require no attention in service other than periodic lubrication and occasional inspection to see that ball studs are tight. Replacement of tie rod ends should be made when excessive up and down motion is evident or if any lost motion or end play at ball end of stud exists.

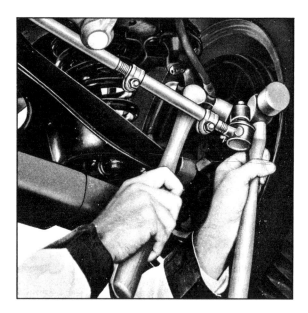

Freeing ball stud

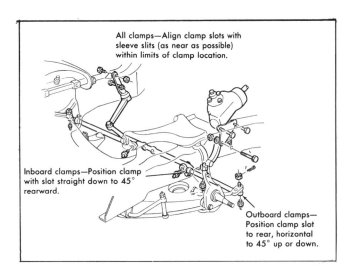

All clamps—Align clamp slots with sleeve slits (as near as possible) within limits of clamp location.

Inboard clamps—Position clamp with slot straight down to 45° rearward.

Outboard clamps—Position clamp slot to rear, horizontal to 45° up or down.

Steering linkage

Removal

Place vehicle on hoist. Remove cotter pins from ball studs and remove castellated nuts. To remove outer ball stud, tap on steering arm at tie rod end with a hammer while using a heavy hammer or similar tool as a backing. If necessary, pull downward on tie rod to remove from steering arm. Remove inner ball stud from relay rod. To remove tie rod ends from tie rods loosen clamp bolts and unscrew end assemblies.

Installation

If the tie rod ends were removed, lubricate the tie rod threads with EP Chassis lube and install ends on tie rod making sure both ends are threaded an equal distance from the tie rod. Make sure that threads on ball stud and in ball stud nuts are perfectly clean and smooth. Install seals on ball studs. If threads are not clean and smooth, ball studs may turn in tie rod ends when attempting to tighten nut. Install ball studs in steering arms and relay rod. Install ball stud nuts and torque to specifications, then advance nuts just enough to insert cotter pins and install cotter pins. Lubricate tie rod ends. Before locking clamp bolts on the rods, make sure that the tie rod ends are in alignment with their ball studs (each ball joint is in the center of its travel). If the tie rod is not in alignment with the studs, binding will result. Remove vehicle from hoist. Adjust toe-in.

RELAY ROD
Removal

Place vehicle on hoist. Remove inner ends of tie rods from relay rod as described under Tie Rod-Removal. Remove cotter pin from end of relay rod at pitman arm ball stud attachment, and remove stud nut. Tap ball stud out of pitman arm and lower relay rod. Remove cotter key and nut from idler arm and remove relay rod from idler arm. Remove washer and seal from idler arm.

Installation

Place relay rod on idler arm stud, making certain idler stud seal and washer are in place, then install and tighten nut. Advance nut just enough to align castellation with cotter pin hole and install pin. Install new seal clamp over ball at end of pitman arm. Install inner spring seat and spring to relay rod. Raise end of rod and install on pitman arm. Install spring seat, spring, and end plug. Tighten end plug until springs are compressed and plug bottoms, then back off 3/4 turn plus amount necessary to insert cotter

206

pin. Insert cotter pin to lock adjustment. Install the rod ends to relay rod as previously described under Tie Rods. Lubricate tie rod ends and pitman arm to relay rod ball joint. Remove vehicle from hoist. Adjust toe-in and align steering wheel as described previously in this section.

IDLER ARM
Removal

Place vehicle on hoist. Remove idler arm to frame nut, washer, and bolt. Remove cotter pin and nut from idler arm to relay rod ball stud. Remove relay rod from idler arm by tapping relay rod with a hammer using a heavy hammer as a backing. Remove idler arm.

Installation

Position idler arm on frame and install mounting bolts, washers and nuts. Install relay rod to idler arm, making certain seal is on stud. Install and tighten nut to specifications. Advance nut just enough to insert cotter pin and install cotter pin. Refer to torque specifications at rear of manual for correct torque values. Remove vehicle from hoist.

PITMAN ARM
Removal

Place vehicle on hoist. Remove cotter pin from pitman arm ball stud and remove nut. Remove relay rod from pitman arm by tapping on side of rod or arm in which the stud mounts with a hammer while using a heavy hammer or similar tool as a backing. Pull down on relay rod to remove from stud. Remove pitman arm nut from sector shaft and mark relation of arm position to shaft. Remove pitman arm with Tool J-6632.

Installation

Install pitman arm on sector shaft, lining up the marks made upon removal. Install sector shaft nut. Position relay rod on to pitman arm. Install nut. Continue to tighten nut enough to align castellation with hole in stud and install cotter pin. Remove vehicle from hoist.

STEERING ARM
Removal

Remove tie rod from steering arm. Remove front wheel, hub and brake drum as a unit by removing hub cap and dust cap, cotter pin from spindle nut and the spindle nut. Pull assembly toward outside of vehicle. If removal is difficult, it may be necessary to back off brake adjustment to increase brake shoe-to-drum clearance. On models with disc brakes, remove caliper and disc. With wheel and drum assembly or caliper and disc removed, steering arm retaining bolt heads are accessible and removal of steering arm from vehicle may be accomplished by removing retaining nuts.

Installation

Place steering arm in position on vehicle and install retaining bolts. Install nuts. Use only the special locknut listed for this use in the Parts Catalog or an equivalent approved part. Pack wheel bearings using a high quality wheel bearing lubricant. Install bearings and wheel-to-hub-brake drum assembly removed previously. On disc brake models, install disc and caliper. Install keyed washer and spindle nut. Proceed as outlined under "Front Wheel Bearing Adjustment". Install tie rod ball stud in steering arm. Be sure that the dust cover is in place on ball stud. Install castellated nut on ball stud, torque to specifica-

tions, advance nut enough to align castellation with hole in stud and install cotter pin. Remove vehicle from hoist. Check cornering wheel relationship and toe-in; correct as required.

POWER STEERING
Description

Hydraulic pressure is provided by an engine-driven vane-type pump. Pressure is delivered through a hose from the pump to a valve which senses the requirement for power assistance and supplies the power cylinder accordingly. The steering gear used with this power steering is the same basic unit used on manually steered vehicles.

Bleeding hydraulic system

Fill oil reservoir to proper level and let oil remain undisturbed for at least two minutes. Start engine and run only for about two seconds. Add oil if necessary. Repeat above procedure until oil level remains constant after running engine. Raise front end of vehicle so that wheels are off the ground.

Increase engine speed to approximately 1500 rpm. Turn the wheels (off ground) right and left, lightly contacting the wheel stops. Add oil if necessary. Lower the car and turn wheels right and left on the ground. Check oil level and refill as required. If oil is extremely foamy, allow vehicle to stand a few minutes with engine off and repeat above procedure.

Check belt tightness and check for a bent or loose pulley. (Pulley should not wobble with engine running.) Check to make sure hoses are not touching any other parts of the car, particularly sheet metal and exhaust manifold. Check oil level, filling to proper level if necessary. Check the presence of air in the oil. If air is present, attempt to bleed system. If it becomes obvious that the pump will not bleed after a few trials, proceed as outlined under Hydraulic System Checks.

Fluid level

Check oil level in the reservoir by checking the dip stick when oil is at operating temperature. On models equipped with remote reservoir, the reservoir should be maintained approximately 3/4 full when oil is at operating temperature. Fill, if necessary, to proper level with GM Power Steering Fluid or, if this is not available, Automatic Transmission Fluid "Type A" bearing the mark "AQ-ATF" followed by a number and the suffix letter "A".

Adjustments

The steering gear used with power steering is adjusted in the same manner as the manual steering gear.

Pump belt tension

Loosen pivot bolt and pump brace adjusting nuts. Do not move pump by prying against reservoir or by pulling on filler neck. Move pump, with belt in place until belt is tensioned to specifications as indicated by Tool J-7316. Tighten pump brace adjusting nut. Then tighten pivot bolt nut.

POWER STEERING PUMP
Removal

Disconnect hoses at pump or steering gear. When hoses are disconnected, secure ends in raised position to prevent

Checking belt tension with Tool J-7316

drainage of oil. Cap or tape the ends of the hoses to prevent entrance of dirt. Install two caps at hose fittings to prevent drainage of oil from pump. Remove pump belt.

On Corvette with MK IV engines, loosen alternator adjustment and remove pump to alternator belt. Remove

pump from attaching parts and remove pump from vehicle. Remove drive pulley attaching nut. Remove pulley from shaft with Tool J-21239 (for stamped pulleys) or Tool J-8433-1 with J-8433-2 adapter (for cast iron pulleys). Do not hammer pulley off shaft as this will damage the pump.

Installation
Install pump pulley. Do not hammer on pump shaft. Use pulley nut to pull pulley on to shaft. Position pump assembly on vehicle and install attaching parts loosely. Connect and tighten hose fittings. Fill reservoir. Bleed pump by turning pulley backward (counter-clockwise as viewed from front) until air bubbles cease to appear. Install pump belt over pulley. Tension belt as outlined under "Pump Belt Tension Adjustment". Bleed as outlined under "Maintenance and Adjustments".

CONTROL VALVE AND ADAPTER ASSEMBLY
Ball stud seal replacement
A ball stud seal is used on the power steering control valve. To replace the seal place vehicle on a hoist. Remove the pitman arm as outlined under "Steering Linkage". Remove clamp by removing nut, bolt and spacer or, if crimped type clamp is used, straighten clamp end and pull clamp and seal off end of stud. Install new seal and clamp over stud so lips on seal mate with clamp. (A nut and bolt attachment type clamp replaces the crimped type for service). Center the ball stud, seal and clamp at opening in adapter housing, then install spacer, bolt and nut. Reinstall the pitman arm as outlined under "Steering Linkage" and remove the vehicle from the hoist.

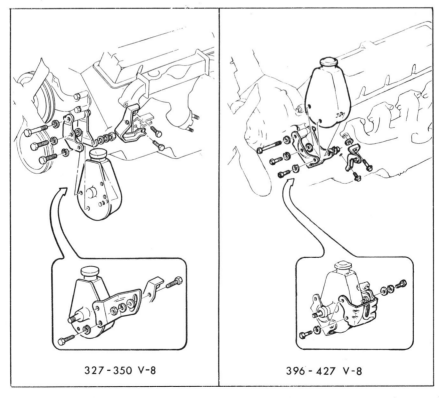

327 - 350 V-8   396 - 427 V-8

Power sterring pump mounting

208

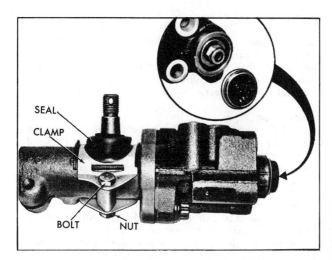

Control valve ball stud seal replacement (service type)

Removal

Raise the front of the vehicle off the floor and place it on stands. Remove the relay rod to control valve clamp bolt. Disconnect the two pump to control valve hose connections and allow fluid to drain into a container, then disconnect the two remaining valve to power cylinder hoses. Remove the retaining nut from the ball stud to pitman arm connection and disconnect the control valve from the pitman arm. Turn the pitman arm to the right to clear the control valve and unscrew the control valve from the relay rod. Remove the control valve from the vehicle.

Installation

Install the control valve on the vehicle by reversing the removal procedure. Reconnect the hydraulic lines, fill the system with fluid and bleed out air using the procedure outlined under "Maintenance and Adjustments." Grease ball joint. The control valve to pitman arm and relay rod fasteners are important attaching parts in that they could affect the performance of vital components and systems, and/or could result in major repair expense. They must be replaced with one of the same part number or with an equivalent part if replacement becomes necessary. Do not use a replacement part of lesser quality or substitute design. Torque values must be used as specified during reassembly to assure proper retention of this part.

Valve balancing

The control valve must be adjusted, after being disassembled. The same procedure may be followed to correct a complaint of harder steering effort required in one direction than the other. Install valve in vehicle. Connect all hoses and fill the pump reservoir with oil. Do not connect the piston rod to the frame bracket. If the vehicle is already in operation, it will be necessary to detach the piston rod from the frame bracket.

With the car on a hoist, start the engine. One of the following two conditions will exist: If the piston rod remains retracted, turn the adjusting nut clockwise until the rod begins to move out. Then turn the nut counterclockwise until the rod just begins to move in. Now turn

the nut clockwise to exactly one-half the rotation needed to change the direction of the piston rod movement. If the rod extends upon starting the pump, move the nut counter-clockwise until the rod begins to retract, then clockwise until the rod begins to move out again. Now position the rod to exactly one-half the rotation needed to change the direction of the piston rod movement. Do not turn the nut back and forth more than is absolutely necessary to balance the valve, as this is a special friction nut.

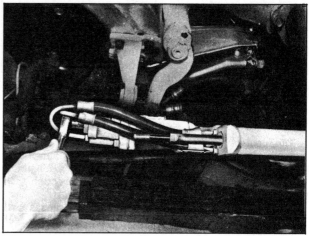

Balancing valve

With the valve balanced, it should be possible to move the rod in and out manually. Turn off the engine and connect the cylinder rod to the frame bracket. The power cylinder to frame bracket fastener is an important attaching part in that it could affect the performance of vital components and systems, and/or could result in major repair expense. It must be replaced with one of the same part number or with an equivalent part if replacement becomes necessary. Do not use a replacement part of lesser quality or substitute design. Torque values must be used as specified during reassembly to assure proper retention of this part.

Restart the engine. If the front wheels (vehicle still on the hoist) do not turn in either direction from center, the valve has been properly balanced. Correct the valve adjustment if necessary. When the valve is properly adjusted, grease end of valve and install the dust cap. Remove vehicle from hoist.

POWER CYLINDER
Removal

Place vehicle on a hoist. Disconnect the two hydraulic lines connected to the power cylinder and drain fluid into a container. Do not reuse. Remove cotter pin, nut, retainer and grommet from power cylinder rod attached to the frame bracket. Also remove grommet and retainer from bracket if replacement parts are required. Remove cotter pin, nut and ball stud at relay rod. Remove the power cylinder from the vehicle.

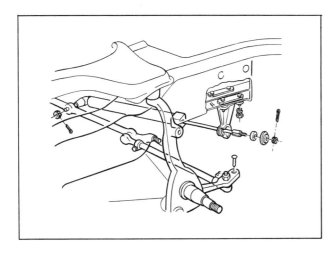

Power cylinder installation

Inspection

Inspect the seals for leaks; if leaks are present, replace the seals. Examine the brass fitted hose connection seats for cracks or damage and replace if necessary. For service other than ball seat or seal replacement and ball stud removal, replace the power cylinder. Check the frame bracket parts for wear.

Installation

Install the power cylinder on the vehicle by reversing the removal procedure. Reconnect the two hydraulic lines, fill the system with fluid and bleed out air using the procedure outlined under "Maintenance and Adjustments." Grease ball joint. Remove vehicle from hoist. The power cylinder to frame bracket fastener is an important attaching part in that it could affect the performance of vital components and systems, and/or could result in major repair expense. It must be replaced with one of the same part number or with an equivalent part if replacement becomes necessary. Do not use a replacement part of lesser quality or substitute design. Torque values must be used as specified during reassembly to assure proper retention of this part.

ELECTRICAL 13

13

BATTERY
Description
The battery is made up of a number of separate elements, each located in an individual cell in a hard rubber case. Each element consists of an assembly of positive plates and negative plates containing dissimilar active materials and kept apart by separators. The elements are immersed in an electrolyte composed of dilute sulfuric acid. Plate straps located on the top of each element connect all the positive plates and all the negative plates into groups. The elements are connected in series electrically by connectors that pass directly through the case partitions between cells. The battery top is a one-piece cover. The cell connectors, by-passing through the cell partitions, connect the elements along the shortest practical path.

PERIODIC SERVICING
Level indicator
The Battery features an electrolyte level indicator, which is a specially designed vent plug with a transparent rod extending through the center. When the electrolyte is at the proper level, the lower tip of the rod is immersed, and the exposed top of the rod will appear very dark; when the level falls below the tip of the rod, the top will glow. The indicator reveals at a glance if water is needed, without the necessity of removing the vent plugs.

Electrolyte level
The electrolyte level in the Battery should be checked regularly. In hot weather, particularly during trip driving, checking should be more frequent because of more rapid loss of water. If the electrolyte level is found to be low, then colorless, odorless, drinking water should be added to each cell until the liquid level rises to the split vent located in the bottom of the vent well. DO NOT OVERFILL because this will cause loss of electrolyte resulting in poor performance, short life, and excessive corrosion.

Cleaning
The external condition of the Battery should be checked periodically for damage such as cracked cover, case and vent plugs or for the presence of dirt and corrosion. The Battery should be kept clean in the area of the terminals. An accumulation of acid film and dirt may permit current to flow between the terminals, which will slowly discharge the Battery. For best results when cleaning batteries, wash first with a diluted ammonia or a soda solution to neutralize any acid present; then flush with clean water. Care must be taken to keep vent plugs tight, so that the neutralizing solution does not enter the cells.

Cables
To insure good electrical contact, the cables should be clean and tight on the Battery. If the posts or cable terminals are corroded, the cables should be disconnected and the terminals cleaned separately with a soda solution and a wire brush. After cleaning and installing clamps, apply a thin coating of petroleum jelly on the cable clamps to retard corrosion.

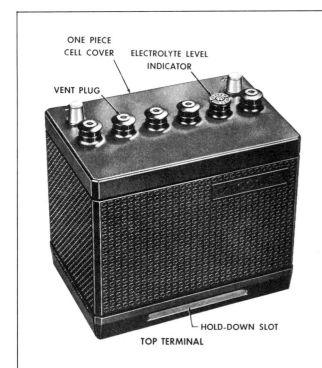

ONE PIECE CELL COVER
ELECTROLYTE LEVEL INDICATOR
VENT PLUG
HOLD-DOWN SLOT
TOP TERMINAL

MANIFOLD AND VENT PLUG
SEALED TERMINAL
HOLD DOWN SLOT
SIDE TERMINAL

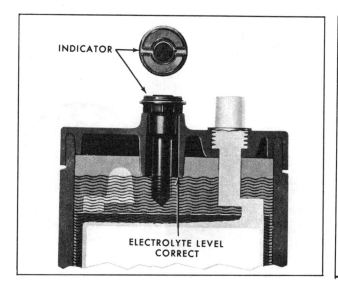

Electrolyte at proper level

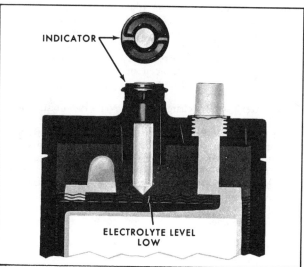

Electrolyte at low level

Carrier and hold-down

The battery carrier and hold-down should be clean and free from corrosion before installing the battery. The carrier should be in a sound mechanical condition so that it will support the battery securely and keep it level. To prevent the battery from shaking in its carrier, the hold-down bolts should be tight (60-80 in. lbs.). However, the bolts should not be tightened to the point where the battery case or cover will be placed under a severe strain.

Safety precautions

When batteries are being charged, an explosive gas mixture forms in each cell. Part of this gas escapes through the holes in the vent plugs and may form an explosive atmosphere around the battery itself if ventilation is poor. This explosive gas may remain in or around the battery for several hours after it has been charged. Sparks or flames can ignite this gas causing an internal explosion which may shatter the battery.

Do not smoke near batteries being charged or which have been very recently charged. Do not break live circuits at the terminals of batteries because a spark usually occurs at the point where a live circuit is broken. Care must always be taken when connecting or disconnecting booster leads or cable clamps on fast chargers. Poor connections are a common cause of electrical arcs which cause explosions.

Charging

Before charging a battery the electrolyte level must be checked and adjusted if needed. Battery charging consists of applying a charge rate in amperes for a period of time in hours. Thus, a 10-ampere charge rate for seven hours would be a 70 ampere-hour (A.H.) charging input to the battery.

Charging rates in the three to 50 ampere range are generally satisfactory. No particular charge rate or time can be specified for a battery because of many variables.

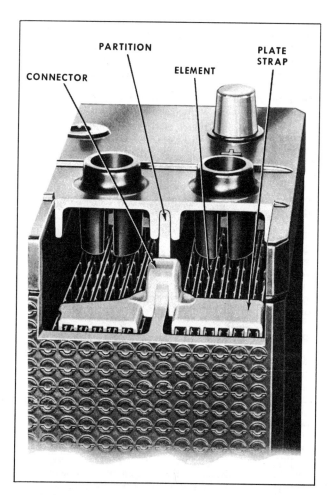

Internal view of two cells showing connector through partition

Any battery may be charged at any rate in amperes for as long as spewing of electrolyte due to violent gassing does not occur, and for as long as electrolyte temperature does not exceed 125°F. If spewing of electrolyte occurs, or if electrolyte temperature exceeds 125°F., the charging rate in amperes must be reduced or temporarily halted to avoid damage to the battery.

The battery is fully charged when over a two-hour period at a low charging rate in amperes all cells are gassing freely (not spewing liquid electrolyte), and no change in specific gravity occurs. The full charge specific gravity is 1.260–1.280, corrected for electrolyte temperature with the electrolyte level at the split ring, unless electrolyte loss has occurred due to age or over-filling in which case the specific gravity reading will be lower. For the most satisfactory charging, the lower charging rates in amperes are recommended.

If after prolonged charging a specific gravity of at least 1.230 on all cells cannot be reached, the battery is not in an optimum condition and will not provide optimum performance; however, it may continue to provide additional service if it has performed satisfactorily in the past.

An "emergency boost charge", consisting of a high charging rate for a short period of time, may be applied as a temporary expedient in order to crank an engine. However, this procedure usually supplies insufficient battery reserve to crank a second and third time. Therefore, the "emergency boost charge" must be followed by a subsequent charging period of sufficient duration to restore the battery to a satisfactory state of charge.

When out of the vehicle, the sealed side terminal battery will require adapters for the terminals to provide a place for attachment of the charging leads. Adapters are available through local parts service. When the side terminal battery is in the car, the studs provided in the wiring harness are suitable for attachment of the charger's leads. Exercise care when attaching charger leads to side terminal studs to avoid contact with vehicle metal components which would result in damage to the energizer.

Visual inspection

Check the outside of the battery for a broken or cracked case or a broken or cracked cover. If any damage is evident, the battery should be replaced. Note the electrolyte level. Levels that are too low or too high may cause poor performance. Check for loose cable connections, and for evidence of corrosion.

Specific gravity readings

A hydrometer can be used to measure the specific gravity of the electrolyte in each cell. The hydrometer measures the percentage of sulphuric acid in the battery electrolyte in terms of specific gravity. As a battery drops from a charged to a discharged condition, the acid leaves the solution and enters the plates, causing a decrease in specific gravity of electrolyte. An indication of the concentration of the electrolyte is obtained with a hydrometer.

When using a hydrometer, the hydrometer must be clean, inside and out, to insure an accurate reading. Hydrometer readings must never be taken immediately after water has

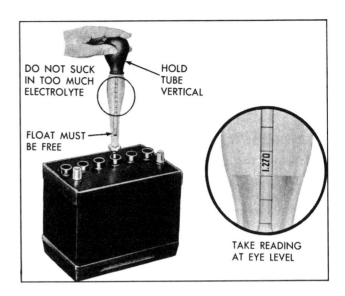

Checking specific gravity

been added. The water must be thoroughly mixed with the electrolyte by charging for at least 15 minutes at a rate high enough to cause vigorous gassing. If hydrometer has built-in thermometer, draw liquid into it several times to insure correct temperature before taking reading. Hold hydrometer vertically and draw in just enough liquid from battery cell so that float is free floating. Hold hydrometer at eye level so that float is vertical and free of outer tube, then take reading at surface of liquid. Disregard the curvature where the liquid rises against float stem due to surface tension. Avoid dropping battery fluid on car or clothing as it is extremely corrosive. Any fluid that drops should be washed off immediately with baking soda solution.

The specific gravity of the electrolyte varies not only with the percentage of acid in the liquid but also with temperature. As temperature increases, the electrolyte expands so that the specific gravity is reduced. As temperature drops, the electrolyte contracts so that the specific gravity increases. Unless these variations in specific gravity are taken into account, the specific gravity obtained by the hydrometer may not give a true indication of the concentration of acid in the electrolyte.

A fully charged battery will have a specific gravity reading of approximately 1.270 at an electrolyte temperature of 80°F. If the electrolyte temperature is above or below 80°F, additions or subtractions must be made in order to obtain a hydrometer reading corrected to the 80°F standard. For every 10° above 80°F, add four specific gravity points (.004) to the hydrometer reading. Example: a hydrometer reading of 1.260 at 110°F would be 1.272 corrected to 80°F, indicating a fully charged battery. For every 10° below 80°F, subtract four points (.004) from the reading. Example: a hydrometer reading of 1.272 at 0°F would be 1.240 corrected to 80°F, indicating a partially charged battery.

Specific Gravity Cell Comparison Test – This test may be used when an instrumental tester is not available. To perform this test measure the specific gravity of each cell, regardless of state of charge. If specific gravity readings show a difference between the highest and lowest cell of .050 (50 points) or more, the battery is defective and should be replaced.

Replacement

Disconnect battery negative cable and then the positive cable. Disconnect battery vent covers from vent plugs. Remove battery hold-down clamp bolt and clamp. Remove battery from tray and transfer vent cover to replacement battery. Position new battery, which has been properly activated, in the battery tray. Install battery clamp and bolt. Recommended hold-down bolt torque is 60–80 in. lbs. Connect battery vent covers to vent plugs. Connect battery positive and negative cables. On side terminal units, tighten terminal studs to 70 in. lbs. torque.

5.5" SERIES 1D DELCOTRON SYSTEM
Description

The Series 1D Delcotron continuous output alternator consists of two major parts, a stator and a rotor. The stator is composed of a large number of windings assembled on the inside of a laminated core that is attached to the generator frame. The rotor revolves within the stator on bearings located in each end frame. Two brushes are required to carry current through the two clip rings to the field coils wound concentric with the shaft of the rotor. Six rectifier diodes are mounted in the slip ring end frame and are joined to the stator windings at three internally located terminals.

Diodes are mounted in heat sinks to provide adequate heat dissipation. The six diodes replace the separately mounted rectifier as used in other types of application. The diodes change the A.C. to D.C.

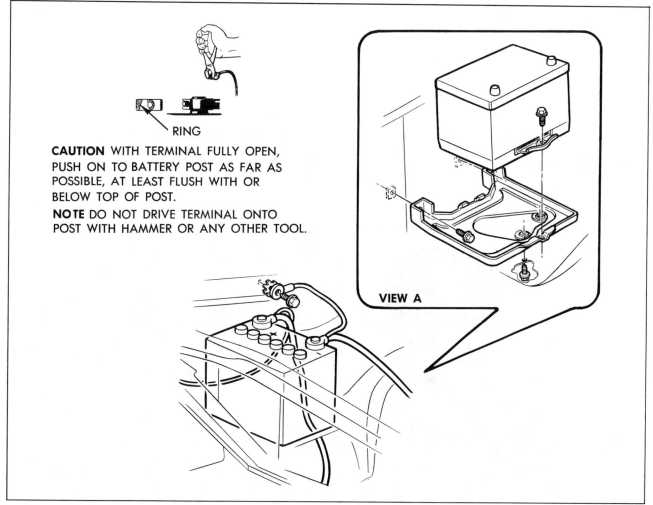

RING

CAUTION WITH TERMINAL FULLY OPEN, PUSH ON TO BATTERY POST AS FAR AS POSSIBLE, AT LEAST FLUSH WITH OR BELOW TOP OF POST.

NOTE DO NOT DRIVE TERMINAL ONTO POST WITH HAMMER OR ANY OTHER TOOL.

VIEW A

Battery installation

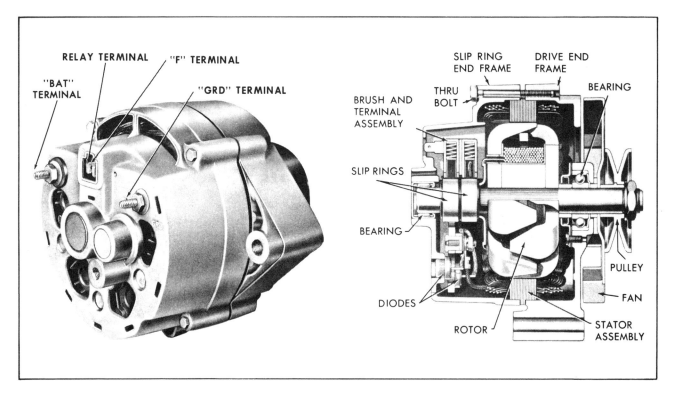

RELAY TERMINAL   "F" TERMINAL

"BAT" TERMINAL   "GRD" TERMINAL

SLIP RING END FRAME   DRIVE END FRAME

THRU BOLT   BEARING

BRUSH AND TERMINAL ASSEMBLY

SLIP RINGS

BEARING

DIODES

ROTOR

PULLEY

FAN

STATOR ASSEMBLY

5.5" series 1D delcotron

The function of this regulator in the charging system is to limit the generator voltage to a pre-set value by controlling the generator field current. The double contact regulator has an internal field relay unit. The relay unit allows the telltale lamp to light (as a bulb check) with the ignition key on and engine not running. When the engine is started and the generator begins to charge, the indicator light goes out indicating that the system is operating normally.

The double-contact regulator assembly consists of a double contact voltage regulator unit and a field relay unit. This unit uses two sets of contact points on the voltage regulator unit to obtain desired field excitation under variable conditions. An internal circuit wiring diagram of the double contact regulator is shown.

Engine compartment wiring harness incorporates several fusible links. Each link is identified with its gauge size. A fusible link is a length of special wire (normally four wire gauges smaller than the circuit it is protecting) used in wiring circuits that are not normally fused, such as the ignition circuit. The same size wire with a hypalon insulation must be used when replacing a fusible link.

The pigtail lead at the battery positive cable is installed as a molded splice at the solenoid "Bat" terminal and servicing requires splicing in a new link.

A 16 gauge black fusible link is located at the horn relay to protect all unfused wiring of 12 gauge or larger. It is installed as a molded splice and servicing requires splicing in a new link.

The generator warning light and field circuitry (16 gauge wire) is protected by a fusible link (20 gauge orange wire)

FIELD RELAY

"LATCH"

"F" TERMINAL

NO. 2 TERMINAL

NO. 3 TERMINAL

NO. 4 TERMINAL

VOLTAGE REGULATOR

Voltage regulator

216

used in the battery feed to voltage regulator #3 terminal wire. The link is installed as a molded splice in the generator and forward lamp harness and is serviced by splicing in a new 20 gauge wire as required.

The ammeter circuit on all models is protected by two orange, 20 gauge wire fusible links installed as molded splices in the cirucit at the junction block or the solenoid "Bat" terminal and at the horn relay. Each link is serviced by splicing in a new 20 gauge wire as required.

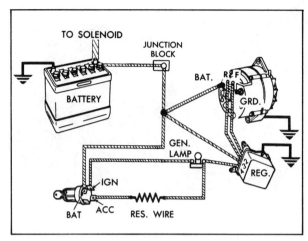

Typical wiring diagram

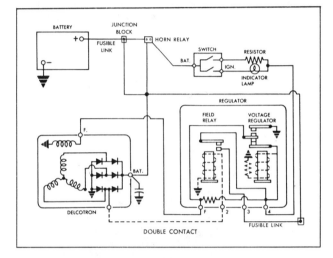

Voltage regulator circuitry

Maintenance and adjustments

At regular intervals, inspect the terminals for corrosion and loose connections, and the wiring for frayed insulation. Check mounting bolts for tightness. Check the drive belt for alignment, proper tension and wear. Because of the higher inertia and load capacity of the rotor used in alternators, proper belt tension is more critical than on D.C. generators.

Since the alternator and its companion regulator are designed for use on negative polarity systems only, the following precautions must be observed. Failure to observe these precautions may result in serious damage to the charging system.

When installing battery always make absolutely sure the ground polarity of the battery, generator and regulator is the same. When connecting a booster battery, make certain to connect the correct battery terminals together.

When connecting a charger to the battery, connect the correct charger leads to the battery terminals. Never operate the generator on an uncontrolled open circuit. Make absolutely certain all connections in the circuit are secure. Do not shore across or ground any of the terminals on the generator or regulator. Do not attempt to polarize the generator. Do not disconnect lead at generator without first disconnecting battery ground cable.

Trouble in the A.C. charging system will usually be indicated by faulty indicator lamp or ammeter operation. An undercharged battery (usually evidenced by slow cranking speeds). An overcharged battery (usually evidenced by excessive battery water usage). Excessive generator noise or vibration.

Static checks

Before making any electrical checks, check for loose fan belt. Check for defective battery. (Refer to Battery.) Inspect all connections, including the slip-on connectors at the regulator and Delcotron. Do not short field to ground to check if generator is charging since this will seriously damage the charging system.

DOUBLE CONTACT REGULATOR
Description

While most regular adjustments are made on the vehicle as outlined under "Maintenance and Adjustments", the regulator may be removed for field relay point and air gap adjustment. However, voltage regulating contacts should be cleaned as they are made of special material that may be destroyed by cleaning with any abrasive material. A sooty or discolored condition of the contacts is normal after a relatively short period of operation.

Removal and installation

To remove the regulator assembly, disconnect the battery ground cable and the wiring harness connector at the regulator, then remove the screws securing the regulator to the vehicle. Electrical settings must be checked and adjusted after making mechanical adjustments. Before installing regulator cover, make sure the rubber gasket is in place on the regulator base.

Mechanical adjustments

Only an approximate voltage regulator air gap setting should be made by the "feeler gauge" method.

Field relay adjustment

Point Opening: Check the point opening as shown. If

Pulley removal

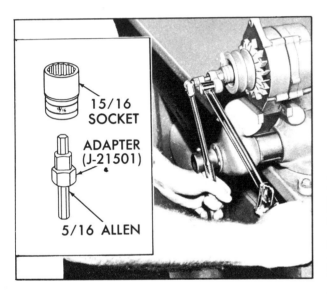

15/16 SOCKET

ADAPTER (J-21501)

5/16 ALLEN

Torquing pulley nut

on retaining nut, insert a 5/16" allen wrench through socket and adapter into hex on shaft to hold the shaft while removing the nut. Remove washer and slide pulley, fan and spacer from shaft. To install, slide spacer, fan, pulley and washer on shaft and start the nut. Use the socket and adapter with a torque wrench and tighten nut to 50 ft. lbs. torque.

ALTERNATOR
Removal and installation
Disconnect the battery cables at battery. Disconnect wiring leads at Delcotron. Remove generator brace bolt, then detach drive belt (belts). Support the generator and remove generator mount bolt and remove from vehicle. Reverse the removal procedure to install then adjust drive belt(s).

Disassembly
Hold generator in a vise, clamping the mounting flange lengthwise. Remove 4 thru bolts then break loose the end

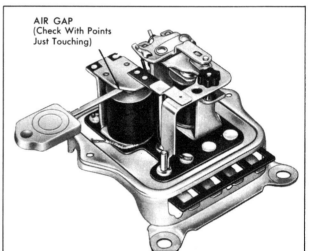

AIR GAP (Check With Points Just Touching)

adjustment is necessary, carefully bend the armature stop. Air Gap: Check the air gap with the points just touching. The air gap normally need not be adjusted. If the point opening and closing voltages are within specifications, the relay will operate satisfactorily even though the air gap may not be exactly according to specifications. If adjustment is necessary, bend the flat contact spring.

PULLEY — SINGLE GROOVE
Removal and installation
Place 15/16" box wrench on retaining nut and insert a 5/16" allen wrench into shaft to hold shaft while removing nut. Remove washer and slide pulley, fan and spacer from shaft. Reverse the procedure to install, use a torque wrench with a crow-foot adapter (instead of box wrench) and torque the nut to 50 ft. lbs.

PULLEY — DOUBLE GROOVE
Removal and installation
Place a 15/16" socket (with wrench flats on the drive end)

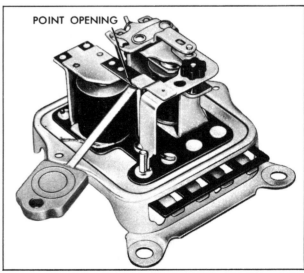

POINT OPENING

218

Checking field relay point opening

frames by prying at bolt locations. Remove the slip ring end frame and stator (as an assembly) from drive end and rotor assembly. Place a piece of tape over the slip ring end frame bearing to prevent entry of dirt or other foreign matter. Brushes may drop onto rotor shaft and become contaminated with bearing lubricant. Clean brushes prior to installing with a non-toxic cleaner such as trichlorethylene. Remove the three stator lead attaching nuts and separate stator from end frame. Remove screws, brushes and holder assembly. Remove heat sink from end frame by removing "BAT" and "GRD" terminals and one attaching screw. Remove slip ring end frame bearing (if necessary). Refer to bearing replacement included in this section.

Remove pulley retaining nut and slide washer, pulley and fan from shaft. Single Groove Pulley — Place 15/16" box wrench on retaining nut and insert a 5/16" allen wrench into shaft to hold shaft while removing nut. Double

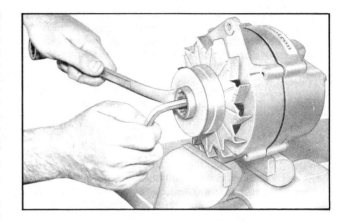

Pulley removal

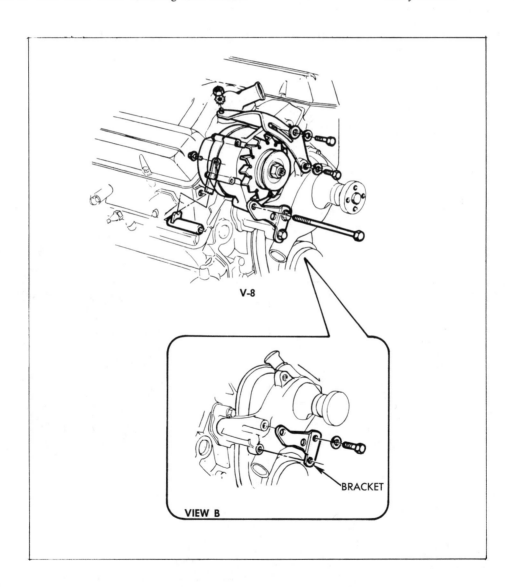

V-8

BRACKET

VIEW B

Delcotron installation

Groove Pulley — Place a 15/16" socket (with wrench flats on the drive end and a box wrench) on retaining nut, insert a 5/16" allen wrench through socket and adapter into hex on shaft to hold the shaft while removing the nut.

Remove rotor and spacers from end frame assembly. Remove drive end frame bearing retainer plate and bearing assembly from frame.

Cleaning and inspection

With Delcotron completely disassembled, except for removal of diodes, the components should be cleaned and inspected. Be sure testing equipment is in good working order before attempting to check the generator. Wash all metal parts except stator and rotor assemblies. Clean bearings and inspect for sealing, pitting or roughness. Inspect rotor slip rings, they may be cleaned with 400 grain polishing cloth. Rotate rotor for this operation to prevent creating flat spots on slip rings. Slip rings which are out of round may be trued in a lathe to .001" maximum indicator reading. Remove only enough material to make the rings smooth and concentric. Finish with 400 grain polishing cloth and blow dry.

Slip rings are not replaceable — excessive damage will require rotor assembly replacement. Inspect brushes for wear. If they are worn halfway, replace. Inspect brush springs for distortion or weakening. If brushes appear satisfactory and move freely in brush holder, springs may be reused.

Testing Rotor — The rotor may be checked electrically with a 110-volt test lamp or an ohmmeter.

Grounds — Connect test lamp or ohmmeter from either slip ring to the rotor shaft or to the rotor poles. If the lamp lights or if the ohmmeter reading is low, the field windings are grounded.

Open Circuit — Connect one test lamp or ohmmeter lead to each slip ring. If the lamp fails to light or if the ohmmeter reading is high, the windings are open.

Short Circuit — The windings are checked for shorts by connecting a 12-volt battery and an ammeter in series with the two slip rings. Note the ammeter reading. An ammeter reading above the specified field amperage draw indicates shorted winding. Refer to Specifications at the end of this manual.

STATOR
Testing

Grounds — Connect a 110-volt test lamp or an ohmmeter from any stator lead to the stator frame. If the test lamp lights or if ohmmeter reads low, the windings are grounded.

Open Circuit — If lamp fails to light or if ohmmeter reads high when successively connected between each pair of stator leads, the windings are open.

Short Circuit — A short in the stator windings is difficult to locate without special test equipment due to the low resistance of the windings. However, if all other electrical checks are normal and the generator fails to supply rated output, shorted stator windings are indicated. Also, look for heat discoloration on the windings.

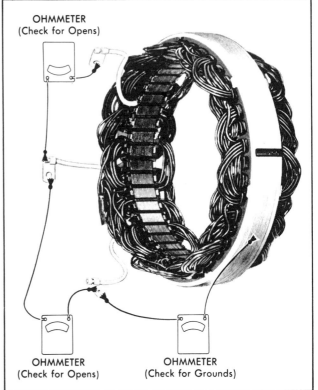

Checking rotor for grounds or opens

Checking stator

220

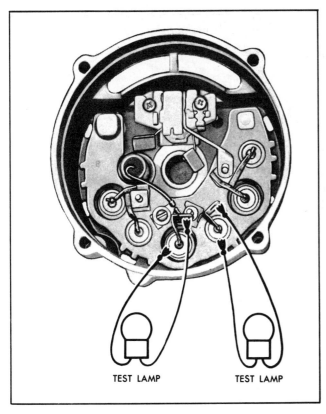

Checking diodes

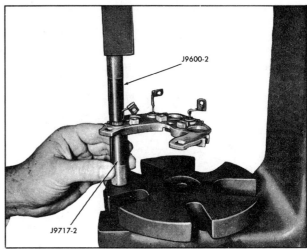

Installing diodes with press tools

TESTING DIODES

Two methods may be used to check diodes for shorts or opens, a test lamp of not more than 12 volts or an ohmmeter. Do not use a 110-volt test lamp to test diodes.

Test lamp method
Diode in heat sink

With the stator previously disconnected connect one of the lamp leads to the heat sink and other lead to the light in only one of the test directions. If lamp lights in both directions or fails to light at all, the diode is defective.

Diode in the end frame

Connect one lamp lead to the end frame and the other lamp lead to the diode lead, and observe lamp condition. Reverse the lamp lead connections and observe the lamp condition. A good diode will allow lamp to light in only one direction. If lamp lights in both directions or fails to light at all, the diode is defective.

Ohmmeter method

Use an ohmmeter with a 1½ volt cell and use the lowest range scale. Connect the ohmmeter leads at each diode as previously described using a test lamp first in one direction and then the other. Note the readings. If both readings are identical (very high or very low), the diode is defective. A good diode will give one high and one low reading.

Diode replacement

Support end frame with support Tool J-9717-2 and press out diode with diode removal Tool J-9717-1 and an arbor press or vise. Do not strike diode as shock may damage other diodes. Select diode with proper color marking.

Diodes in the heat sink are positive (red markings) and those in the end frame are negative (black markings).

Support outside end frame around diode hole on a flat, smooth surface and press diode into position with J-9600-2 and an arbor press or vise. Make sure diode is square with end frame and started straight. Avoid bending or moving diode stem as excessive movement can cause internal damage and result in diode failure.

Heat sink replacement

Detach heat sink from end frame by removing the two attaching bolts. Note carefully the proper stack up of parts so the "BAT" and "GRD" terminal bolts can be reassembled in the same manner. Replace diodes, if necessary, as outlined in Diode Replacement. Assemble heat sink to the end frame, following carefully the proper stack up of parts.

Brush replacement

After through bolt removal and Delcotron separation, remove stator lead nut that also holds relay terminal connector. Remove 2 mounting screws and brush holder assembly. Position new brush holder assembly and install retaining screws. Connect relay terminal wire lead and install stator lead nut.

End frame replacement

Remove heat sink. Attach brush holder assembly to the new end frame and insert pin or wire through the hole to hold the brushes in the holder. After the Delcotron unit has been completely assembled, withdraw the pin or wire from the end frame hole to allow the brushes to drop down onto the slip rings. Replace heat sinks to end frame.

221

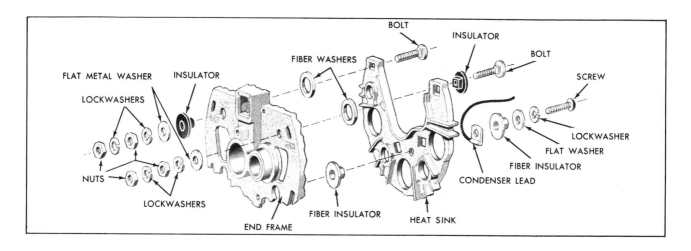

Heat sink — parts location 5.5" delcotron

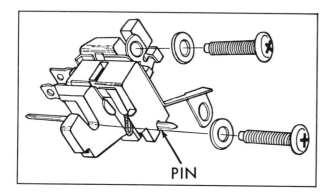

Brush assembly — 5.5" delcotron

ALTERNATOR
Assembly
Install stator assembly in slip ring end frame and locate diode connectors over the relay, diode and stator leads, and tighten terminal nuts. Install rotor in drive end frame. Install fan, spacer, pulley washer and nut. Place torque wrench and adapter on shaft nut and insert allen wrench into opening at end of drive shaft. Tighten shaft nut to 40—50 ft. lbs. Assemble slip ring end frame and stator assembly to drive end frame and rotor assembly. Install 4 through bolts in the end frame assemblies.

Bearing replacement — Drive end frame
The drive end frame bearing can be removed by detaching the retainer plate bolts and separating retainer plate and seal assembly from end frame, and then pressing bearing out using suitable tube or pipe on inner race. Refill bearing one-quarter full with Delco-Remy No. 1948791 grease or equivalent. Do not overfill. Press bearing into end frame using tube or pipe that fits over outer race. Install retainer plate. Use new retainer plate if felt seal is hardened or excessively worn. Stake retainer plate bolts to plate.

Bearing replacement — Slip ring end frame
Replace the bearing if the grease supply is exhausted. Make no attempt to relubricate and reuse the bearings. Press bearing from outside of housing toward inside using suitable tool that just fits inside the end frame. To install, place a flat plate over the bearing and press in from outside of housing until bearing is flush with the outside of the end frame. Support inside of end frame around bearing bore with a suitable tool to prevent distortion. Use extreme care to avoid misalignment. Saturate felt seal with S.A.E. 20 engine oil and install seal and steel retainer at inner end of bearing assembly.

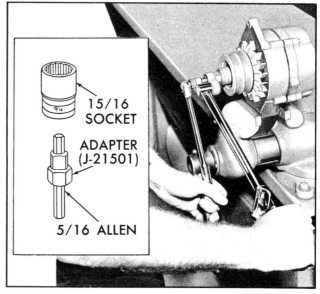

222

Torquing pulley nut

10-S1 SERIES DELCOTRON SYSTEM
Description
The 10-S1 Series Delcotron generator features a solid state regulator that is mounted inside the generator slip ring end frame. All regulator components are enclosed into a solid mold, and this unit along with the brush holder assembly is attached to the slip ring end frame. The regulator voltage setting lever needs adjusting, and no provision for adjustment is provided.

No periodic adjustments or maintenance of any kind are required on the entire generator assembly. The generator rotor bearings contain a supply of lubricant sufficiently adequate to eliminate the need for periodic lubrication. Two brushes carry current through the two slip rings to the field coil mounted on the rotor, and under normal conditions will provide long periods of attention-free service.

The stator windings are assembled on the inside of a laminated core that forms part of the generator frame. A rectifier bridge connected to the stator windings contains six diodes, and electrically changes the stator a.c. voltages to a d.c. voltage which appears at the generator output terminal. Generator field current is supplied through a diode trio which also is connected to the stator windings. A capacitor or condenser, mounted in the end frame protects the rectifier bridge and diode trio from high voltages, and suppresses radio noise.

Operating principles
When the switch is closed, current from the energizer flows to the generator No. 1 terminal, through resistor R1, diode D1, and the base-emitter of transistor TR1 to ground, and then back to the battery. This turns on transistor TR1, and current flows through the generator field coil and TR1 back to the energizer.

With the generator operating, a.c. voltage is generated in the stator windings, and the stator supplies d.c. field current through the diode trio, the field, TR1, and then through the grounded diodes in the rectifier bridge back to the stator. Also, the six diodes in the rectifier bridge change the stator c.v. voltages to a d.c. voltage which appears between ground and the generator "BAT" terminal. As generator speed increases, current is provided for charging the energizer and operating electrical accessories. The No. 2 terminal on the generator is always connected to the energizer, but the discharge current is limited to a negligible value by the high resistances of R2 and R3. As the generator speed and voltage increase, the voltage between R2 and R3 increases to the point where zener diode D2 conducts. Transistor TR2 then turns on and TR1 turns off. With TR1 off, the field current and system voltage decrease, then D2 then blocks current flow, causing TR1 to turn back on. The field current and system voltage increase, and this cycle then repeats many times per second to limit the generator voltage to a pre-set value. Capacitor C1 smooths out the voltage across R3,

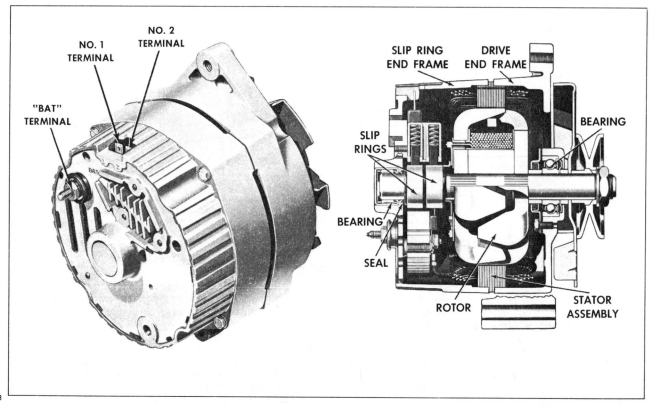

10-SI series delcotron

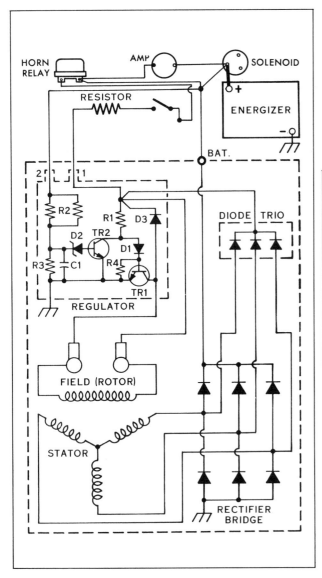

Integral charging system circuitry

The following precautions must be observed when working on the charging circuit. Failure to observe these precautions will result in serious damage to the electrical equipment. Do not polarize the generator. Do not short across or ground any of the terminals in the charging circuit except as specifically instructed in these procedures. Never operate the generator with the output terminal open circuited. Make sure the generator and Energizer are of the same ground polarity. When connecting a charger or a booster Energizer to the vehicle Energizer, connect negative to negative and positive to positive.

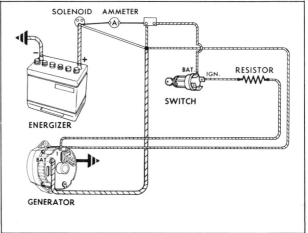

Basic wiring diagram

Static check
Before making any electrical checks, visually inspect all connections, including slip-on connectors, to make sure they are clean and tight. Inspect all wiring for cracked, frayed or broken insulation. Be sure generator mounting bolts are tight and unit is properly grounded. Check for loose fan belt.

Undercharged battery condition check
This condition, as evidenced by slow cranking and low specific gravity readings, can be caused by one or more of the following conditions even though the ammeter may be operating normally.

Insure that the undercharged condition has not been caused by accessories having been left on for extended periods. Check the drive belt for proper tension. Check energizer. Test is not valid unless energizer is good and fully charged. Inspect the wiring for defects. Check all connections for tightness and cleanliness, including the slip connectors at the generator and firewall, and the cable clamps and battery posts.

resistor R4 prevents excessive current through TR1 at high temperatures, and diode D3 prevents high-induced-voltages in the field windings when TR1 turns off.

Charging circuit checks
Most charging system troubles show up as an undercharged or overcharged battery. Since the battery itself may be defective, it should be checked first to determine its condition. Also, in the case of an undercharged battery, check for battery drain caused by grounds or by accessories being left on.

With ignition switch "on" connect a voltmeter from generator "BAT" terminal to ground, generator No. 1 terminal to ground and generator No. 2 terminal to ground. A zero reading indicates an open between voltmeter connection and Energizer.

If the previous check is satisfactory, connect a voltmeter in the circuit at the "BAT" terminal of the generator. Operate engine at moderate speed (approximately 1500-2000 rpm) and turn on electrical loads (high beam headlights, windshield wiper, heater or A/C blower, radio, etc.). Without sufficient electrical load to demand maximum Delcotron output the following voltage check is invalid.

Observe Voltmeter reading: If reading is 12.8 volts or more, generator is not defective. Turn off electrical loads, stop engine and disconnect test equipment. Recheck steps 1 thru 5. If reading is less than 12.8 volts, ground field winding by inserting a screwdriver into the test hole in the end frame. TAB IS WITHIN 3/4 INCH OF CASTING SURFACE. DO NOT FORCE SCREWDRIVER DEEPER THAN ONE INCH INTO END FRAME.

If voltage increases (13 volts and above) regulator unit is defective. Replace regulator. If voltage does not increase significantly, generator is defective. Check the field winding, diode trio, rectifier bridge and stator and repair as required. Turn off electrical loads, shut off engine and disconnect all test equipment, if not previously done.

Overcharged battery condition check

Determine battery condition. Test is not valid if battery is not good and fully charged. Connect a voltmeter from generator No. 2 terminal to ground. If reading is zero, No. 2 lead circuit is open. If Energizer and No. 2 lead circuit check good, but an obvious overcharge condition exists as evidenced by excessive battery water usage, separate end frames as covered in Delcotron "Disassembly" section.

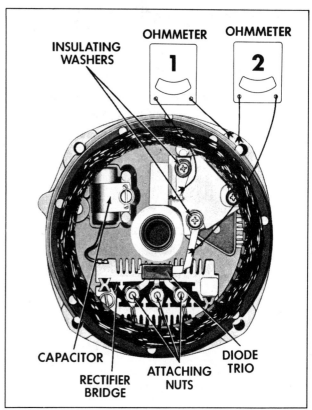

Slip ring end frame

Connect ohmmeter using lowest range scale from brush lead clip to end frame, then reverse lead connections. If both readings are zero, either the brush lead clip is grounded, or regulator is defective. A grounded brush lead clip can result from omission of insulating washer, omission of insulating sleeve over screw, or damaged insulating sleeve. Remove screw to inspect sleeve. If satisfactory, replace regulator.

Generator output test

To check the generator in a test stand, make connections as shown, except leave the carbon pile disconnected. Use a fully charged battery, and a 10 ohm resistor rated at six watts or more between the generator No. 1 terminal and

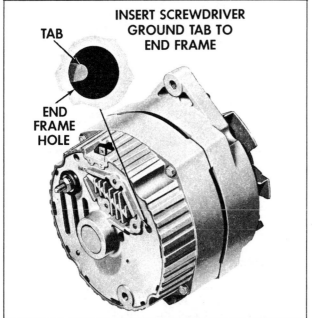

Delcotron end view

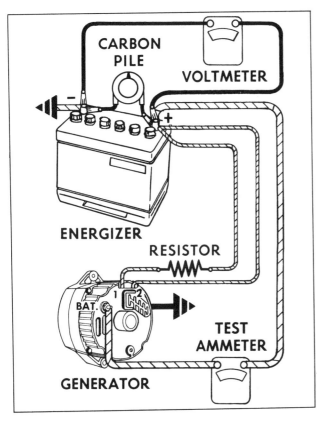

Generator output test

the battery. Slowly increase the generator speed and observe the voltage. If the voltage is uncontrolled with speed and increases above 16 volts, check for a grounded brush lead clip. If not grounded, replace the regulator. The Energizer must be fully charged when making this check.

Connect the carbon pile. Operate the generator at moderate speed as required and adjust the carbon pile as required to obtain maximum current output. If output is within ten percent of rated output as stamped on generator frame, generator is good. If output is not within ten percent of rated output, ground generator field. Operate generator at a moderate speed and adjust carbon pile as required to obtain maximum output. If output is within ten percent of rated output, replace regulator as covered in "Regulator Replacement" section. If output is not within ten percent of rated output, check the field winding, diode trio, rectifier bridge and stator as previously covered.

DIODE TRIO
Check
With the diode trio unit removed from the end frame, connect an ohmmeter to the single connect and to one of

the three connectors. Observe the reading. Then reverse the ohmmeter leads to the same connectors. A good diode trio will give one high and one low reading. If both readings are the same, replace the diode trio. Repeat this test between the single connector and each of the other two connectors. There are two diode trio units differing in appearances used in the generator and are completely interchangeable. The diode trio may also be checked for a grounded brush lead while still installed in the end frame. Connect an ohmmeter from the brush lead clip to the end frame, then reverse the lead connection. If both readings are zero, check for a grounded brush lead clip caused by omission of the insulating washer, of the insulating sleeve over the screw, or damaged insulating sleeve. Remove the screw to inspect the sleeve. If sleeve or screw are not grounded, replace regulator. Do not use high voltage to check the diode trio, such as 110-volt test lamp.

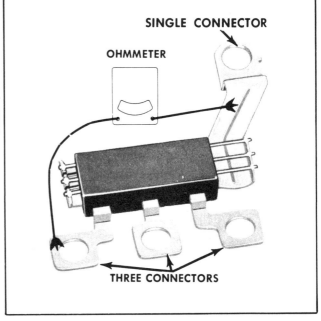

Diode trio checks

RECTIFIER BRIDGE
Check
Connect an ohmmeter to the grounded heat sink and one of the three terminals. Then reverse the lead connections to the grounded heat sink and same terminal. If both readings are the same, replace the bridge. A good rectifier bridge will give one high and one low reading. Repeat this same test between the grounded heat sink and the other two terminals, and between the insulated heat sink and

each of the three terminals. When this is done all six diodes are checked with two readings taken for each diode. The diodes are not replaced individually. The entire rectifier bridge is replaced if one or more diodes are defective. Do not use high voltage to check the rectifier bridge, such as a 110-volt test lamp.

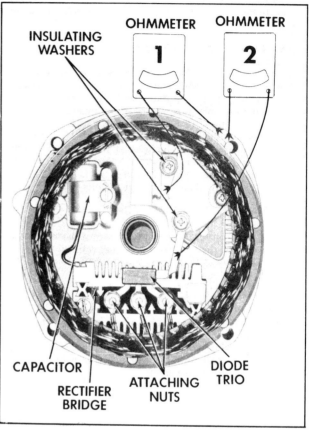

Brush lead clip checks

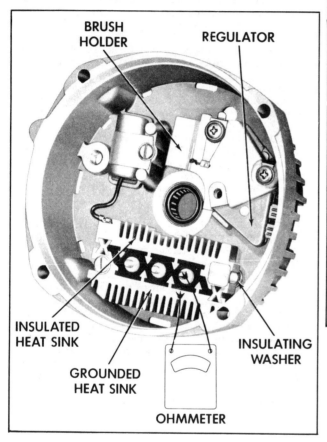

Rectifier bridge checks

VOLTAGE REG./BRUSH LEAD CLIP
Check
Connect an ohmmeter from the brush lead clip to the end frame as shown in Step 1, fig. 57F. Then reverse lead connections. If both readings are zero, either the brush lead clip is grounded or the regulator is defective.

A grounded brush lead clip can result from omission of the insulating washer, omission of the insulating sleeve on the screw, or a damaged insulating sleeve. Remove the screw and inspect the sleeve. If it is satisfactory, replace the regulator unit.

BRUSH HOLDER AND REGULATOR
Replacement
If not previously removed, remove the three stator lead attaching nuts, the stator, diode trio brush lead screw and diode trio from the end frame. If the brush lead screw does not have a sleeve it must not be interchanged with either one of the other two screws. Remove the remaining two screws from the brush holder and regulator and remove these units from the end frame. These two screws have special insulating sleeves over the screw body above the threads. If they are damaged or missing a ground will result causing uncontrolled or no output.

Slip ring servicing
If the slip rings are dirty, they may be cleaned and finished with 400 grain or finer polishing cloth. Spin the rotor, and hold the polishing cloth against the slip rings until they are clean. The rotor must be rotated in order that the slip rings will be cleaned evenly. Cleaning the slip rings by hand without spinning the rotor may result in flat spots on the slip rings, causing brush noise.

Slip rings which are rough or out of round should be trued in a lathe to .002 inch maximum indicator reading. Remove only enough material to make the rings smooth and round. Finish with 400 grain or finer polishing cloth and blow away all dust.

BEARING — DRIVE END FRAME
Replacement
The drive end frame bearing can be removed by detaching the retainer plate bolts and separating retainer plate and seal assembly from end frame, and then pressing bearing out using suitable tube or pipe on inner race. Refill bearing one-quarter full with Delco-Remy No. 1948791 grease or equivalent. Do not overfill. Press bearing into end frame using tube or pipe that fits over outer race, with bearing and slinger assembled. Install retainer plate. Use new retainer plate if felt seal is hardened or excessively worn. Stake retainer plate bolts to plate.

BEARING — SLIP RING END FRAME
Replacement
Replace the bearing if the grease supply is exhausted. Make no attempt to re-lubricate and reuse the bearing. Press out from outside of housing toward, inside, using suitable tool that just fits inside end frame. To install, place a flat plate over the bearing and press in from outside of housing until bearing is flush with the outside of the end frame. Support inside of end frame around bearing bore with a suitable tool to prevent distortion. Use extreme care to avoid misalignment. Install new seal whenever bearing is replaced. Lightly coat the seal lip with oil and press seal into the end frame with the seal lip toward the inside of the end frame.

Assembly
Install rotor in drive end frame and attach spacer, fan, pulley, washer, and nut. Using adapter J-21501, insert an allen wrench into hex shaped hole at end of shaft and torque the shaft nut to 40—50 ft. lbs. Install capacitor and retaining screw in slip ring end frame. Position brush holder and regulator assemblies in end frame and install two retaining screws. The two screws retaining the brush clips have insulating washers over the top of the brush clips and special insulating sleeves over the screw body above the threads. If the third screw does not have an insulating sleeve, it must not be interchanged with either one of the other two screws.

Position rectifier bridge to end frame. Install attaching screw and the "BAT" terminal screw. Connect capacitor lead to bridge. Position diode trio on rectifier bridge terminal and install screw attaching brush lead clip to brush holder. Insulating washer on the screw must be assembled over top of the connector.

Position stator in end frame. Connect stator leads to rectifier bridge terminals and install attaching nuts. Position slip ring end frame to drive end frame and install four thru bolts. Remove tooth pick from brush holder at opening in slip ring end frame before operating machine on vehicle.

STARTER CIRCUIT
Description
The function of the starting system, composed of the starting motor, solenoid and battery, is to crank the engine. The battery supplies the electrical energy, the solenoid completes the circuit to the starting motor, and the motor then does the actual work of cranking the engine. The starting motor consists primarily of the drive mechanism, frame, armature, brushes, and field winding. The starting motor is a pad mounted 12-volt extruded frame type, having four pole shoes and four fields, connected with the armature. The aluminum drive end housing is extended to enclose the entire shift lever and plunger mechanism, protecting them from dirt, splash, and icing. The drive end frame also includes a grease

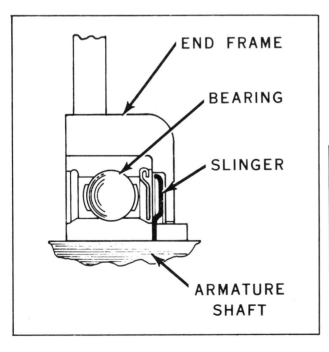

Drive end bearing cross section

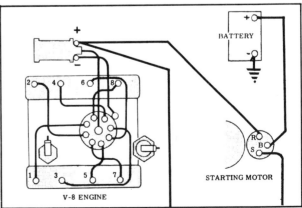

228

Starting circuit diagram

reservoir to provide improved lubrication of the drive end bearing. The flange mounted solenoid switch operates the overrunning clutch drive by means of a linkage to the shaft lever.

Maintenance and adjustments

No periodic lubrication of the steering motor solenoid is required. Since the starting motor and brushes cannot be inspected without disassembling the unit, no service is required on these units between overhaul periods.

Resistance checks

Although the starting motor cannot be checked against specifications on the car, a check can be made for excessive resistance in the starting circuit. Place a voltmeter across points in the cranking circuit as outlined below and observe the reading with the starting switch closed and the motor cranking (distributor primary lead grounded to prevent engine firing).

From battery positive post to solenoid battery terminal. From battery negative post to starting motor housing. From solenoid battery terminal to solenoid motor terminal. If voltage drop in any of above check exceeds 0.2 volts, excessive resistance is indicated in that portion of starting circuit and the cause of the excessive resistance should be located and corrected in order to obtain maximum efficiency in the circuit. Do not operate the starting motor continuously for more than 30 seconds in order to avoid overheating.

When the solenoid fails to pull in, the trouble may be due to excessive voltage drop in the solenoid control circuit. To check for this condition, close the starting switch and measure the voltage drop between the BATTERY terminal of the solenoid and the SWITCH(S) terminal of the solenoid. If this voltage drop exceeds 3.5 volts, excessive resistance in the solenoid control circuit is indicated and should be corrected. If the voltage drop does not exceed 3.5 volts and the solenoid does not pull in, measure the voltage available at the SWITCH terminal of the solenoid. If the solenoid does not feel warm, it should pull in whenever the voltage available at the SWITCH terminal is 7.7 volts or more. When the solenoid feels warm, it will require a somewhat higher voltage to pull in.

Starting motor and solenoid check

The following checks may be made if the specific gravity of the battery is 1.215 or higher. If the solenoid does not pull in, measure the voltage between the switch(s) terminal of the solenoid and ground with the starting switch cclosed. If the solenoid feels warm, allow to cool before checking.

If the voltage is less than 7.7 volts, check for excessive resistance in the solenoid control circuit. If the voltage exceeds 7.7 volts, remove the starting motor and check (1) solenoid current draw, (2) starting motor pinion clearance, and (3) freedom of shift lever linkage.

If the solenoid "chatters" but does not hold in, check the

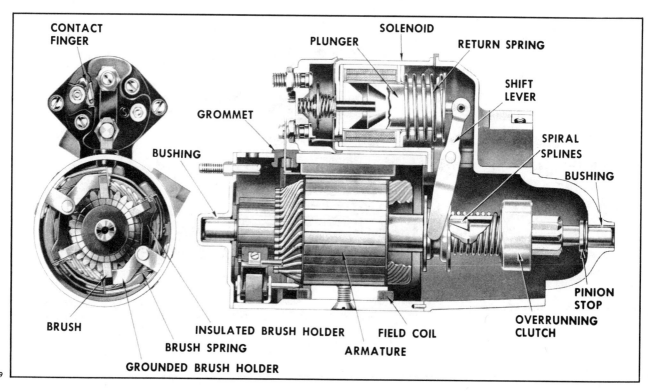

Starting motor cross section (typical)

solenoid for an open "hold-in" winding. Whenever it is necessary to replace a starting motor solenoid, always check starting motor pinion clearance.

If motor engages but does not crank or cranks slowly, check for excessive resistance in the external starting circuit, trouble within the starting motor, or excessive engine resistance to cranking.

STARTING MOTOR
Removal and installation
The following procedure is a general guide for all models and will vary slightly depending on series and model.

Disconnect battery ground cable at battery. Raise vehicle to a good working height. Disconnect all wires at solenoid terminals. Reinstall the nuts as each wire is disconnected as thread size is different but may be mixed and stripped. Loosen starter front bracket nut then remove two mount bolts. On V-8 Engines incorporating solenoid heat shield, remove front bracket upper bolt and detach bracket from starter motor. Remove the front bracket bolt or nut and rotate bracket clear of work area, then lower starter from vehicle by lowering front end first — (hold starter against bell housing and sort of roll end-over-end). Reverse the removal procedure to install. Torque the mount bolts to 25—35 ft. lbs. first, then torque brace bolt. Check operation of starter on vehicle.

Disassembly — light duty models
Disconnect the field coil connector(s) from the motor solenoid Terminal. Remove through bolts. Remove Commutator end frame, field frame assembly and armature assembly from drive housing.

Remove overrunning clutch from armature shaft as follows: Slide two piece thrust collar off end of armature shaft. Slide a standard half-inch pipe coupling or other metal cylinder of suitable size (an old pinion of suitable size can be used if available) onto shaft so end of coupling or cylinder butts against edge of retainer. Tap end of coupling with hammer, driving retainer towards armature end of snap ring. Remove snap ring from groove in shaft using pliers or other suitable tool. If the snap ring is too badly distorted during removal, it may be necessary to use a new one when reassembling clutch. Slide retainer and clutch from armature shaft.

Disassemble brush rigging from field frame. Release "V" spring from slot in brush holder support. Remove support pin. Lift brush holders, brushes and spring upward as a unit. Disconnect leads from each brush. Repeat operation for other set of brushes.

Disassembly — Intermediate models
Note relative position of solenoid, lever housing, and nose housing so starter can be reassembled in same manner.

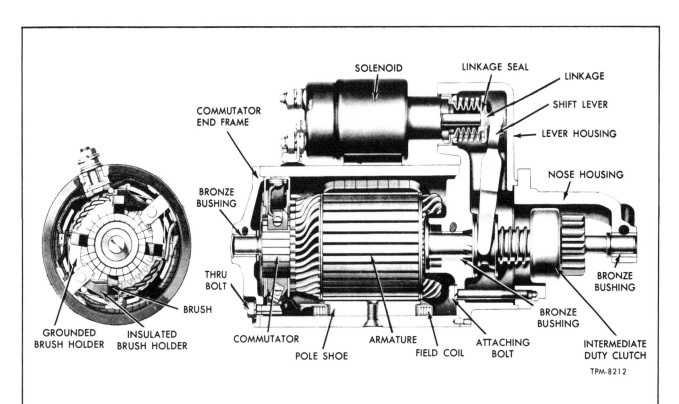

Intermediate duty overrunning clutch type starting motor

230

Disconnect field coil connector from solenoid "motor" terminal, and remove solenoid mounting screws. Remove two through-bolts and lock washers from commutator end frame, then remove commutator end frame from field frame and field frame from lever housing. Remove nose housing attaching bolts and separate nose housing from lever housing.

Slide a standard 1/2 inch pipe coupling or other metallic cylinder of correct size onto shaft so end of coupling or cylinder butts against edge of retainer. Tap coupling with hammer to drive retainer toward armature and off snap ring. Remove snap ring from groove in shaft, using pliers or other suitable tool. If snap ring is distorted during removal, it must be discarded and a new one obtained for assembly. Remove armature and clutch from lever housing. Separate solenoid from lever housing.

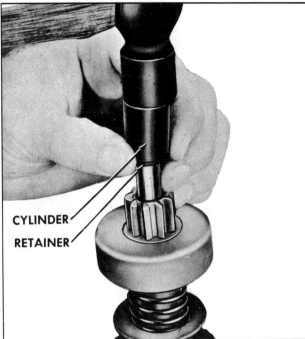

Driving retainer off snap ring

Cleaning and inspection
With the starting motor completely disassembled except for removal of field coils, the component parts should be cleaned and inspected as described below. Field coils need be removed only where defects in the coils are indicated by the tests described in this section, in which case the pole shoe screws should be removed and the pole shoes and field coils disassembled. Any defective parts should be replaced or repaired.

Clean all starting motor parts, but do not use grease dissolving solvent for cleaning the overrunning clutch, armature, and field coils since such a solvent would dissolve the grease packed in the clutch mechanism and would damage armature and field coil insulation. Test overrunning clutch action. The pinion should turn freely in the overrunning direction and must not slip in the cranking direction. Check pinion teeth to see that they have not been chipped, cracked, or excessively worn. Check the spring for normal tension and the drive collar for wear. If necessary the spring or collar can be replaced by forcing the collar toward the clutch and removing lock ring from end of tube.

Check brush holders to see that they are not deformed or bent, but will properly hold brushes against the commutator. Check the condition of the brushes and if pitted or worn to one-half their original length, they should be replaced. Check fit of armature shaft in bushing of drive housing. Shaft should fit snugly in the bushing. If the bushing is worn, it should be replaced. Apply a silicone lubricant to this bushing before reassembly. Avoid excessive lubrication.

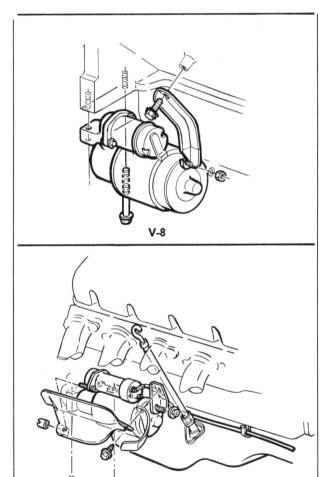

V-8

V-8 WITH SOLENOID HEAT SHIELD

231

Starter mounting installation

Starting motor parts layout (light duty)

1. Drive housing
2. Shift lever bolt
3. Shift lever nut and lock washer
4. Pin
5. Shift lever
6. Solenoid plunger
6A. Solenoid return spring
7. Solenoid case

8. Screw and lock washer
9. Grommet
10. Field frame
11. Through bolts
12. Thrust collar
13. Snap ring
14. Retainer

15. Overrunning clutch assembly
16. Armature
17. Braking washer
18. Commutator end frame
19. Brush springs
20. Washer
21. Insulated brush holders

22. Grounded brush holders
23. Brushes
24. Screws
25. Field coils
26. Insulators
27. Pole shoes
28. Screws

Check fit of bushing in commutator end frame. If this bushing is damaged or worn excessively, the end frame assembly must be replaced. Apply a silicone lubricant to this bushing before reassembly. Avoid excessive lubrication. Lubricant forced onto the commutator would gum and cause poor commutation with a resulting decrease in cranking motor performance.

Inspect armature commutator. If commutator is rough or out of round, it should be turned down and undercut.

Inspect the points where the armature conductors join the commutator bars to make sure that it is a good firm connection. A burned commutator bar is usually evidence of a poor connection. Some starter motor models use a molded armature commutator design and no attempt to undercut the insulation should be made or serious damage may result to the commutator. Undercutting reduces the bonding of the molding material which holds the commutator bars and since the molding material is softer than the copper bars, it is not necessary to undercut the material between the bars of the molded commutator.

Armature test for shorts

Check the armature for short circuit by placing on growler and holding hack saw blade over armature core while armature is rotated. If saw blade vibrates, armature is shorted. Recheck after cleaning between the commutator bars. If saw blade still vibrates, replace the armature.

Field coil open circuit test

Armature short circuit text

Armature test for ground

Place one lead on the armature core or shaft and the other on the commutator. If the lamp lights, the armature is grounded and must be replaced.

Field coil test for ground

Place one lead on the connector bar and the other on the grounded brush. If the lamp lights, the field coils are grounded. Be sure to disconnect the shunt coil before performing this test (when applicable).

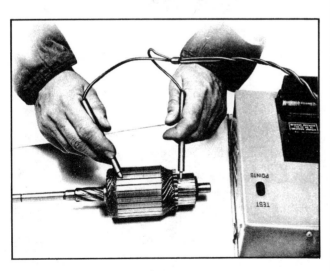

Armature ground test

Field coil test for open circuit

233 Place one lead on the insulated brush and the other to the field connector bar. If the lamp does not light, the field coils are open and will require replacement.

Field coil ground test

Loose electrical connections

When an open soldered connection of the armature to commutator leads is found during inspection, it may be resoldered provided resin flux is used for soldering. Acid flux must never be used on electrical connections.

Turning the commutator

When inspection shows commutator roughness, it should be cleaned by turning down commutator in a lathe until it is thoroughly cleaned. Do not cut beyond section previously turned.

Undercut insulation between commutator bars 1/32". This undercut must be the full width of insulation and flat at the bottom; a triangular groove will not be satisfactory. After undercutting, the slots should be cleaned out carefully to remove any dirt and copper dust. Sand the commutator lightly with No. 00 sandpaper to remove any slight burrs left from undercutting. Recheck armature on growler for short circuits.

Brush holder replacement

If brush holders are damaged, they can be replaced by special service units which are attached with screws and nuts.

Overrunning clutches

Roll clutches are designed to be serviced as a complete unit. Intermediate duty-sprag clutches are used on diesel engine starting motors. Service the unit as follows:

Remove and disassemble the starting motor. Using a snap ring pliers and prying tool, remove the snap ring retaining the pinion to the sleeve.

Remove the large snap ring retaining the sprags and pinion sleeve in the shell. Remove the flatwasher, feltwasher and stepped flatwasher from the shell. Stepped flatwasher is held in place by a rubber "O" ring. Hold clutch with pinion sleeve downward, rapping sleeve sharply to remove this washer. Use care not to remove sprags.

Removing tru-arc retaining ring

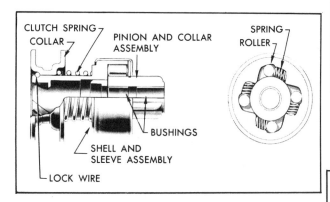

Roll type clutch cross-section

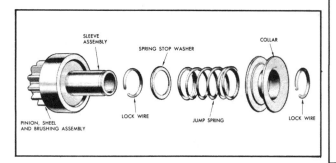

Intermediate duty sprag clutch

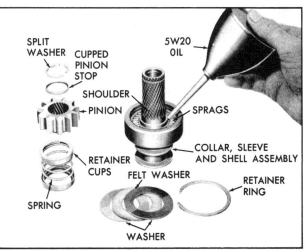

Lubricating sprag clutch

Lubricate the sprags and the feltwasher with No. 5W20 oil. Heavier oil must not be used. Reassemble the unit by reversing the above steps. The clutch should not slip when 2200 to 2400 inch lbs. are applied.

Assembly light duty models

After all parts have been thoroughly tested and inspected and worn or damaged parts replaced, the starter should be reassembled. Assemble brushes to brush holders. Assemble insulated and grounded brush holder together with the "V" spring and position as unit on the support pin. Push holders and spring to bottom of support and rotate spring to engage the "V" in slot in support. Attach ground wire to grounded brush and field lead wire to insulated brush. Repeat for other set of brushes.

Assemble overrunning clutch assembly to armature shaft. Lubricate drive end of armature shaft with silicone lubricant. Slide clutch assembly onto armature shaft with pinion outward. Slide retainer onto shaft with cupped surface facing end of shaft (away from pinion). Stand armature on end of wood surface with commutator down. Position snap ring on upper end of shaft and hold in place with a block of wood. Hit wood block a blow with hammer forcing snap ring over end of shaft. Slide snap ring down into groove. Assemble thrust collar on shaft with shoulder next to snap ring. Place armature flat on work bench, and position retainer and thrust collar next to snap ring. Then using two pair of pliers at the same time (one pair on either side of shaft), grip retainer and thrust collar and squeeze until snap ring is forced into retainer.

Lubricate the drive housing bushing with a silicone lubricant. Make sure thrust collar is in place against snap ring and retainer and slide armature and clutch assembly into place in drive housing engaging shift lever with

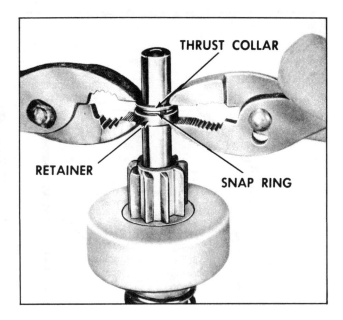

Forcing snap ring into retainer

clutch. Position field frame over armature and apply special sealing compound between frame and solenoid case. Position frame against drive housing using care to prevent damage to the brushes.

Lubricate the bushing in the commutator end frame with a silicone lubricant. Place leather brake washer on armature shaft and slide commutator end frame onto shaft. Reconnect the field coil connectors to the "motor" solenoid terminal. After overhaul is completed, perform "Pinion Clearance Check".

Assembly Intermediate models

Install solenoid to lever housing. Insert armature shaft through lever housing and install clutch assembly on shaft. Place retainer over armature shaft with cupped surface facing end of shaft. Install snap ring in groove of shaft. Place a suitable washer on top of snap ring and using two pair of pliers, force retainer over snap ring. REMOVE WASHER AFTER OPERATION. Assemble nose housing to lever housing in same position as before disassembling. Install field frame to lever housing and commutator end frame to field frame, using thru bolts. Install solenoid mounting screws and connect field coil connector to solenoid "motor" terminal. After overhaul is completed, perform "Pinion Clearance Check".

Pinion clearance check

Connect a battery, of the same voltage as the solenoid, from the solenoid switch terminal to the solenoid frame or ground terminal. Disconnect the motor field coil connector for this test. Momentarily flash a jumper lead from the solenoid motor terminal to the solenoid frame or ground terminal. The pinion will now shift into cranking position and will remain there until the battery is disconnected. Push the pinion back towards the commutator end to eliminate slack movement. Measure the distance between the pinion and pinion stop. If clearance

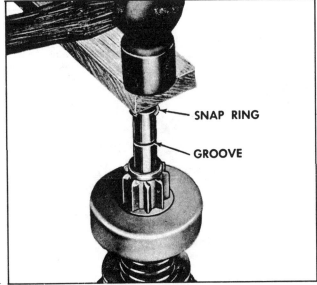

Forcing snap ring over shaft

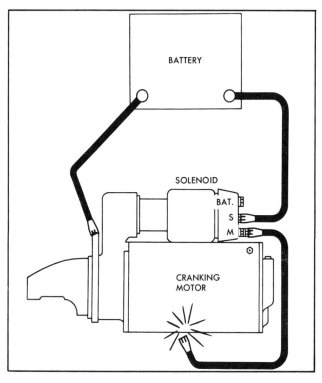

Circuit for checking pinion clearance

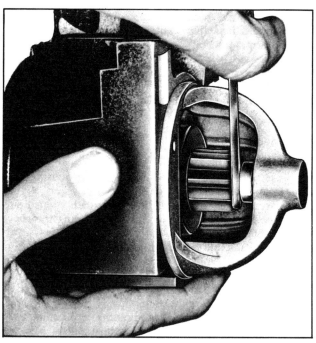

Checking pinion clearance

is not within specified limits (.010—.140) it may indicate excessive wear of solenoid linkage shift lever yoke buttons or improper assembly of the shift lever mechanism. Worn or defective parts should be replaced.

STARTING SOLENOID
Removal
Remove the outer screw and washer from the motor connector strap terminal. Remove the two screws retaining solenoid housing to end frame assembly. Twist solenoid clockwise to remove flange key from keyway slot in housing; then remove solenoid assembly.

Replacement of contacts
With solenoid removed from motor, remove nuts and washers from switch and motor connector strap terminals. Remove the two solenoid end cover retaining screws and washers and remove end cover from solenoid body. Remove nut and washer from battery terminal on end cover and remove battery terminal. Remove resistor by-pass terminal and contactor. Remove motor connector strap terminal and solder new terminal in position. Using a new battery terminal, install terminal washer and retaining nut to end cover. Install by-pass terminal and contactor.

Position end cover over switch and motor terminals and install end cover retaining screws. Also install washers and nuts on the solenoid switch and starting motor terminals. Bench test solenoid for proper operation.

Installation
With solenoid return spring installed on plunger, position solenoid body to drive housing and turn counterclockwise to engage the flange key in the keyway slot. Install two screws retaining solenoid housing to end frame. Install outer screw and washer securing motor connector strap terminal. Install starter motor as previously described.

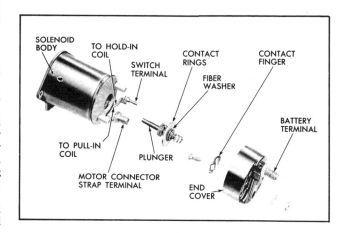

Exploded view of solenoid

BODY ELECTRICAL 14

INDEX

14

ELECTRICAL BODY AND CHASSIS
Lighting system
Description

It is recommended that the battery be disconnected before performing any electrical service other than bulb or fuse replacement.

The lighting system includes the main lighting switch, stop light, dimmer and back lamp switches, headlamps, parking, side marker, stop, tail and directional lamps. It also includes instrument illumination, indicator lamps and the necessary wiring to complete the circuits. Vacuum system components for headlamp operation are now included. A fuse panel provides convenient power taps and fuse clips for the appropriate circuits. Vacuum actuated headlamp covers are standard.

The side marker lamps have clear bulbs (with amber lenses) in the front fenders and with red lenses in the quarter panels. In addition the parking and marker lamps will operate in the headlamp switch "park" or "headlights" position the same as license and tail lamps. Marker lamps are part of the front and rear lighting harnesses. Power windows can be opened or closed with the ignition switch in the "ACC", accessory, position or the "ON" position.

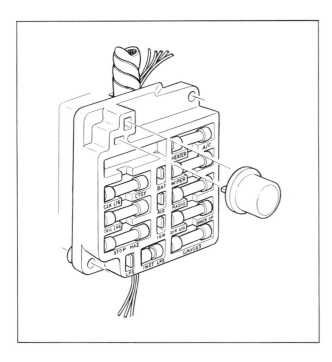

Fuse panel assembly

The fiber optic monitoring system is standard on Corvettes. Fiber optic conductors are attached to light source housings remotely located from the viewing lens. The monitoring lens will glow when the EXTERIOR lights are lit. Backup and marker lamps are not monitored. The fiber optic conductors are multi strand coated plastic bundles, of a precision length having a crimped ferrule on each end. In addition to fuses, the wiring harness incorporates fusible links to protect the wiring. Links are used rather than a fuse in wiring circuits that are not normally fused, such as the ignition circuit. Fusible links in the Chevrolet wiring are four gauge sizes smaller than the cable it is designed to protect. The links are marked on the insulation with wire gauge size because of the heavy insulation which makes the link appear a heavier gauge than it actually is.

Engine compartment wiring harnesses incorporate several fusible links. The same size wire with special hypalon insulation must be used when replacing a fusible link. The links are: The pigtail lead at the battery positive cable is installed as a molded splice at the solenoid "Bat" terminal and servicing requires splicing in a new link. A 16 gauge black fusible link is located at horn relay to protect all unfused wiring of 12 gauge or larger. It is a serviceable piece with an in-line connector and is not integral with the wiring harness. The generator warning light and field circuitry (16 gauge wire) is protected by a fusible link (20 gauge orange wire) used in the "battery feed to voltage regulator #3 terminal" wire. The link is installed as a molded splice in the circuit at the junction block or the solenoid "Bat" terminal and at the horn relay. Each link is serviced by splicing in a new 20 gauge wire as required.

The ammeter circuit on all models is protected by two orange, 20 gauge wire fusible links installed as molded splices in the circuit at the junction block or the solenoid "Bat" terminal and at the horn relay. Each link is serviced by splicing in a new 20 gauge wire as required.

The wiring harnesses use a standardized color code common to all Chevrolet vehicles. Under the color code, the color of the wire designates a particular circuit. The harness title indicates the type of harness, single or multiple wire, and also describes the location of the harness. The body harness is routed through the vehicle near the center of the body. Composite wiring diagrams are included at the end of this section.

Maintenance and adjustments

Maintenance of the lighting units and wiring system consists of an occasional check to see that all wiring connections are tight and clean, that the lighting units are tightly mounted to provide good ground and that the headlamps are properly adjusted. Loose or corroded connections may cause a discharged battery, difficult starting, dim lights, and possible damage to generator and regulator. Wire harnesses must be replaced if insulation becomes burned, cracked, or deteriorated. Whenever it is necessary to splice a wire or repair one that is broken, always use solder to bond the splice. Always use rosin flux solder on electrical connections. Use insulating tape to cover all splices or bare wires.

When replacing wires, it is important that the correct size be used. Never replace a wire with one of a smaller gauge size. Each harness and wire must be held securely in place by clips or other holding devices to prevent chafing or wearing away the insulation due to vibration. By referring to the wiring diagrams, circuits may be tested for continuous circuit or shorts with a conventional test lamp or low reading voltmeter.

HEADLAMP ADJUSTMENT – T-3 HEADLAMPS
Headlamp aiming
The headlamps must be properly aimed in order to obtain maximum road illumination and safety that has been built into the headlighting equipment. With the Guide T-3 type sealed beam units, proper aiming is even more important because the increased range and power of this lamp make even slight variations from recommended aiming hazardous to approaching motorists. The headlamps must be checked for proper aim whenever a sealed beam unit is replaced and after an adjustment or repairs of the front end sheet metal assembly.

Regardless of method used for checking headlamp aim, car must be at normal weight, that is, with gas, oil, water, and spare tire. Tires must be inflated to specified pressures. If car will regularly carry an unusual load in rear compartment, or a trailer, these loads should be on car when headlamps are checked. Some states have special requirements for headlamp aiming adjustment and these requirements should be known and observed. Horizontal and vertical aiming of each sealed beam unit is provided by two adjusting screws which move the mounting ring in the body against the tension of the coil spring. There is no adjustment for focus since the sealed beam unit is set for proper focus during manufacturing assembly.

HEADLAMP PANEL (66-67)
Alignment procedures
The headlamp panel travel is limited by two adjusting screws which are located on the arms of the shaft mounted stop. Raise hood and as a safety precaution install a bolt through the hole in the hood support – secure bolt with a nut. Adjusting screw (A) limits headlamp panel travel in open position – adjust this screw so that mounting face of panel is within 2 degrees of vertical in the fully open position. Each headlamp operates independently of the other, therefore individual adjustment is required for each panel.

Adjusting screw (B) limits headlamp panel travel in closed position – adjust this screw so that panel is flush to upper body panel in the fully closed position. Lock both screws by tightening lock nut against stop. Remove safety bolt as installed in Step 1 and close the hood.

Removal
Remove engine compartment hood. Actuate headlamp panel to the open position. In the event headlamp motor is inoperative, manual positioning of the panel can be accomplished by turning the knurled knob at inboard end of motor. As an assist in manual operation of panel, apply light hand pressure to panel in desired direction of rotation.

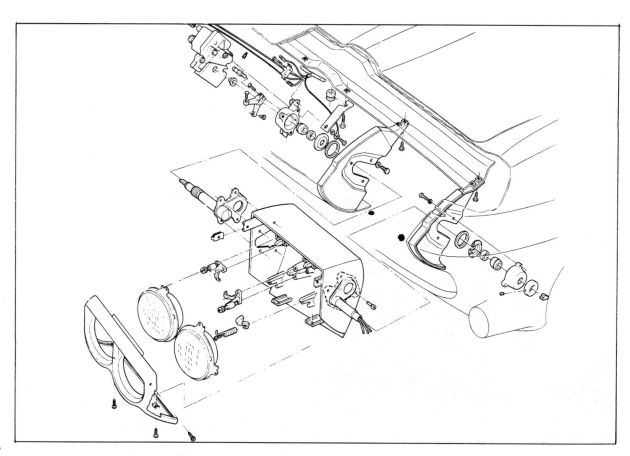

239

Headlamp, headlamp panel, and motor details

Remove positive lead from battery terminal. Remove headlamp bezel. Remove the sealed beam housing unit as an assembly. Remove motor from panel pivot shaft. Then remove the panel stop from panel inboard pivot shaft, and disconnect switch lead wires from panel, support-mounted motor switch.

Remove the panel retaining bolt access hole plugs from inside the panel then rotate panel as required and remove the retaining bolts and slide supports from ends of panel pivot shaft. Remove bearing, felt seal, retainer and washer from inboard pivot shaft. Remove panel from its location by alternately disengaging pivots from their retaining slots and withdrawing unit forward through opening in body

Installation
Loosen allen screw in spacer and disassemble parts from the panel outboard pivot shaft. Install washer, retainer, felt seal, bearing and spacer, in that order, to panel outboard pivot shaft. Do not tighten spacer on shaft at this time. Position panel in body opening and index panel pivot shafts in retaining slots; then loosely install support retaining bolts. Install washer, retainer, felt seal, bearing and support, in that order, to panel inboard pivot shaft, and loosely install support retaining bolts.

Check side-to-side alignment of panel, making sure that there is no panel-to-body contact; position spacer snugly against bearing; then tighten spacer seat screw to 30-50 in. lbs. Install stop on panel inboard shaft so that it rests against bearing, being sure that index mark on stop is aligned with flat on pivot shaft. Install stop lock bolt and torque bolt to 45-60 in. lbs. Make sure that side-to-side panel alignment is not changed when installing and tightening the stop.

Tighten panel pivot support bolts snugly but still allowing panel to be moved by applying hand pressure. Position panel to the closed position and align with body so that all surfaces are flush. With the panel in the closed position

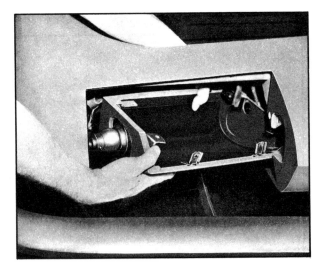

Headlamp panel removal

the outboard access hole is aligned with the forward bolt head. Tighten this bolt with the panel in the closed position – access to bolt head can be obtained by working through opening between the hood and body. Rotate panel to obtain access to each bolt head and torque bolts to 100-140 in. lbs. Connect switch lead wires to support-mounted limit switch.

Install sealed beam housing unit and at the same time position sealed beam lead wires through panel outboard pivot shaft. Install housing unit retaining screws and install lead wires in connector, making sure to match colors between harness and connector. Install motor assembly. Connect positive lead to battery terminal. Adjust headlamp panel. Adjust headlamp aiming. Install headlamp bezel and engine compartment hood.

HEADLAMP PANEL MOTOR
Removal
Raise hood and as a safety precaution install a bolt through the hole in the hood support – secure bolt with a nut. Remove positive lead from battery terminal. Disconnect motor lead wire and position it out of the way. Turn the knurled knob at inboard end of motor until the gear seems to turn freely – turn the knob in one direction until a definite drag is experienced then rotate knob approximately six complete turns in the opposite direction. This is necessary to produce a no-load condition on the drive gear and to permit separation of motor from the panel pivot shaft. Remove the retainer from groove in motor locating stud then remove the motor-to-support retaining screw and remove motor assembly from the vehicle.

Installation
Rotate motor and headlamp panel as required to align slot in motor with headlamp panel pivot shaft, and install motor on shaft so that the bracket is aligned with the locating stud. It may be necessary to turn knurled-knob on end of motor to permit alignment of motor with shaft. Slide motor onto shaft until it seats against shaft shoulder; then install retainer in groove on locating stud. Install motor-to-support retaining screw making sure ground wire is installed between screw head and bracket. Connect motor lead wires, making sure that contacts are clean and that connection is secure. Connect positive lead battery terminal, remove bolt from hood support, close hood and check operation of headlamp panel.

HEADLAMP PANEL (68 – 72)
Alignment procedure
"In-out" – loosen screws fastening slotted bracket to underside of headlamp housing assembly.

"Closed" – lamp cover top to opening; by turning hex head screw fastened to top of pivot link.

"Open" fully extended actuator with rod. Remove spring from actuator rod pin. Remove cotter pin from rod pin. Turn actuator rod until bushing hole aligns with forward end of slot in connecting link extended position, with engine idling for vacuum. Shut off engine, retract actuator rod and unscrew rod ½ turn to preload actuator rod in link.

"Up" (bezel to opening alignment) loosen jam nut and turning bumper covered screw up or down to touch then up 1½ turns more. Micro switch on linkage must shut off warning lamp when lights are fully extended. The headlamp housing must be properly aligned before headlamps are aimed.

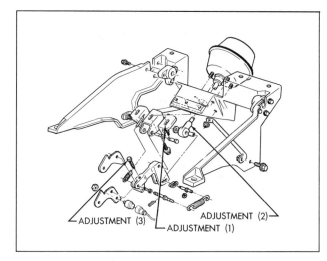

Headlamp panel adjustment

SERVICE OPERATIONS
Front lighting
Headlamp replacement (68 — 72)
Open the headlamp panel to the open position. Remove headlamp bezel retaining screws and remove bezel. Disengage spring from the retaining ring and remove two attaching screws. Remove retaining ring, disconnect sealed beam unit at wiring connector and remove the unit. Attach connector to replacement unit and position unit in

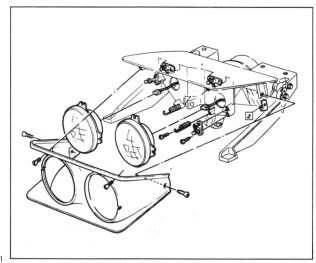

Headlamp assembly

place making sure the number molded into the lens face is at the top. In the dual headlamp installation the inboard unit (No. 1) takes a double connector plug, the outboard unit (No. 2) takes a triple connector plug. Position retaining ring into place and install the retaining ring attaching screws and spring. Check operation of unit and install the headlamp bezel.

VACUUM OPERATED HEADLAMP SERVICE
Headlamp door
Replacement
Actuate door open. Remove radiator decorative grille and attaching screws. Remove headlamp bezel and apply protective tape to headlamp door opening. Remove headlamp assemblies inner and outer by removing springs with a hooked wire or small vise grips. Remove front screw and bushing attaching headlamp housing to support assembly link. Remove "J" bar from side of headlamp housing extending downward. Reaching in through grille opening, remove three (3) cap screws retaining bearing and headlamp housing assemblies to support assembly.

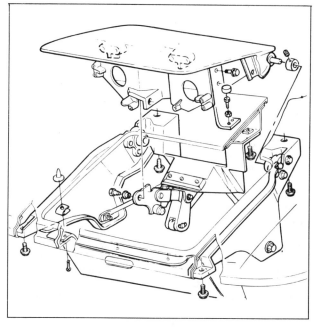

Headlamp door

Lift headlamp door assembly forward and out of opening. Transfer remaining door attachments to new door assembly and install assembly in the reverse order of removal.

ACTUATOR ASSEMBLY
Replacement
Remove radiator decorative radiator grille and screws. Actuate headlamps partially open and remove long spring on either side of pivot link pin. Remove two (2) vacuum hoses attached to actuator. Remove cotter pin and slide out pivot pin, freeing actuator rod. Remove four (4) nuts

from actuator studs and slide actuator down and out grille opening. Reverse removal procedure when replacing actuator.

VACUUM RELAY
Replacement

Raise hood. Gently remove four (4) hoses from relay. Remove two (2) screws attaching vacuum relay and lift relay from engine compartment. Install vacuum relay in reverse order of removal and check operation of headlamp covers with applied vacuum. Reinstall color coded hoses on proper fittings. Connectors have molded R-Y-G letters and color striped hoses.

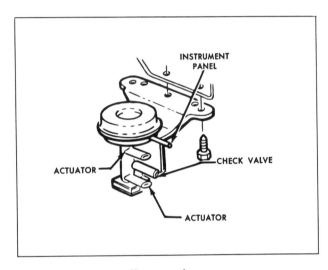

Vacuum relay

VACUUM TANK
Replacement

Raise hood. Remove power brake booster and master cylinder. Remove vacuum hoses from tank noting color stripe and location of hoses. Tape hose ends to keep out contaminants. From underside of left fender skirt remove screw assemblies retaining vacuum tank. Reaching inside engine compartment, lift out tank with integral brackets. Install tank in the reverse order of removal procedure. Reinstall power brake assembly.

Vacuum hose routing

Many vacuum hoses are color stipped and bundled on Corvettes for ease of assembly. Replacement hoses must not be pinched in routing. Actuator to relay valve hoses are an exact length for each side. Vacuum hoses connect the following headlamp system components: manifold vacuum to check valve, check valve to vacuum tank, check valve to headlamp switch, headlamp switch to relay valve, vacuum tank to relay valve, relay valve to closed side of actuator, relay valve to open side of actuator and manual control valve to relay valve.

Lamp housing

Remove the lamp assembly retaining screws and remove lamp assembly from mounting location (rear bumper

center face bar most models.) Disconnect lamp wiring from chassis harness connector (inside trunk on some models). Force grommet and wiring down through body opening (where applicable) and remove assembly from vehicle. To install replacement unit, insert grommet and wiring through body opening where applicable and connect wiring to chassis harness. Position assembly and install retaining screws. Make sure ground wire is properly installed. Check operation of lamp assembly.

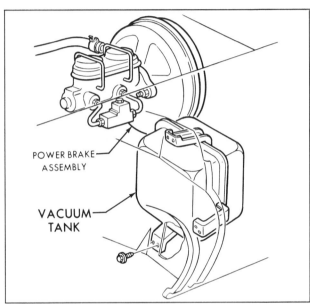

Vacuum tank

Ignition switch

Due to the integral design relationship of the ignition lock and switch with the energy absorbing steering column service procedures for the ignition lock and switch are covered in Section 9, Steering.

Lighting switch — replacement

Disconnect battery ground cable at battery. Remove screws securing mast jacket trim covers and remove covers for access to steering column to instrument panel mounting bracket. Unclip and remove the left side console forward trim panel so that left side instrument cluster may be lowered for access to the light switch in latter steps. Lower steering column. Remove screws and washers securing left instrument panel to door opening, top of dash, and left side of center instrument panel. Pull cluster assembly down and then tip forward for access to the light switch. Depress the switch shaft retainer and remove the knob and shaft assembly. Remove the switch retaining bezel. Disconnect vacuum hoses from the switch, tagging them for assembly. Pry the connector from the switch and remove the lighting switch.

Wiper-washer control

Disconnect battery ground cable. Remove screws from upper portion of center console labeled "Corvette". Remove wiring connector from switch. Remove switch and plate from center console. Grasping the switch arm with a small pair of lock pliers, gently pry the knob off switch then remove screws securing the switch. The first step in replacing the switch is to insert a small rod in the switch arm before attempting to push the knob on the arm outside of the trim plate then follow steps in the reverse order of removal.

STOP LIGHT SWITCH REPLACEMENT
Removal

Disconnect wiring harness connector at switch. Remove retaining nut and unscrew switch from bracket.

Installation

Install stop light switch through bracket and install retaining nut. Plug connector onto switch. Check switch for proper operation. Electrical contact should be made when the brake pedal is depressed 1/4" to 5/8" from fully released position. The stop light switch bracket has a slotted screw hole for switch adjustment.

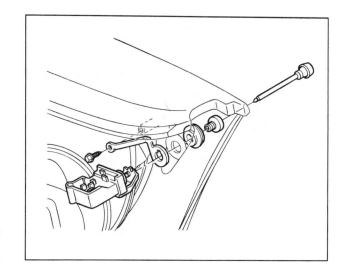

lighting switch

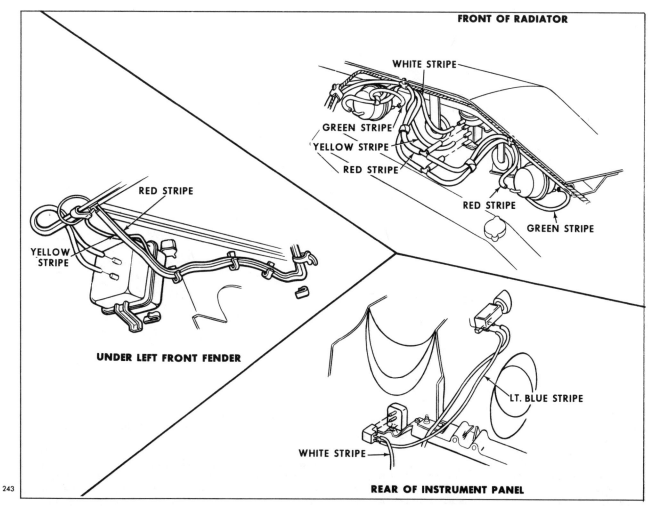

Vacuum hose routing

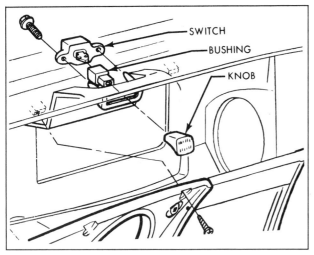

Wiper washer control

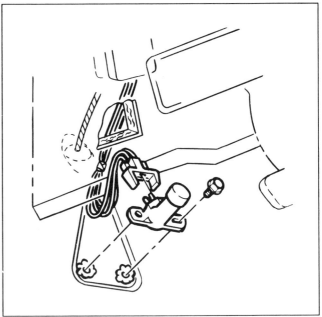

Dimmer switch (typical)

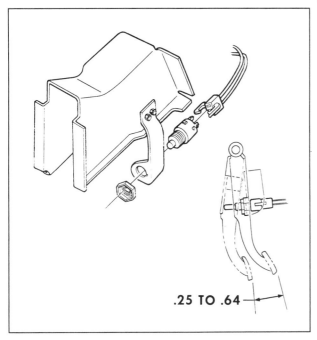

Light switch

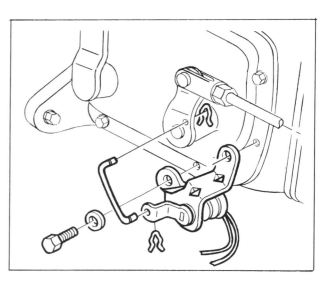

Back-up switch

Dimmer switch replacement

Fold back upper left corner of front floor mat, disengage connector lock fingers, and disconnect multi-plug connector from dimmer switch.

Remove two screws retaining dimmer switch to toe pan. Connect multi-plug connector to new switch and check operation. Install new switch to toe pan with two screws. Replace floor mat.

BACKING LAMP SWITCH REPLACEMENT
Located on transmission

Place vehicle on hoist. Disconnect switch wiring from harness wiring at in-line connector. Remove bolt retaining wiring attaching clip to transmission. Remove wire clip retaining reverse lever rod to switch. Remove screws retaining switch and shield assembly to transmission, remove switch. Do not remove transmission-to-bracket retaining bolts. To install, reverse removal procedure and check switch for operation in transmission reverse range. Remove vehicle from hoist.

NEUTRAL SAFETY SWITCH
Replacement

Disconnect shift control lever arm from the control rod. Remove the shift control knob. Remove Trim Plate retaining screws and trim plate assembly. Remove control assembly retaining screws. Remove control assembly from the seal. Remove switch retaining nuts and switch from the control assembly. To install, position gearshift in "Drive" position align hole in contact support with hole in switch and insert pin (3/32" dia.) to hold support in place. Place the contact support drive slot over the drive tang and tighten switch mounting screws, then remove pin. Connect wiring harness to switch wiring, and check operation of switch. Install trim plate assembly and shift lever control knob. Connect shift lever arm to the transmission control rod.

Clutch operated neutral start switch

The clutch pedal must be fully depressed and the ignition switch in thhe START position for vehicle starting. The switch mounts with a tang and screw to the pedal mounting bracket. The switch lever is moved by a wire link attached to the clutch pedal arm.

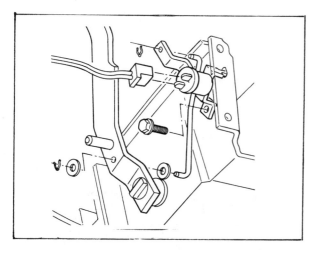

Clutch operated neutral start switches

Removal

Unplug electrical connector from switch. Remove retainer from link on clutch pedal arm. Remove screw fastening switch to pedal support and lift off switch.

Installation

Slide switch tang, into pedal support. Secure switch to pedal support with screw, switch lever over link in pedal arm. Install link in switch lever then fasten with retainer. Install electrical connector onto switch.

Parking brake alarm switch replacement

Remove parking brake cover assembly from rear center console. Disconnect wire from brake alarm switch. Remove retaining screw and switch from reinforcement.

Position new switch to reinforcement and install retaining screws. Connect switch wire and check operation of switch. Install cover assembly.

Parking brake switch

INSTRUMENT PANEL COMPARTMENT
Lamp/switch replacement

Disconnect battery ground cable. Reach into glove box, depress bulb in end of switch and turn counter-clockwise to remove bulb. Remove switch from socket. Carefully detach wire and terminal from switch. On some model switches, the wire and terminal cannot be detached from the switch making it necessary to cut and splice the switch wire.

Insert wire and terminal into new switch. Push switch into place and install bulb by setting it in place, depressing and turning it clockwise.

Cigarette lighter replacement

Disconnect ground cable from battery. Remove vent control set screws and gear shift trim plate secured by screws. Compress clip for lighter lamp and unplug connector from beneath lighter housing. Unscrew retainer from rear of housing assembly and disengage lighter unit from trim plate. To install, reverse removal procedure.

LEFT HAND SIDE
Removal and installation

Disconnect ground cable from battery. Lower steering column. Remove screws and washers securing left instrument panel to door opening, top of dash and left-side of center instrument panel. Unclip and remove floor console left forward trim panel. Pull cluster assembly slightly forward to obtain clearance for removal or speedometer

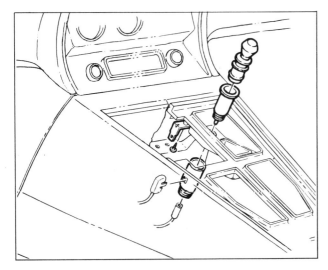

Cigar lighter installation

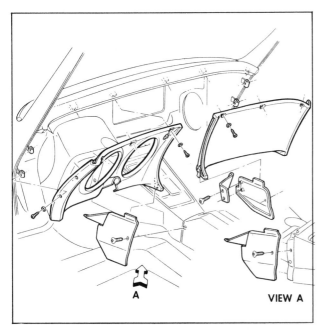

Instrument panels – lower and trim

cable housing nut, tachometer cable housing nut, head-lamp and ignition switch connectors and panel illuminating lamps. Install by reversing removal procedure, being careful not to kink the speedometer or tachometer cable housings.

Center instrument cluster

Disconnect ground cable from battery. Remove wiper switch trim plate screws and tip plate forward for access to remove switch connector. Lift trim plate out from cluster assembly. Unclip and remove right and left console

forward trim pads to gain access to studs at lower edge of cluster assembly. Remove nuts from studs at lower edge of cluster. Remove remaining screws retaining cluster assembly to instrument panel. Remove right instrument panel pad. Remove radio knobs, bezel retaining nuts and one radio support bolt (from behind cluster assembly). Slide radio back towards firewall and pull cluster forward. Reaching behind cluster, disconnect oil pressure line, wiring harness, and bulbs. The center instrument cluster trim plate is designed to collapse under impact. Consequently, do not try to deflect the cluster plate forward to gain more access to back of gauges. Lift cluster assembly up and forward for removal. To install cluster assembly in the dash, reverse "Removal" procedure.

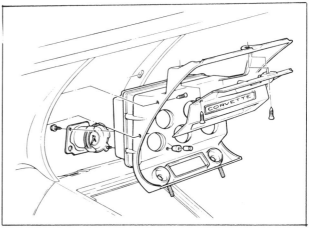

Center instrument cluster

SPEEDOMETER AND/OR TACHOMETER
Removal and installation

The left-hand cluster must be removed from the vehicle to service the speedometer or tachometer head assembly. With the cluster removed from the vehicle, remove odometer reset connector and speedometer connector from rear of speedometer head. Remove the three screws fastening the speedometer or tachometer head to back of cluster and carefully remove speedometer or tachometer head. Transfer wiper solenoid to rear of tachometer and or speedminder buzzer to the rear speedometer head and fasten appropriate head to cluster bezel. Connect odometer and/or speedminder reset cable to rear of speedometer head. Connect the left-hand cluster as previously described in this section.

Cable replacement or lubrication

Disconnect the cable from the speedometer or tachometer head. Remove the old cable by pulling it out from speedometer end of conduit. If old cable is broken it may be necessary to remove lower piece from transmission or distributor end of conduit as applicable. Lubricate the lower 3/4 of cable with AC speedometer cable lubricant and push the cable into the conduit. Connect the upper end to the speedometer or tachometer head and road test

246

vehicle for proper operation. DO NOT KINK CABLE HOUSINGS. It will be necessary to remove the center instrument panel as previously outlined before further proceeding with instrument removals.

Fuel gauge

The gasoline fuel gauge circuit consists of an electrical indicator in the instrument cluster and a float-controlled rheostat in the fuel tank. Since the fuel gauge consists of two remotely located units and connecting wires, it is sometimes difficult to determine which unit is at fault when the gauge fails to operate properly. Center gauge pack removal is necessary to replace the fuel gauge. Be sure to check gas gauge fuse in fuse panel before attempting to trouble shoot for inoperative gauge or tank sending unit.

Temperature gauge

The temperature indicator requires very little service other than testing for malfunctioning and replacing defective units. Cluster must be removed to service temperature gauge. Do not attempt to repair either the engine unit or the gauge. When installing new engine unit, do not use thread compound on unit threads, as this will increase electrical resistance of unit and cause faulty reading on gauge.

Ammeter or oil pressure gauge

The ammeter or oil pressure gauge requires very little attention other than keeping ammeter terminals clean and tight. If the oil pressure control line should become restricted it should be blown out or replaced. The cluster must be removed to service these gauges.

CLOCK
Replacement

Remove clock set shaft knob retaining screw and knob from set shaft. Remove screws attaching clock to rear of housing and remove clock from rear of cluster housing. Install clock in the reverse order of removal.

Seat belt and door ajar indicators

Seat belt indicator is a thermal switch with a push reset. The switch is powered from the fuel gauge terminal. The switch is replaced as a unit. Door ajar indicator is a bulb with a remote door jamb switch located on the lock pillar.

Headlamp up indicator

An indicator in the center cluster will glow when the headlamps are turned on, but not fully extended into the locked position. The indicator is signaled by a switch on each of the headlamp linkages.

Instrument cluster lamp replacement

Loosen either left hand or center instrument cluster as previously described in this section. Reach behind cluster and grasp lamp socket and wiggle socket from cluster. Remove and replace illumination bulb, push lamp socket back into cluster and reinstall instrument cluster in place.

WINDSHIELD WIPERS AND WASHERS
Headlamp washers

The headlamp washer system (standard on Corvette) only operates with the headlamps turned on and the windshield washer button held in the depressed position. A solenoid mounted to the wiper motor diverts washer fluid through hoses to the nozzles mounted on the low beam headlamp bezel.

WIPER TRANSMISSION ASSEMBLY
Removal and installation

Make sure wiper motor is in Park position. Open hood and disconnect battery ground cable. Remove rubber plug from front of wiper door actuator then insert a screw driver, pushing internal piston rearward to actuate wiper door open. Remove wiper arm and blade assemblies from the transmission. On articulated left-hand arm assemblies, remove carburetor type clip retaining pinned arm to blade arm. Remove plenum chamber air intake grille, or screen, if so equipped.

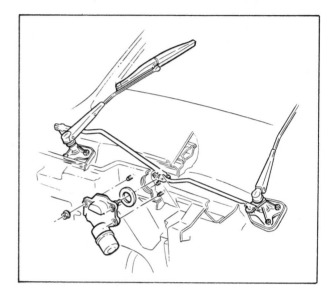

Wiper and motor linkage

Loosen nuts retaining drive rod ball stud to crank arm and detach drive rod from crank arm. Remove transmission retaining screws, or nuts, then lower and drive rod assemblies into plenum chamber. Remove transmission and linakge from plenum chamber through cowl opening. To install, reverse removal procedure. Make sure wiper blade assemblies are installed in the Park position (+ 3/8" from stop of reveal moulding on recessed wiper arms).

WIPER MOTOR ASSEMBLY
Disassembly

Motor section may be disassembled independently of the gear box.

Removal

Make sure that the wiper motor is in the park position. Disconnect washer hoses and electrical connectors from the wiper motor assembly. Remove the plenum chamber grille. Remove the nut which retains the crank arm to the motor assembly. Remove the ignition shield and distributor cap to gain access to the motor retaining screw assemblies or nuts. Remove left bank secondary leads from the cap and mark both cap and leads for aid in reinstallation. Remove the three motor retaining screw assemblies or nuts and remove the motor.

Installation
Check sealing gaskets at motor and retaining screws. Replace if necessary. Make sure that the wiper motor is in the park position prior to installation. Reverse "Removal"

Depressed park 2-speed wiper
The Type "C" two speed electric wiper assembly incorporates a depressed park type (blades park against step on windshield lower molding when the motor is turned off) motor and gear train. The wiper has a compound wound 12 volt motor and a gear box section containing the gear mechanism and relay control. The motor armature is fitted with a worm gear which drives the main gear assembly and crank arm. The relay control, consisting of a relay coil, relay armature and switch assembly, is located in the gear box section and controls the starting and stopping of the wiper through a latching mechanism. An electric washer pump is mounted on the gear box section of the wiper and is driven by the wiper unit gear assembly.

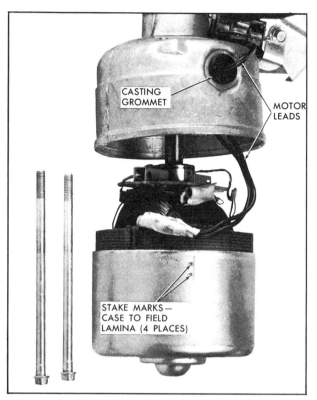

Wiper motor separation

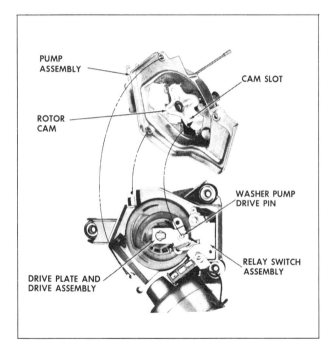

Depressed park wiper and washer mechanism

Brush plate and circuit breaker removal
Scribe a reference line along the side of the casting and end cap to insure proper reassembly. Remove the two motor thru bolts. Feed exposed excess length of motor leads thru the casting grommet and carefully jack the case and field assembly plus the armature away from the casting. It may be necessary to remove the armature end play adjusting screw and insert a rod thru the opening in order to apply pressure against the end of the armature. Unsolder the black lead from circuit breaker. Straighten out the 4 tabs that secure the brush plate to the field coil retainers. Be careful not to break any of the retainer tabs.

Install "U" shaped brush retainer clip over brush holder that has brush lead attached to circuit breaker. Holding the opposite brush from that retained in Step 6, carefully lift the brush holder off the mounting tabs far enough to clear the armature commutator. Allow the brush to move out of its holder. Remove the brush spring and lift the brush holder off the armature shaft.

Armature removal
Follow brush plate removal. Lift armature out of case and field assembly. Remove thrust ball from end of armature shaft as required and save for reassembly. Thrust ball may be easily removed with a magnet.

Case and field assembly removal
Remove brush plate and armature. The end case and field assembly is serviced as a unit. To free the field and case assembly, cut the solid black and black with pink stripe leads in a location convenient for splicing — preferably near the wiper terminal board. Refer to figure 25. Remove steel thrust plate and rubber disc from case bearing as required.

Assembly
If new field and case assembly is being installed, splice the black and black with pink stripe leads of the new field with the corresponding leads of the wiper. Install the rubber thrust disc, steel thrust disc and felt lubricating washer in the case assembly bearing in the order indicated. Lubricate end of armature shaft that fits in case bearing.

248

Next, install thrust ball in end of shaft. Assemble armature in the case and field assembly. Position the partially assembled brush plate over the armature shaft far enough to allow re-assembly of the remaining brush in its brush holder; then position the brush plate assembly on the mounting tabs in the position shown in figure 26. Circuit breaker ground lead will not reach circuit breaker terminal if brush plate is positioned wrong.

Center the brush plate mounting holes over the mounting tabs and bend the tabs toward the brush holders as required to secure the brush plate in position. Be sure tabs are centered in brush plate mounting holes. Remove brush retainer clips and resolder circuit breaker ground lead to circuit breaker. If new case and field assembly is used, scribe a line on it in the same location as the one scribed on the old case. This will insure proper alignment of the new case with the scribed line made on the housing. Position armature worm shaft inside the housing and, using the scribed reference marks, line up as near as possible the case and field assembly with the housing. Maintaining the armature in its assembled position in the case, start the armature worm shaft through the field and housing bearing until it starts to mesh with the worm gear. At the same time carefully pull the excess black and black with pink stripe leads thru the housing grommet. It may be necessary at this point to rotate armature slightly before the armature worm will engage with worm gear teeth. Rotate the case as required to align the bolt holes in the case with those in the housing. Secure the case to the housing with the two tie bolts. Adjust armature end play.

GEAR BOX SECTION

The gear box section is subdivided into two areas, (A) the relay control and latching mechanism and (B) the drive gear mechanism.

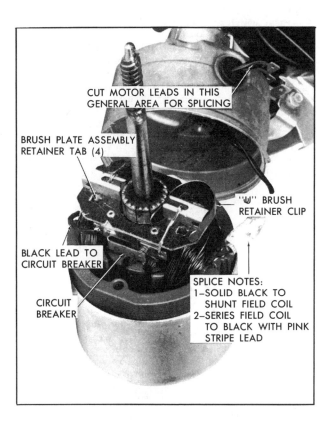

Circuit breaker

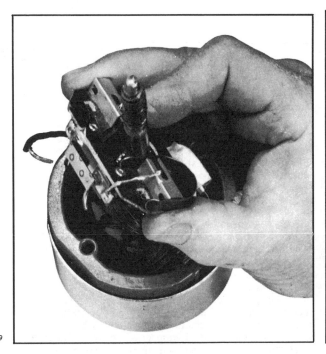

Removing brush holder

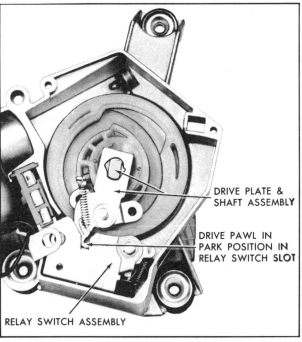

Drive pawl in full park position

A – Relay switch and latch assembly
Terminal board removal

Remove screws retaining washer pump assembly to wiper unit. If wiper pawl is in full park position (drive pawl located in magnetic switch assembly slot). It becomes necessary to remove gear assembly first (see Gear Assembly Removal). If wiper gear mechanism is not in park position (drive pawl away from latch arm), proceed. Remove relay-switch attaching screw and carefully lift the relay-switch assembly out of the gear box. Unsolder leads from switch terminals as required. To remove terminal board assembly simply slide it out of housing and unsolder leads as required.

Relay switch – latch and terminal board assembly

Resolder leads to wiper terminal board as required. Slide terminal board into wiper housing being careful to position the terminal board resistor lead.

Resolder leads to relay switch assembly as required. Position relay-switch assembly in housing. Be very careful to route leads in such a manner as to avoid having them pinched between relay and wiper housing. Install relay-switch mounting screw. Assemble gear box cover and washer pump assembly to wiper being careful that the ground strap is properly connected.

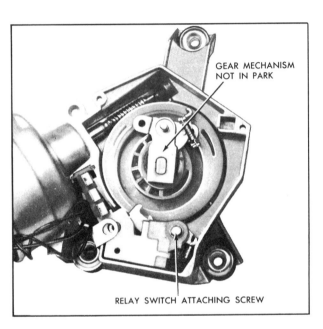

Drive gear NOT in park position

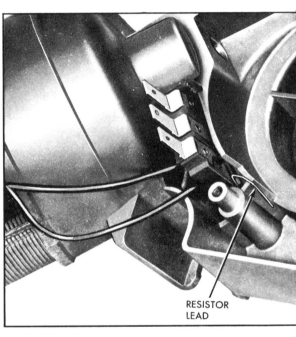

Terminal board resistor

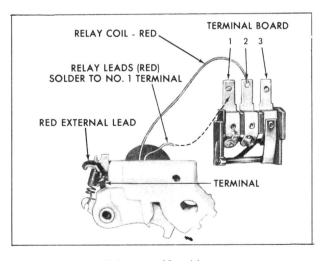

Relay assembly wiring

B – Drive gear disassembly

Remove washer pump assembly. Remove crank arm retaining nut, crank arm, rubber seal cap, retaining ring, shim washers, shield and spacer washer in the order indicated. Slide gear assembly out of housing. Slide drive plate and shaft out of gear and remove the drive pawl, lock pawl and coil spring as required.

Drive gear – assembly

Position drive pawl on drive plate. Assemble lock pawl over drive pawl. Slide gear and tube over the drive shaft. Move drive and lock pawls as required to allow their respective pins to fit in the gear guide channel. Holding the gear, manually rotate the drive plate until the drive and lock pawl guide pins snap into their respective pockets in the gear. Reinstall coil spring between lock and drive pawls. Be very careful to maintain lock and drive pawl guide pins in their respective pockets. Assemble inner spacer washer over gear shaft and assemble gear mechanism in housing so that it is positioned with respect

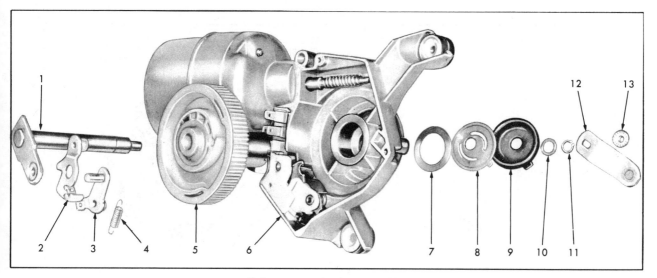

Drive gear mechanism assembly

1. Drive gear shaft
2. Drive pawl
3. Lock pawl
4. Coil spring
5. Drive gear
6. Relay assembly
7. Spacer
8. Shield
9. Seal
10. Washer
11. Snap ring
12. Crank arm
13. Nut

to the housing in the approximate location. Reassemble the outer spacer washer, shield, shim washers as required to obtain .005" max. end play, snap ring and rubber seal cap in the order indicated. Operate wiper to "park" or "off" position and install crank arm. Reassemble washer pump to wiper.

Washer pump unit
The washer pump and/or valve assembly may be removed from the wiper assembly as a unit; therefore, it is not necessary to remove the wiper assembly from the vehicle if only the washer pump and/or valve assembly requires service. When the pump is removed from the wiper assembly, all working parts are readily accessible and may easily be serviced as necessary.

Removal of pump assembly
Raise vehicle hood. Disconnect washer hoses and electrical connections from assembly. Remove 3 screws securing washer pump and cover to wiper assembly. Remove pump from wiper gear box.

VENT GRILLE AND WINDSHIELD WIPER COVER PANEL
Removal
Raise hood then remove ground cable from battery behind driver's seat. Remove eight screws located in the vent grille. Move the vent grille forward and up to remove. If vacuum is present in system, open cover panel by pulling downward wiper cover panel switch. If panels do not open, remove plug at front of vacuum actuator and insert a screw driver into the hole. Press in on actuator piston until cover panel raises to the opened position. The actuator rubber plug is functional in the vacuum system. It must be replaced immediately. Remove four bolts with bushings retaining cover panel to control assembly arms and lift panel from vehicle.

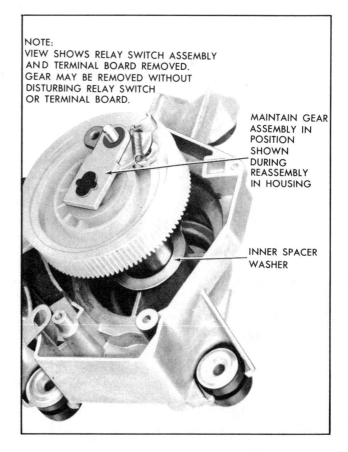

NOTE:
VIEW SHOWS RELAY SWITCH ASSEMBLY AND TERMINAL BOARD REMOVED. GEAR MAY BE REMOVED WITHOUT DISTURBING RELAY SWITCH OR TERMINAL BOARD.

MAINTAIN GEAR ASSEMBLY IN POSITION SHOWN DURING REASSEMBLY IN HOUSING

INNER SPACER WASHER

Removing gear assembly

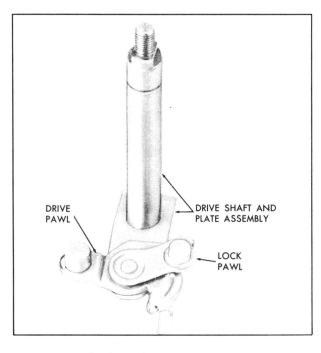

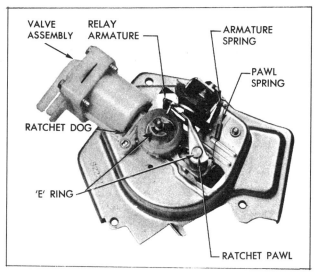

Washer pump assembly

Lock and drive pawl assembly

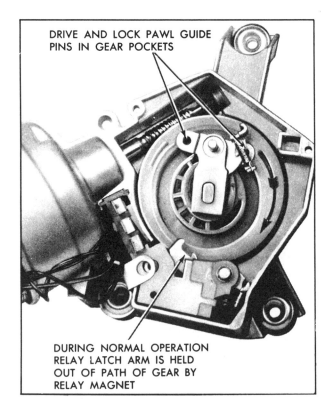

Drive and lock pawl guide pins in pockets

Installation
With actuator rod in the EXTENDED position place cover panel over actuator arms and install bushings and bolts. With actuator in the closed position align vent grille to the mounting brackets and install screws retaining vent to the brackets. Connect ground cable to battery.

Actuate system and check both panels for alignment. The cover panel to windshield adjustment is made at two stop screws "D" located on the rear outboard links of the control assembly under the cover panel. Loosen locking nuts and turn screws to adjust underside of cover panel (at reveal moulding clip) to a .060" gap at the windshield center. The adjustment stop screws seat against plastic pads mounted on top of the plenum. The grille panel has elongated screw holes allowing fore and aft adjustment. Up and down alignment of the entire control assembly with vent grille and cover panel attached is possible at the firewall mounting of the control assembly nuts "E".

WIPER COVER PANEL BUMPERS
Replacement
Remove eight screws securing vent grille and lift grille off. Reaching beneath wiper cover panel, with an offset phillips head screw driver, remove and replace soft bumpers if less than 3/16" thick. Install and align vent grille.

ACTUATOR ASSEMBLY (VIEW C)
Replacement
Remove vacuum hoses from the actuator assembly. Loosen locking nut from turnbuckle rod nut. Remove two (2) nuts, bolts, washers, and bushings attaching actuator assembly to bracket. Rotate actuator assembly to remove from engine compartment. Reverse removal procedures for installation.

LIMIT SWITCH (VIEW B)
Replacement
Remove eight screws located in the vent grille and lift off grille. Remove bolt and bushing rod. Remove two bolts retaining bracket-to-plenum panel and remove assembly. Remove bolt retaining switch to assembly and remove switch. Reverse procedure for installation.

CONTROL ASSEMBLY
Replacement
Remove eight screws securing vent grille and lift grille off. Remove screw and bushings (View B) securing limit switch link to control assembly. Remove wiper cover panel actuator as previously described. Remove four nuts and reinforcements from outboard studs on control assembly mounted to firewall. Rotate entire control assembly, including wiper cover panel, forward and up for removal and further bench disassembly. Reverse removal procedure for installation. Vent grille and cover panel vertical alignment to fenders is possible at nuts "E". Cover panel to windshield gap of .060" is adjusted at screws "D".

VACUUM RELAY (VIEW A)
Replacement
Remove screws securing wiper vacuum relay to inner fender panel. Transfer four hoses from removed relay to new relay. Note hose color strip and match letter on pipe of relay (G-Green, Y-Yellow, R-Red).

Manual overide vacuum switch
For opening wiper cover panel located beneath steering column.

Replacement
Remove screw retaining switch and push switch rearward to release from bracket. Transfer vacuum hoses to new switch. Place new switch in position and secure with screw.

INTERLOCK VALVE (VIEW F)
Replacement
Raise cover panel by vacuum or remove eight (8) screws securing vent grille. Remove 3 hoses and 2 screws from interlock switch and lift off. Install new switch, secure

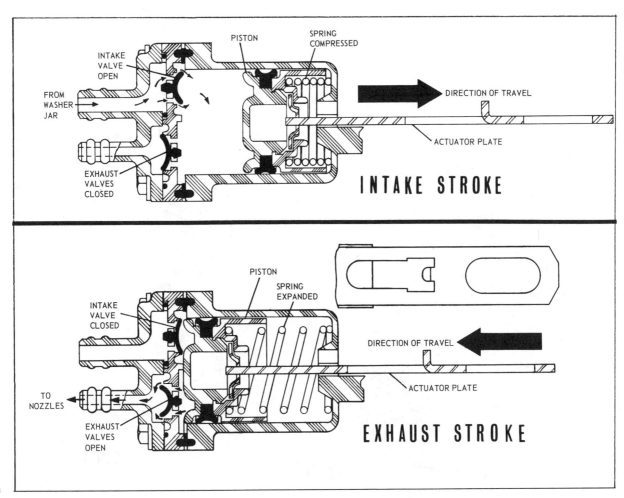

Cross section of windshiled wahser pump valve

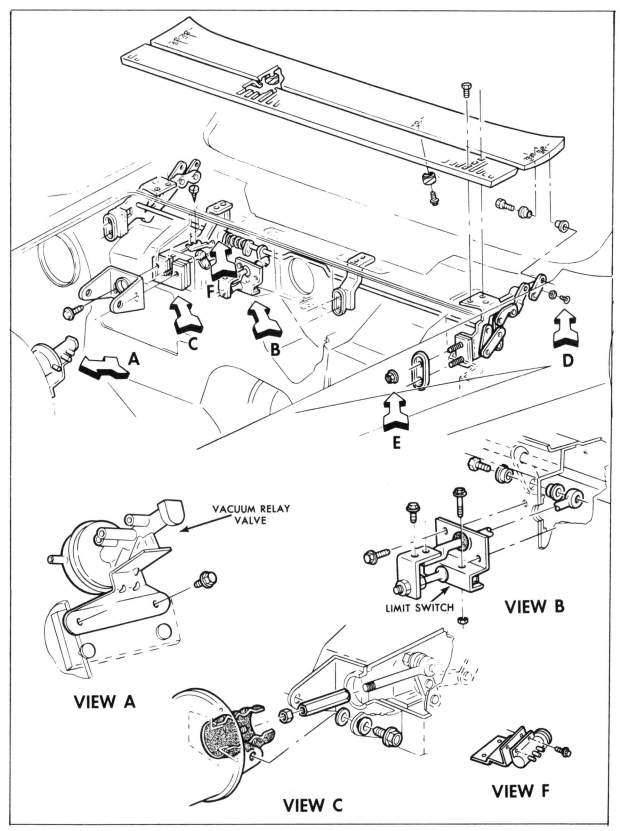

VACUUM RELAY VALVE

LIMIT SWITCH

VIEW A

VIEW B

VIEW C

VIEW F

254

Wiper cover panel system

with washers and screws and attach hoses. Reinstall grille or actuate cover closed.

VACUUM CONTROL SOLENOID
Replacement

Disconnect ground cable from battery. Lower steering column. Remove screws and washers securing left instrument panel to door, opening, top of dash and left side of center instrument panel. Unclip and remove floor console left forward trim panel. Pull cluster slightly forward for access to solenoid valve mounted on rear of tachometer housing. Remove and replace solenoid valve and reinstall cluster in reverse order of removal.

HORN RELAY-BUZZER
(Theft deterrent system)

The horn relay-buzzer is constructed to incorporate both the horn relay and reminder buzzer into one assembly.

Buzzer operation

The horn relay-buzzer operates when the driver's door is opened to remind the vehicle driver that the ignition key has been left in the switch. With the key fully inserted into the ignition switch, the No. 4 terminal on the horn relay-buzzer is connected to ground through the door switch when the driver's door is opened. Current then flows from the battery through the coil winding, the buzzer contacts, the ignition switch, and door switch to ground. The winding magnetism causes the buzzer contacts to open, which opens the winding circuit, and the contacts then re-close. This cycle then repeats many times per second to give the buzzing sound. Closing the door, or removing the key, will stop the buzzer action.

Horn operation

When the horn switch is closed, the coil winding is connected to ground, and the armature moves toward the core to close the horn relay contacts. The horns are then connected to the battery, and operate accordingly. With the horn switch closed, the buzzer contacts remain separated. The horn relay-buzzer can be checked. Although the terminal type and location may vary between models, the numerical designation on all models is standardized. When making electrical checks for improper operation, proceed as follows:

Buzzer check

Insure that the key is fully inserted into the ignition switch. Open the driver's door and observe the dome lamp. If both the dome lamp and buzzer fail to operate, check the door switch for defects. If the dome lamp is on, but the buzzer fails to operate, remove the horn relay buzzer from its mounting and identify the No. 1 and No. 4 terminals. Connect a jumper lead from the No. 4 terminal to ground. Slide a prod into the wiring harness connector to make contact if the terminals are of the slip-on-type. If now the buzzer operates, check the ignition switch wiring and ignition switch for defects. If the buzzer does not operate, connect a voltmeter from the No. 1 terminal to ground. If the reading is zero, the circuit is open between this point and the battery. If a voltage reading is obtained, replace the horn relay-buzzer.

Horn relay check:

Remove the horn relay-buzzer from its mounting, and identify the No. 2 and No. 3 terminals. Connect a jumper from the No. 2 terminal to ground. Slide a prod into the wiring harness connector to make contact if the terminals are of the slip-on type. If the horns operate, check the No. 2 terminal wire and horn switch for defects. If the horns do not operate, leave the No. 2 terminal connected to ground, and connect a voltmeter from the No. 3 terminal to ground. If a reading is obtained, check the horn wiring and horns for defects. If no reading is obtained, replace the horn relay-buzzer.

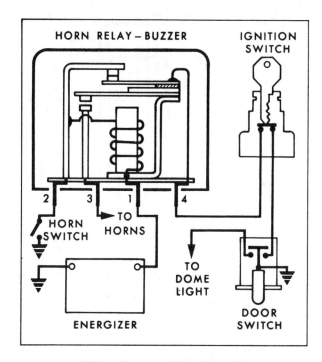

Wiring schematic – Horn relay buzzer

ANTI-THEFT ALARM

When the oval head key is inserted in the door lock cylinder and turned clockwise 90%, it will set or turn off the alarm. Plungers located in both door jambs and under the right side of the hood will trigger the alarm if the hood or doors are opened. This will complete the ground circuit and cause the warning horn to sound.

The roof panels are also protected by anti-theft switches which trigger the alarm if the roof panels are opened.

The only way to shut off the alarm is to insert the oval head key in the door lock and turn it 90% counterclockwise.

To replace the warning horn, remove the electrical connector from the horn, then remove the horn.

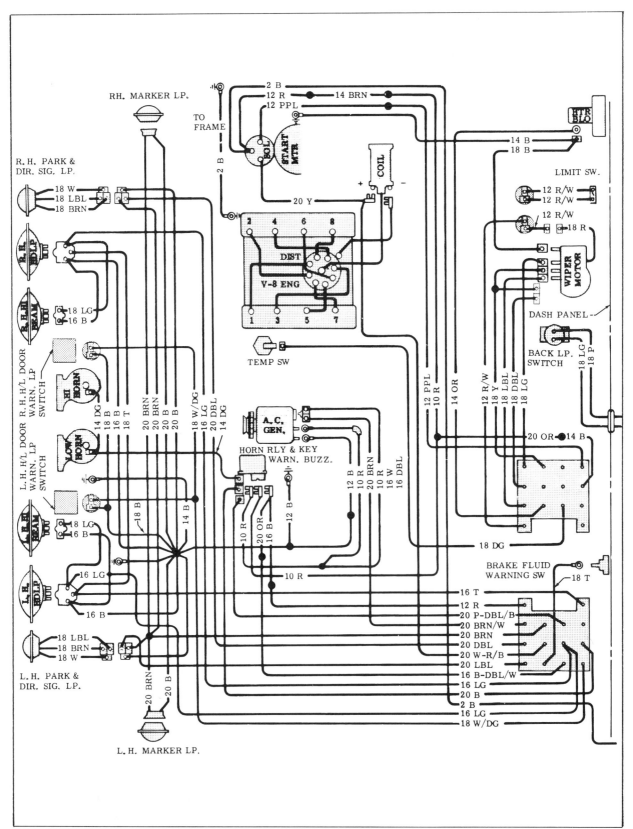

Front lighting and engine compartment

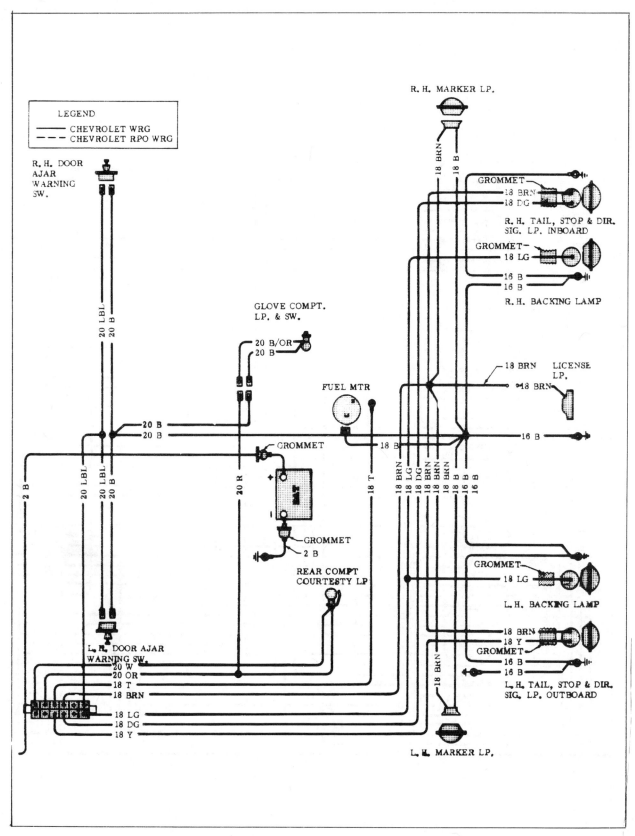

Body and rear lighting

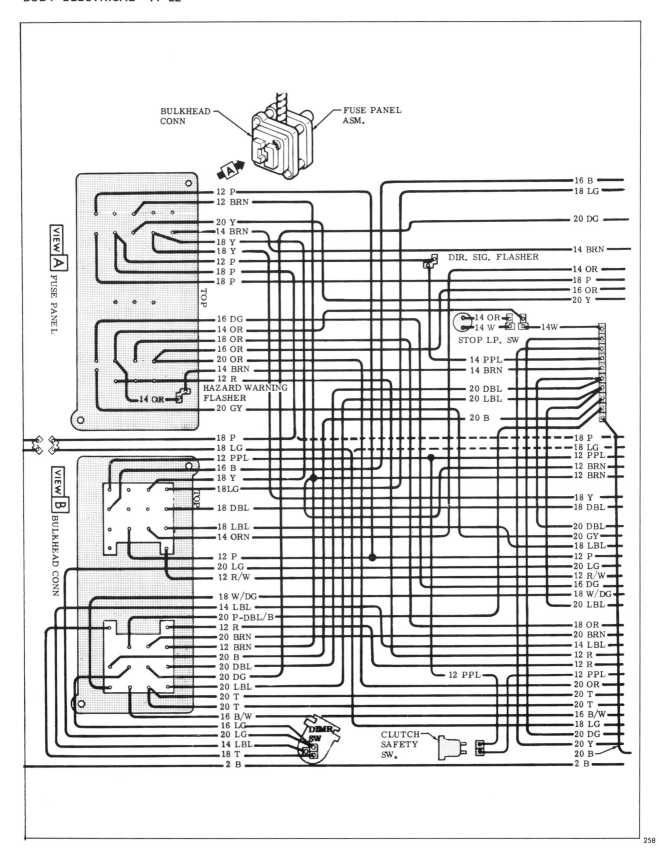

Fuse panel

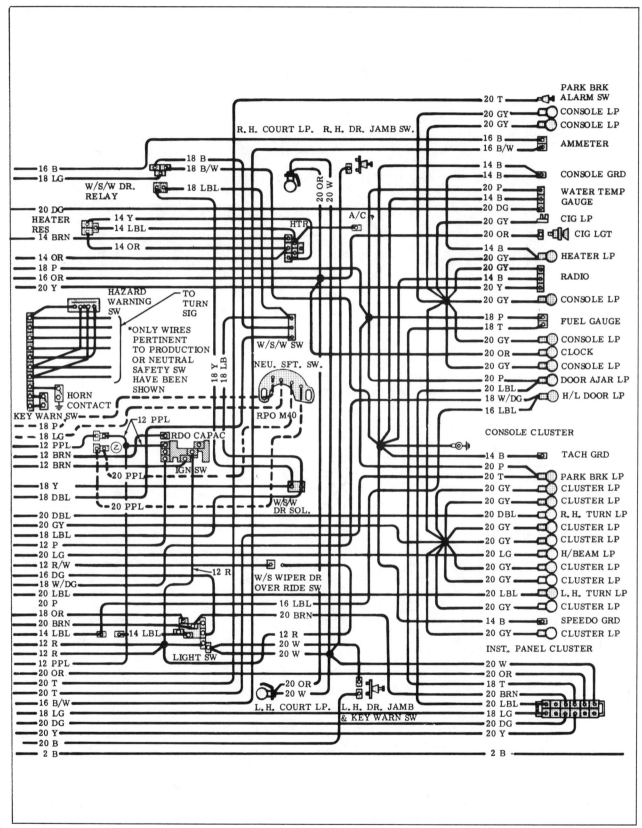

Instrument panel

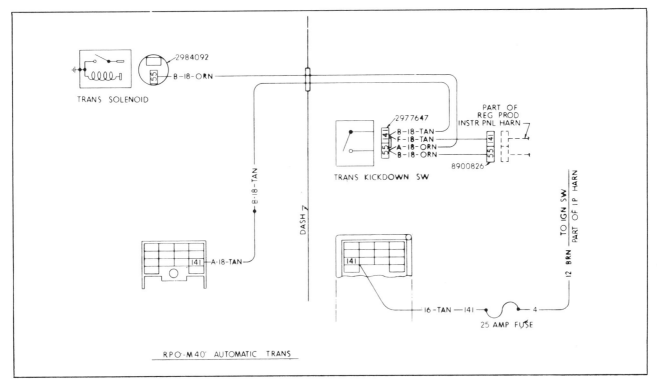

Automatic transmission kickdown switch

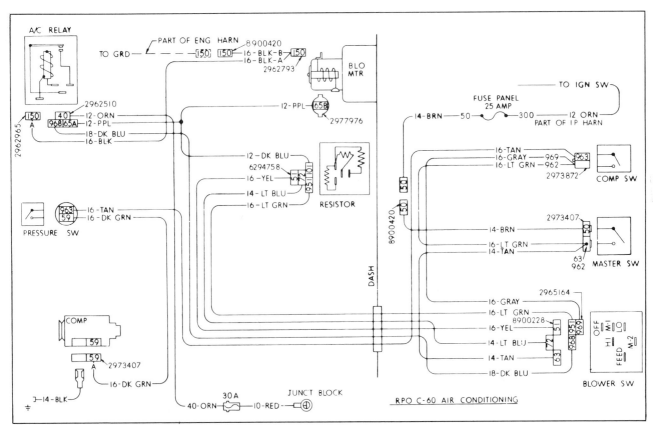

Air conditioning

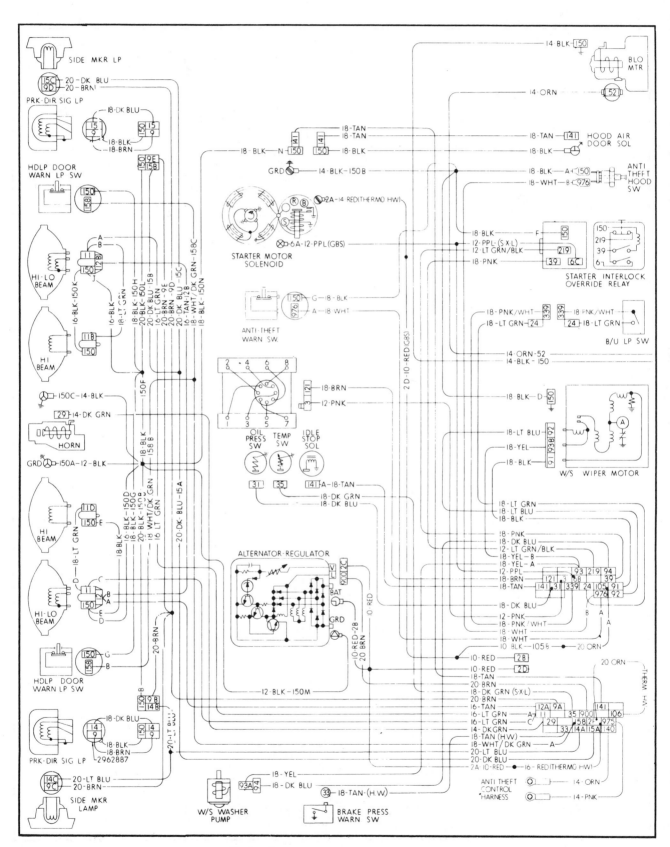

Engine Compartment

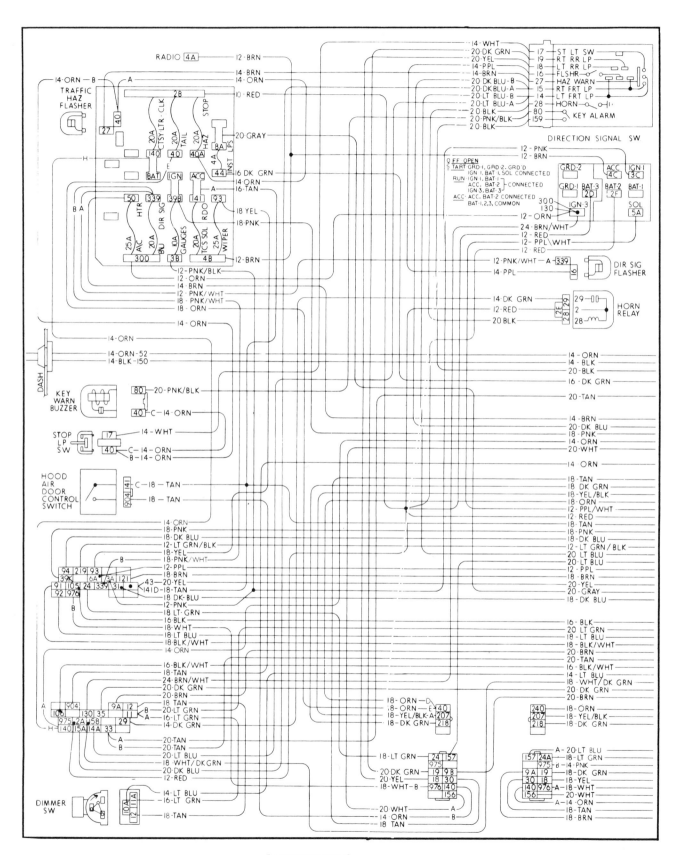

Instrument panel

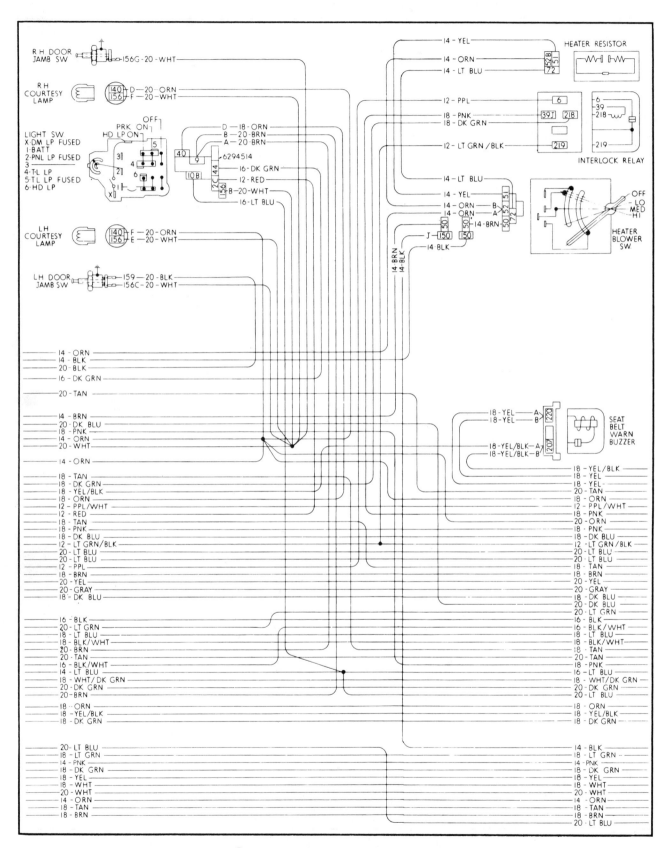

Instrument panel

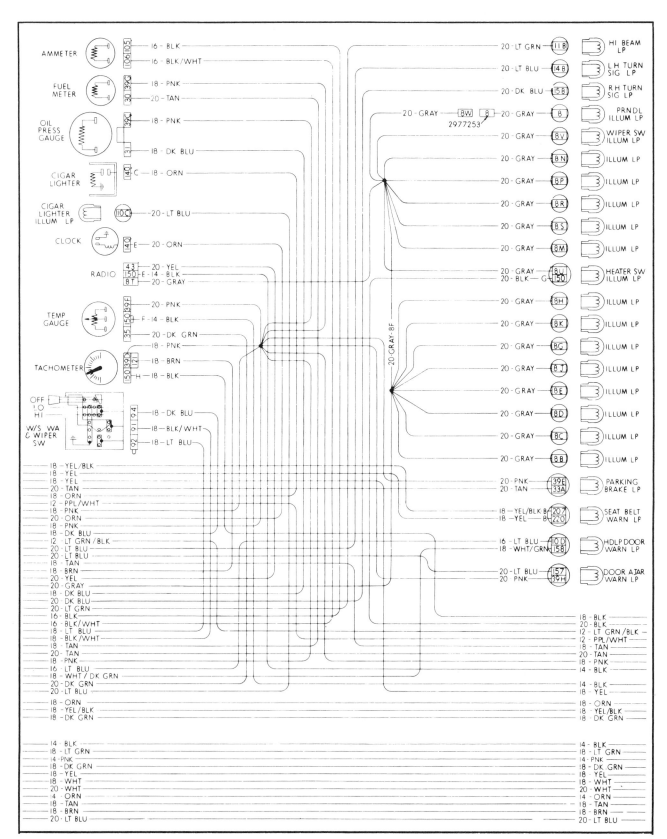

Instrument cluster

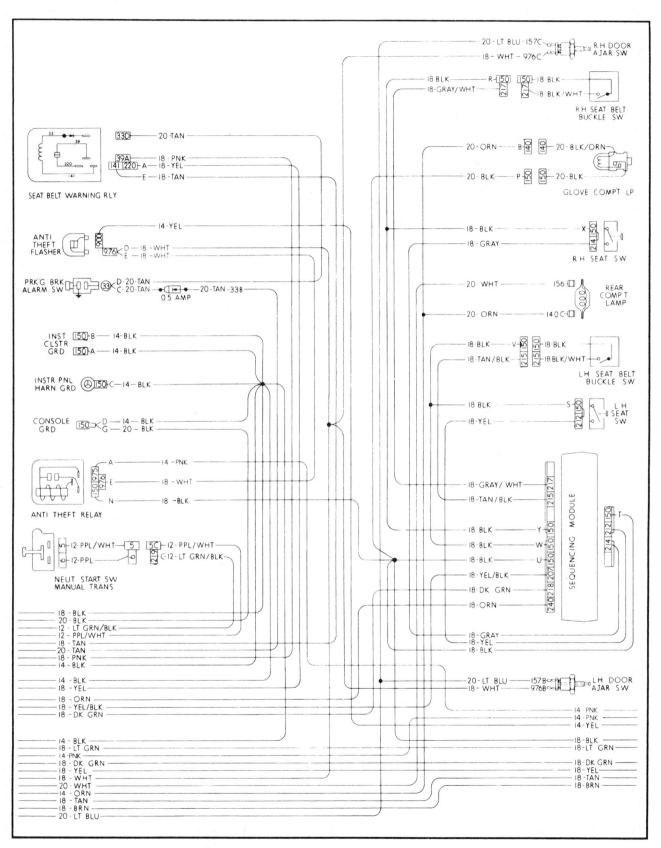

Body

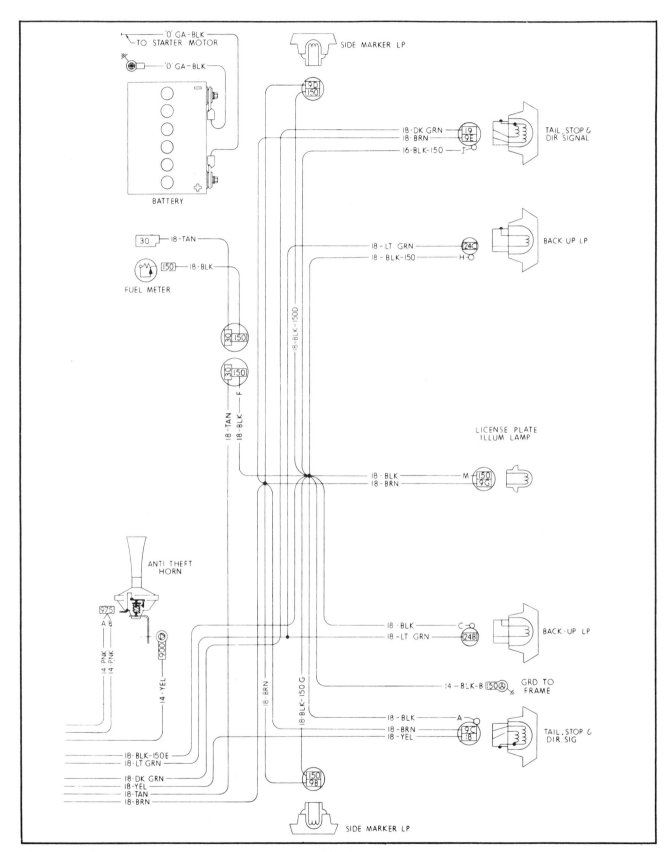

Rear Lighting

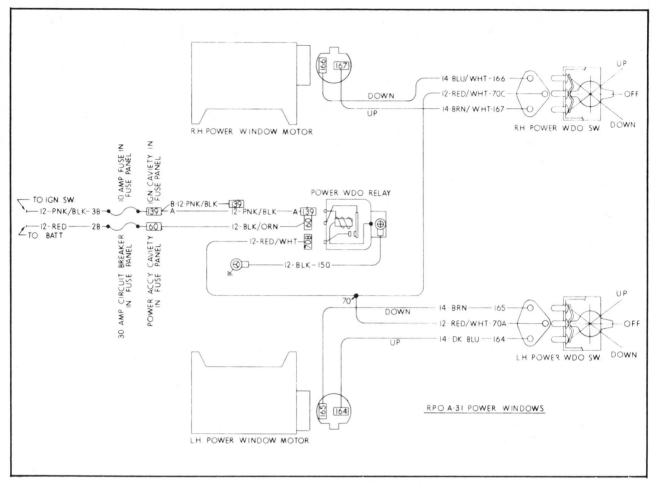

Power windows 1976 shown

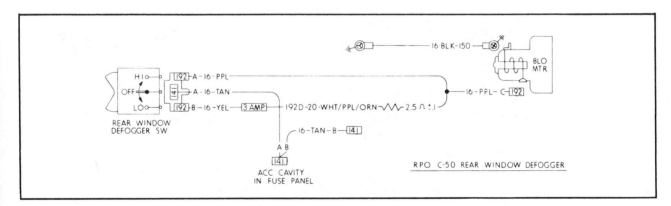

Rear window defogger 1976 shown

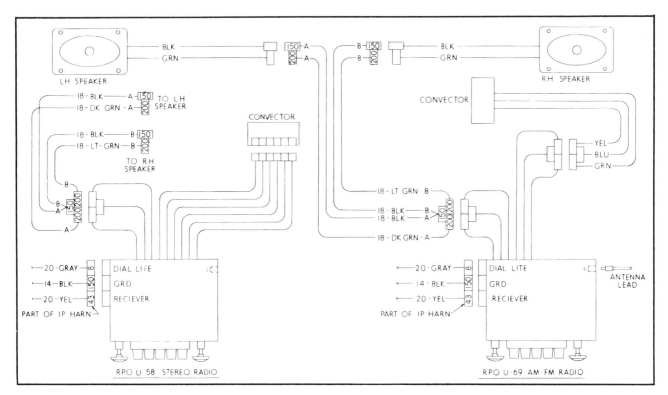

Mono & stereo radios

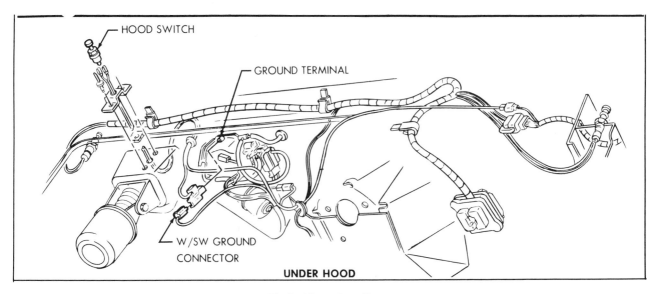

Anti-Theft Alarm Installation

IDENTIFICATIONS & SPECIFICATIONS 15

HOLLEY

CARBURETOR IDENTIFICATION

	w/o AIR		w/AIR	
1966 BASE	**Manual**	**Automatic**	**Manual**	**Automatic**
327/330	R3367A	R3367A	R3416A	R3416A
327/350				
427/390	R3370A	R3370A	R3606A	R3606A
427/425	R3247A			
1967				
327/300	R3810A	R3810A	R3814A	R3814A
327/350				
427/390	R3811A		R3815A	R3815A
1968				
427/430	R4054A			
427/435	R4055A (Prim.)	R4056A		
	R3659Á (Sec.)	R3659A		
1969				
350/370	R4346			
427/430	R4296A			
427/435	R4055-1A (Prim.)	R4056A		
	R3659A (Sec.)	R3659A		

	w/Exhaust Emission		w/Exhaust emission	Fuel Evaporation
1970				
350/370	R-4555A	R-4555A	R-4489A	
454/450	R-4559A	R-4559A	R-4493A	R-4493A
1971				
350/330	R-4801A	R-4800A		
459/425	R-4803A	R-4802A		
1972				
350/225	R-6239A			

ROCHESTER

1966	**Manual**	**Automatic**
327/275	7026203	202
427/390	7026205	204
1967		
327/275	7027203	202
	7027213	212
1968		
327/300	7028207	208
327/350	7028219	
427/390	7028209	216
1969		
350/300	7029203	202
350/350	7029207	
427/390	7029215	204
	7029201	200
1970		
350/300	7040203,213	202
350/300 Calif.	7040903,513	502
454/390	7040205	202

15

454/390 Calif.	7040505	502

1971

350/270	7041213	212
454/365	7041205	204

1972

350 Non-Calif.	7042203	202
350 Calif.	7042903	902
454	7042217	216

1973

350/190	7043203	118
350/250	7043213	212
454/275	7043201	200

	Federal		California	
1974	**Manual**	**Automatic**	**Manual**	**Automatic**
350/185 L-48	7044207	7044206	7044507	7044506
350/245 L-82	7044211	7044210	7044211	7044210
454/235 LS-4	7044221	7044225	7044221	7044505
1975				
350 L-48	7045223	7045222	7045223	7045222
350 L-82	7045211	7045210	7045211	7045210
1976				
350 L-48	17056207	17056206	17056507	17056506
350 L-82	17056211	17056210		
350 L-82 w A/C		17056226		

REAR AXLE IDENTIFICATION

Axle identification is stamped on differential carrier along with date of manufacturer.

STANDARD
1966, 69 — 70 — 71 — 72
3.36   AK   CAK

POSITRACTION		4.11	CAP, CFB
1966—69		4.56	CFC
2.73	AY		
3.08	AL, AR, AT, AU, AW	**1971**	
3.36	AM, AU	3.08	AW
3.55	AN, AZ	3.36	AX, LR
3.70	AO, AS, FA	3.55	AA
4.11	AP, FB	3.70	AB
4.56	FC	4.11	AL
		4.56	AD
1970			
2.73	CAY	**1972**	
3.08	CAL, CAT, CAV, CAW	3.36	AX, LR
3.36	CAM, CAU, CAX, CLR	3.55	AA
3.55	CAN, CAZ	3.70	AB
3.70	CAO, CAS, CFA	4.11	AL

IDLE SPEEDS

	Curb Idle		Fast Idle	
Engine	Man.(N)	Auto (DR)	Man.(N)	Auto.(N)
327/300	700	600		2200
327/350 (1968)	750 (A/C on)			2400
327/350	700			2400
350/165/205 (1975)	800	600		1600 with Vac. Adv.
350/190	900	600		1600 with Vac. Adv.
350/195/270 (1976)	1000	700 (A/C on)		1600 with Vac. Adv.

350/200	800	600	1500	
350/250	900	700	1600 with Vac. Adv.	
350/255	900		2350 with Vac. Adv.	
350/270	900	750		
350/300	700	600	2400	
350/330	700		2200	
350/350	750		2400	
350/370	900		2200	
427/425	800		2200	
427/430	1000 (A/C off)		2200	
427/435	750 (A/C off)			
454/270	750	600	1350	1500
454/275	900	600	1600 W/O Vac.	1600 with Vac. Adv.
454/365	600	600	1350	1500
454/390	700	600	1350	1500
454/425	700	700	2200	

Year	Engine	Dis pl/H.P. @ rpm	C.R.:1	Carb.*
SMALL BLOCK				
66–68	Base	327/300 @ 5000	10.25	4160
66–67	L-79	327/350 @ 5800	11.0	4160
68	L-79	327/350 @ 5800	11.0	4MV
69–71	Base	350/300 @ 4800	10.25	4MV
69–71	LT-46	350/350 @ 5600	11.0	4MV
69–71	LT-1	350/370 @ 5800	11.0	4150
71	Base	350/270 @ 4800	8.5	4MV
71	LT-1	350/330 @ 5600	9.0	4150
72	Base	350/200 @ 4400	8.5	4MV
72	LT-1	350/255 @ 5600	9.0	4150
73	L-48	350/190 @ 4400	8.5	4MV
73-80	L-82	See Specifications 15-7		
MARK IV BIG BLOCK				
66	L-72	427/425 @ 5600	11.0	4150
66–67	L-36	427/390 @ 5400	10.25	4160
68–69	L-36	427/390 @ 5400	10.25	4MV
67–69	L-71	427/435 @ 5800	11.0	2300
67–69	L-68	427/400 @ 5400	10.25	2300
69	L-88	427/430 @ 5200	12.0	4150
70	LS-5	454/390 @ 5400	10.25	4MV
70	LS-7	454/465 @ 5200	12.25	4MV
71	LS-5	454/365 @ 4800	8.5	4MV
71	LS-6	454/425 @ 5600	9.0	4150
72	LS-5	454/270 @ 4000	8.5	4MV
73	LS-4	454/275 @ 4000	8.5	4MV
74	LS-4	454/270 @ 4400	6.25	4MV

*Note: 4150, 4160, 2300 are Holley carburetors. 4MV is Rochester Quadrajet.

Bore & stroke engine

327	4.00 x 3.25
350	4.00 x 3.48
427	4.25 x 3.76
454	4.25 x 4.00

Firing order

All engines 1-8-4-3-6-5-7-2

Compression pressure

Normal at cranking speed . . . . 150 to 170 psi

Minimum . . . . 130 psi

Tappet clearance, engine hot

		All	except LT-1	L-88
Solid lifters	intake	.024	.030	.022
	exhaust	.028	.030	.024
Hydraulic lifters	One turn down from zero lash			

Main bearing clearance (brg. #)

327 & 350 (#1, 2, 3, 4)	.008—.0020 in.
(#5)	.0018—.0034 in.
427 & 454 (#1, 2, 3, 4)	.0013—.0025 in.
(#5)	.0015—.0031 in.

CRANKSHAFT

	327 & 350	427 & 454
Journal dia.	2.448—2.449 in.	2.748—2.749 in.
Service clear., brg. No. 1, 2, 3, 4	.001—.0025 in.	
brg. No. 5	.002—.0035 in.	
End play	.003—.011 in.	.006—.010 in.
Crankpin dia.	2.099—2.100 in.	2.199—2.200 in.
Rod bearing clear.	.001—.004 in. max.	
Rod side clear.	.009—.013 in.	.019—.021 in.
Max. taper & out of round	.001 in. all journals	

CAMSHAFT

Journal dia., 327	1.8682—1.8692 in.
350, 427	1.9482—1.9492 in.

Lobe lift, inches	intake	exhaust
Base	.260	.2733
L-46	.300	.306
LT-1	.3234	.3234
L-71	.3057	.3057
L-36, L-68	.2714	.2714
LS-5, LS-6	.2714	.2824
L-88	.3286	.3412
LT-1 (71, 72)	.3057	.3234

VALVES

Lifters, hydraulic	Base, L-36, L-68, LS-5, L-82
Mechanical	LT-1, L-88, L-71, LS-6
Face angle, int. & exh.	45 deg.
Seat angle, int. & exh.	46 deg. (45 deg. Alum. Hd.)
Seat width, int.	1/32—1/16 in.
exh.	1/16—3/32 in.
Stem clearance, int.	.0010—.0035 in.
exh.	.0012—.0047 in.

Spring, free length	
All 327, 350	2.03
except Base 66	1.94
All 427, 454	2.12
except L-36, L-68	2.094
L-72 outer	2.203
inner	2.109
L-88 outer	2.21
inner	2.12

PISTONS

Clearance	
Base, L-79	.0007—.0013 in.
L-82 LT-1	.0036—.0042 in.
L-72, L-88	.0054—.0063 in.
L-36, L-68	.0012—.0025 in.
L-71, LS-7, LS-6	.0040—.0050 in.
LT-46, LS-5	.0024—.0034 in.

Ring groove clearance	
Compression	.0017—.0027 in.
Oil, 327 & 350	to .006 in.
427 & 454	to .0075 in.
Ring gap	
Compression	.010—.020 in.
Oil, 327 & 350	to .065 in.
427 & 454	to .040 in.
Wrist pin dia.	
327, 350	.9270—.9273 in.
427, 454	.9895—.9898 in.

Wrist pin clearance in piston
All 327, 350 . . . . .00015–.00025 in.
    except LT-1, LT-46 . . . . .00045–.00055 in.
All 427, 454 . . . . .00025–.00035 in.
    except L-88 . . . . .00045–.00055 in.
Wrist pin fit in rod,
All . . . . .0008–.0016 interference (press fit)
    except L-88 . . . . .0001–.0008 in. loose.

CARBURETORS

	327/350	350/300/350
Engine	327/350	350/300/350
Model	Rochester 4MV	
Throttle bore, main	1-3/8 in.	1-3/8 in.
secondary	2-1/4 in.	2-1/4 in.
Main metering jet	.071 in.	.076 in.
Float level	9/32 in.	1/4 in.
Fast idle	2400 rpm	2400 rpm (2 turns)
Idle vent	.375 in.	—
Choke vac. break, man. trans.	.245 in.	.275 in.
auto. trans.	.160 in.	.245 in.
Choke unloader	.300 in.	.450 in.
Air valve spring	3/8 in. 7/8, 1968	7/16 in.
Air valve dashpot	.015 in.	.020 in.
Thermostat choke rod interference		rod top even with hole bottom

Engine	327/300/350
Model	Holley 4160
Throttle bore . . . . main	1-9/16 in.
secondary	1-9/16 in.
Main metering jet . . . . primary	#65 (#63 AIR eng.)
secondary	.076 in.
Float level . . . . primary	.170 in.
secondary	.300 in.
Fast idle	2200 rpm (.035 in.)
Idle vent	.065 in.
Choke vacuum break	.190 in. (.175 AIR eng.)
Choke unloader	.265 in.
Thermostat choke rod interference	1/2 to 1 rod dia.
Secondary stop	1/2 turn after contact

Engine	327/300
Model	4160
Throttle bore . . . . main	1-9/16 in.
secondary	1-9/16 in.
Main metering jet . . . . primary	#65
secondary	.076
Float level . . . . primary	.170
secondary	.300
Fast idle	.035 (2200 rpm)
Choke vacuum break	.170
Choke unloader	.260
Choke	1/2 to 1 rod dia. interference
Secondary stop	1/2 turn after contact
Accelerator pump	.015

Engine	327/350 (66–67) same as 327/300

Engine	327/350 (1968)
Model	4MV
Throttle bore . . . . main	1-3/8
secondary	2-1/4
Main metering jet	.071
Float level, primary	9/32

Fast idle	2 turns (2400 rpm)		
Choke vacuum break	.245		
Choke unloader	.300		
Thermostat choke rod	1 rod dia. interference		
Secondary . . . . closing	.020		
opening	.070		
lockout	.010		
Air valve spring.	7/8		
Accelerator pump	9/32		
Idle vent	3/8		
Carburetor choke rod	.100		
Air valve dashpot	.015		
Secondary metering rod	27/32		
Engine (1975) L-48	350/165	350/185	350/220
Carburetor model . . . . Rochester	M4MC	M4MC	M4MC
Float level	15/32	15/32	15/32
Inner pump rod	.275	.281	.281
Choke coil lever	.120	.120	.120
Choke rod (fast idle cam)	.300	.314 or 46°	.314 or 46°
Air valve dash pot	.015	.015	.015
Front vacuum break	.180	.157	.203
Rear vacuum break	.170	.170	.170
Spring wind-up	7/8"	7/8"	7/8"
Unloader	.325	.277 or 42°	.277 or 42°
Curb idle speed (rpm) . . . . manual	800 (N)	900 (N)	900 (N)
automatic	600 (D)	700 (D)	700 (D).
Fast idle speed	1600 (N)	1600 (N)	1600 (N)
Engine	350/190		
Same as 350/255 except			
Float level	7/32"		
Pump rod location – inner	13/32"		
Choke rod (fast idle cam)	.430		
Air valve wind-up	1/2"		
Vacuum break	.250		
Engine (1974) L-48	350/195		
Carburetor model . . . . Rochester	4QJ		
Float level	1/4"		
Inner pump rod	13/32		
Choke rod (fast idle cam)	.430		
Air valve wind-up	7/8"		
Vacuum break	.230		
Unloader	.450		
Curb idle speed . . . . manual	900 (Neutral)		
automatic	700 (Drive)		
Fast idle speed, with vacuum advance	1600 (Neutral)		
Engine (1976) L-48	350/195		
same as 1975 350 except			
Float level	13/32		
Choke rod (fast idle cam)	.325		
Front vacuum break . . . . manual	.170		
automatic	.185		
Spring wind-up	1"		
Curb idle speed . . . . manual	1000 (Neutral)		
automatic	700 (Drive)		
Engine	350/200		
Model	4MV		
Float level	1/4		
Fast idle . . . . manual	1350 rpm		
automatic	1500 rpm		
Choke vacuum break	.215		
Choke unloader	.450		
Choke rod	.100		

Air valve dashpot 020
Accelerator pump 3/8
Slow curb idle speed . . . . .manual 900
                             automatic 600

Engine (1975) L-82 . . . . 350/205
same as 1975 L-48

Engine 350/250
Same as 350/270 except
Float level . . . . 7/32"
Pump rod location − inner . . . . 13/32"
Choke rod (fast idle cam) 430
Air valve wind-up . . . . 3/4"
Vacuum break 250
Unloader 450

Engine (1974) L-82 . . . . 350/250
same as 1974 L-48 350/195 except
Air valve wind-up . . . . One turn

Engine 350/255
Same as 350/330 except:
Main metering jet . . . . primary No. 68
                           secondary No. 73
Fast idle . . . . .025 (2350 rpm with
vac. spark adv.)

Engine 350/270
Model 4MV
Float level . . . . 1/4"
Choke vacuum break . . . . .manual .275
                           automatic .260
Choke rod 100
Air valve 020
C.E.C. valve idle speed . . . . .manual 900
                           automatic 750

Engine 350/300
Model 4MV
Throttle bore . . . . main 1-3/8"
                         secondary 2-1/4"
Main metering jet . . . . primary .067
Float level 7/32
Fast idle . . . . 2 turns (2400 rpm)
Choke vacuum break . . . . .manual .245
                           automatic .180
Choke unloader 450
Thermostat choke rod 3
Secondary . . . . closing .020
                           opening .070
                         lockout .015
Air valve spring 7/16
Accelerator pump 5/16
Idle vent 3/8
Carburetor choke 100
Air valve dashpot .015

Engine (1977-80) L-82 . . . . (1976) 350/270, (1977) 350/210, (1978) 350/185
same as (1975) L-48 350/165 . . . . 350/220, (1979) 350/225

Engine 350/330
Model 4150
Throttle bore . . . . main 1-11/16"
                         secondary 1-11/16"
Main metering jet . . . . primary No. 70
                           secondary No. 76
Fuel level, (on engine) . . . . Fuel level with bottom of
sight plug hole

Engine	....	350/350
Model	....	4MV
Throttle bore	.... main	1-3/8"
	secondary	2-1/4"
Main metering jet, primary	....	.066
Float level	....	3/16
Fast idle	....	Two turns (2400 rpm)
Choke vacuum break	....	.245
Choke unloader	....	.450
Thermostat choke rod	....	3
Secondary	.... closing	.020
	opening	.070
	lockout	.015
Air valve spring	....	13/16
Accelerator pump	....	5/16
Idle vent	....	5/8
Carburetor choke	....	.100
Air valve dashpot	....	.015
Engine (1977 and later)	....	350
Model	....	M4MC
Throttle bore	.... main	1⅜"
	secondary	2¼"
Float level	....	15/32"
Secondary	.... closing	.020"
	opening	Link in center slot
	lockout	.015"
Air valve spring	....	7/8"
Accelerator pump	....	9/32"
Carburetor choke	....	.120"
Air valve dashpot	....	7/8"

For all other adjustments see emission control label.

Engine	....	350/370
Model	....	4150
Throttle bore	.... main	1-9/16"
	secondary	1-9/16"
Main metering jet	.... primary	No. 68
	secondary	No. 76
Float level	....	.50
Fast idle	....	.025 (2200 rpm)
Choke vacuum break	....	.300
Choke unloader	....	.350
Thermostat choke rod	....	3
Secondary stop	....	1/2 turn open
Accelerator pump	....	.015

Engine	....	427/390
		(same as 327/350 except)
Float level	....	3/16"

Engine	....	427/400 — Same as 427/435

Engine	....	427/425
Model	....	Holley 4150
Throttle bore	.... main	1-3/4" (67), 1-9/16" (68)
	secondary	1-3/4" (67), 1-9/16" (68)
Main metering jet	.... primary	No. 74 choke side (67)
		No. 78 throttle lever side (67)
		No. 68 (68)
	secondary	No. 82 choke side (67)
		No. 80 throttle lever side (67)
		No. 76 (68)
Float level	.... primary	.350"
	secondary	.450"
Fast idle	....	2200 rpm (.025")
Choke vacuum break	....	.350" (67), .300" (68)
Choke unloader	....	.350"
Thermostat choke rod interference	....	1/2 to 1 rod dia.
Secondary stop	....	1/2 turn after contact
Air valve spring	....	7/8"

Engine		427/430
Model		Holley 4150
Throttle bore	main	1-3/4 in.
	secondary	1-3/4 in.
Main metering jet	primary	No. 78 choke side
		No. 82 throttle lever side
	secondary	No. 82 choke side
		No. 80 throttle lever side
Float level	primary	.350 in.
	secondary	.500 in.
Fast idle		2200 rpm (.025 in.)
Choke vacuum break		.350 in.
Choke unloader		.350 in.
Thermostat choke rod interference		rod top even with hole bottom
Secondary stop		1/2 turn open

Engine		427/435
Model	primary	Holley 2300C
Throttle bore		1-1/2 in.
Main metering jet, manual		No. 64
	automatic	No. 62
Float level		.350 in.
Choke vacuum break		.250 in.
Choke unloader		.250 in.
Thermostat choke rod interference		1/2 to 1 rod dia.
Secondary stop		closed

Model	secondary	Holley 2300 (same as 2300C except)
Throttle bore		1-3/4 in.
Main metering jet		No. 76

Engine (1974) LS-4		454/270
Carburetor model	Rochester	4QJ
Float level		3/8"
Inner pump rod		13/32
Choke rod (fast idle cam)		.430
Air valve wind-up		7/16
Vacuum break		.250
	Federal Emiss. Automatic	.220
Unloader		.450
Curb idle speed	manual	800 (Neutral)
	automatic	600 (Drive)
Fast idle speed		1600 (Neutral)

Engine		454/270
Model		4MV
Float level		1/4"
Fast idle	manual	1350 rpm
	automatic	1500 rpm
Choke vacuum break		.250
Choke unloader		.450
Choke rod		.100
Air valve dashpot		.020
Accelerator pump		3/8
Slow curb idle speed	manual	750/450
	automatic	600/450

Engine		454/275
Same as 454/270 except		
Pump rod location – inner		13/32"
Choke rod (fast idle cam)		.430
Air valve wind-up		11/16"
Slow curb idle speed	Manual	900 rpm
	Automatic	600 rpm

Engine		454/365
Model		Rochester 4MV
Main metering jet		
Float level		1/4"
Fast idle	manual trans.	1350 rpm
	automatic trans.	1500 rpm
Choke vacuum break	manual	.275
	Automatic	.260

Choke rod 100
Air valve dashpot 020
Thermostat choke rod interference . . . . Rod top even with bottom
of choke lever hole

Engine 454/390
Model 4MV
Float level . . . . primary 1/4
Fast idle . . . . .manual 1350 rpm
automatic 1500 rpm
Choke rod 100
Air valve dashpot 020
C.E.C. valve idle speed . . . . .manual .900
automatic .650

Engine . . . . 454/465 (Same as 454/390)

Engine 454/425
Model . . . . Holley 4150
Throttle bore . . . . main 1-11/16"
secondary 1-11/16"
Main metering jet . . . . primary No. 70
secondary No. 76
Float level . . . . float centered in inverted bowl
Float level, on engine . . . . gas level with bottom of sight plug hole
Fast idle . . . . .025 (2200 rpm)
Choke vacuum break 350
Choke unloader 350
Choke rod adjustment . . . . 1.320 ± .015

FUEL PUMP
Type . . . . Mechanical, diaphragm
Pressure, all 66–69 . . . . 5–6½ psi
except 427 (68–69) 5–8½
all 1970 and later 7½–9

DISTRIBUTOR TIMING AND SPARK PLUGS

Engine year	Timing deg. BTDC man.	auto	trans.	Spark plugs (AC)
Base 66	10	10		R44
Base 67–69	4	4		R44
Base 70	0	4		R44
Base 71	8	8		R44TS
Base 72	8	8		R44T
L-79 66–68	4	4		R44
L-79 67	10			R44
LT-46 69	4	—		R44
LT-46 70				R43
LT-1 69,70	4	—		R43
LT-1 71	8	12		R44TS
LT-1 72	4	12		R44T
L48	12	12		R44T
L-82 73	8	8		R44T
L-82 73–74	8 (4-Calif.)	8		R44T
L-82, L-48 75		6		R44TX
L-82, L-48 76		6		R45TS
L-48 77		8		R45TS
L-48 78, 79		6		R45TS
Big block Mark IV				
L-72 66	8	8		R43N
L-36 66-69	4	4		R43N
L-71 67	5	—		R43N
L-71 68, 69	4	—		R43N
L-68 67, 69	4	4		R43N
L-88 69	12	—		R43XL
LS-5 70	6	6		R43T
LS-5 71	8	8		R43TS
LS-5 72	8	8		R44T
LS-7 70	4	4		R43T
LS-6 71	8	12		R43TS
LS-4 73	10	10		R44T

Distributor dwell angle .... 29–31 deg.

Point gap .... new points .019 in.

                    used points .016 in.

Spark plug gap .... .035, .060 (75), .045 (76)

Spark plug torque .... 25 lb./ft., 15 lb./ft. (72-76)

CLUTCH

Type .... single dry plate, semi-centrifugal

Diameter	10.4 in.	11.0 in. (Mark IV)
Total friction area, sq. in.	103.5	123.7
Thickness of facing	.135 in ea.	.140 in. ea.

Springs .... diaphragm bent finger design

Pedal travel .... free

                    normal 1-1/4 to 1-3/4 in.

              ·heavy duty 2 to 2-1/2 in.

TRANSMISSION

Type .... Three speed forward and one reverse synchro on all forward gears

Ratios .... first 2.54:1

                second 1.50:1

                third 1:1

                reverse 2.63:1

Type .... Four speed forward and one reverse synchro on all forward gears

Ratio	Regular	Close ratio
first	2.52:1	2.2:1
second	1.88:1	1.64:1
third	1.47:1	1.27:1
fourth	1:1	1:1
reverse	2.59:1	2.26:1

Type .... Hydra-matic 400

Ratio .... first 2.5:1

                second 1.5:1

                third 1:1

                reverse 3.70:2

                low 3.70:2

REAR AXLE

Type .... Fixed differential carrier, hypoid gears

Ratios, transmissions ....

                three speed 3.36:1

            four speed, regular 3.36:1

          four speed, close ratio 3.70:1

      Hydra-matic (positraction) 3.08:1

Positraction ratios ....

                three speed 3.08:1, 3.36:1

            four speed, regular 3.08:1, 3.36:1, 3.55:1

         four speed, close ratio 3.08:1, 3.36:1, 3.55:1, 3.70:1, 4.11:1, 4.56:1 (L-88 only)

              Hydra-matic 400 2.73:1 (427 only), 3.08:1, 3.36:1 (L-71, 88, 89)

Ring gear to pinion backlash .... .005-.008

Pinion bearing preload ....new 20-25 inch, pounds, rotating

                used 5-10

STEERING

Type .... Recirculating ball (power assist optional)

Turns, lock to lock 3.40

Steering gear ratio 16.0:1

Overall ratio 20.2:1

(with fast steering adjustment) 17.6:1

SUSPENSION

Front

Type .... unequal length A-frames

Springs .... independent coil

Shock absorber type .... telescopic double action

Rear

Type .... trailing arm, three link independent

Springs . . . . transversely mounted multi-leaf
Shock absorber type . . . . telescopic double action

BRAKES

Type . . . . four wheel disc, dual hydraulic
circuit
Disc diameter . . . . 11.75 in.
Limit of disc runout . . . . .002 in.
Minimum disc thickness . . . . .065 in (1 in disc),
l.215 in. (1-1/4 in disc)
Parallelism . . . . within .001 in.
Parking brake . . . . drum dia. 6.5 in.
lining width 1.25 in.
thickness .175 in.

WHEEL ALIGNMENT

Front wheels
Castor . . . . manual steering   0-2 deg.
power steering   2-1/4 deg. + 1 pos.
camber . . . . 0°-2¼ deg.
steering axis inclination . . . . 7 deg. ± 1/2
toe-in . . . . 3/16 to 5/16 in.
Rear wheels
Camber . . . . −7/8 deg. ± 1/4
Toe-in (total) . . . . 1/16 to 3/16 in.

ELECTRICAL

Type system . . . . 12 volt
Ground . . . . Negative
Generator type . . . . Alternator
Rating . . . . with air conditioner   61 amps
w/o air conditioner   37 amps (42 amps, 1969)
427 w/o air conditioner   10-S1 (42 amps)
Regulator
Field relay air gap . . . . .015 in.
point opening   .030 in.
closing voltage   1.5−3.2
Voltage regulator air gap . . . . .067 in.
point opening   .014 in.
setting   13.8−14.8 volts @ 85 deg. F
Battery
Capacity . . . . 62 amp/hrs.
No. plates per cell 13
cranking power @ 0 deg. F. . . . . 3250 watts

TORQUE VALUES

ENGINE

Crankcase front cover . . . . 80 lb-in
Flywheel housing pans . . . . 80 lb-in
Oil filter bypass valve . . . . 80 lb-in
Oilpan (to crankcase) . . . . 80 lb-in (327, 350)
Oilpan (to crankcase) . . . . 135 lb-in (427, 454)
Oilpan (to front cover) . . . . 80 lb-in
Oil pump cover . . . . 80 lb-in
Rocker arm cover . . . . 50-55 lb-in
Connecting rod cap . . . . 30-35 lb-ft (327, 350)
Connecting rod cap . . . . 50 lb-ft (427, 454)
Camshaft sprocket . . . . 20 lb-ft
Clutch pressure plate . . . . 35 lb-ft
Flywheel housing . . . . 30 lb-ft
Exhaust manifold . . . . 20 lb-ft
Inlet manifold . . . . 30 lb-ft
Water outlet . . . . 20 lb-ft
Water pump . . . . 30 lb-ft
Cylinder head . . . . 80 lb-ft (396-427)
Cylinder head, L-88, aluminum . . . . short   65 lb-ft
long   75 lb-ft
Cylinder head . . . . 65 lb-ft (327, 350)
Oil pump . . . . 65 lb-ft
Rocker arm stud . . . . 50 lb-ft (427)
Flywheel . . . . 60 lb-ft
Torsional damper . . . . 60 lb-ft (327)
Torsional damper . . . . 85 lb-ft (427, 454)

Main bearing cap . . . . 80 lb-ft (327, 350)
                                         two bolt 95 lb-ft (427, 454)
                                     four bolt 105 lb-ft (427, 454)
Oil pan drain plug . . . . 20 lb-ft

MANUAL TRANSMISSION—THREE & FOUR SPEED SAGINAW

Clutch Gear Retainer to Case Bolts . . . . 15 lb-ft
Side Cover to Case Bolts . . . . 15 lb-ft
Extension to Case Bolts . . . . 45 lb-ft
Shift Lever to Shifter Shaft Bolts . . . . 25 lb-ft
Lubrication Filler Plug . . . . 13 lb-ft
Transmission Case to Clutch Housing Bolts . . . . 75 lb-ft
Crossmember to Frame Nuts . . . . 25 lb-ft
Crossmember to Mount Bolts . . . . 40 lb-ft
Mount to Transmission Bolts . . . . 32 lb-ft

MANUAL TRANSMISSION—FOUR SPEED WARNER

Clutch Gear Retainer to Case Bolts . . . . 18 lb-ft
Side Cover to Case Bolts . . . . 18 lb-ft
Extension to Case Bolts . . . . 40 lb-ft
Shift Lever to Shifter Shaft Bolts . . . . 20 lb-ft
Lubrication Filler Plug . . . . 15 lb-ft
Transmission Case to Clutch Housing Bolts . . . . 52 lb-ft
Crossmember to Mount and Mount to Extension Bolts . . . . 25 lb-ft
Rear Bearing Retainer to Case Bolts . . . . 25 lb-ft
Extension to Rear Bearing Retainer Bolts (Short) . . . . 25 lb-ft
Retainer to Case Bolt . . . . 35 lb-ft
Transmission Drain Plug . . . . 20 lb-ft

TURBO HYDRA-MATIC

Pump cover . . . . 18 lb-ft
Parking pawl bracket . . . . 18 lb-ft
Pump to case . . . . 18 lb-ft
Extension housing to case . . . . 23 lb-ft
Rear servo cover . . . . 18 lb-ft
Modulator retainer . . . . 18 lb-ft
Governor cover . . . . 18 lb-ft
Manual lever to manual shaft nut . . . . 8 lb-ft
Manual shaft to inside detent lever . . . . 18 lb-ft
Transmission to engine mounting . . . . 35 lb-ft
Converter to flywheel . . . . 33 lb-ft
Oil cooler line . . . . 113 lb-in
Line pressure take-off plug . . . . 13 lb-ft
Strainer retainer plug . . . . 10 lb-ft
Pump Cover to Pump Body . . . . 17 ft. lbs.
Pump Assembly to Case . . . . 18-1/2 ft. lbs.
Valve Body and Support Plate. . . . . 130 in. lbs.
Parking Lock Bracket . . . . 29 ft. lbs
Oil Suction Screen . . . . 40 in. lbs.
Oil Pan to Case . . . . 130 in. lbs.
Extension to Case. . . . . 25 ft. lbs.
Modulator Retainer to Case. . . . . 130 in. lbs.
Inner Selector Lever to Shaft . . . . 25 ft. lbs.
Detent Valve Actuating Bracket . . . . 52 ft. lbs.
Converter to Flywheel Bolts . . . . 35 ft. lbs.
Under Pan to Transmission Case . . . . 110 in. lbs.
Transmission Case to Engine . . . . 35 ft. lbs.
Oil Cooler Pipe Connectors to Transmission
   Case (Straight Pipe Fitting) . . . . 25 ft. lbs.
         (Tapered Pipe Fitting) . . . . 15 ft. lbs.
Oil Cooler Pipe to Conectors . . . . 10 ft. lbs.
Gearshift Bracket to Frame . . . . 15 ft. lbs.
Gearshift Shaft to Swivel . . . . 20 ft. lbs.
Manual Shaft to Bracket . . . . 20 ft. lbs.
Detent Cable to Transmission . . . . 75 in. lbs.
Intermediate Band Adjust Nut. . . . . 15 ft. lbs.

STEERING

Steering gear mounting . . . . 70 lb-ft
Pitman shaft nut . . . . 185 lb-ft
Idler arm to frame nuts . . . . 35 lb-ft
Tie rod clamp nuts . . . . 132 lb-in

Steering coupling to shaft flange nuts	20 lb-ft
Steering wheel nut	30 lb-ft
Power steering pump pulley nut	58 lb-ft
Power steering pump mounting	20-25 lb-ft
Power cylinder to relay rod nut	45-60 lb-ft
Control valve to pitman arm	45 lb-ft

FRONT SUSPENSION

Ball joint	upper stud	50 lb-ft
	lower stud	80 lb-ft
Stud to control arm nuts		20 lb-ft
Lower forging nuts		40 lb-ft
Upper control arm attaching nuts		50 lb-ft
Collar bolts		40 lb-ft
Brake anchor bolts		120 lb-ft
Lower control arm cross shaft bolts	front	70 lb-ft
	rear	95 lb-ft
Shock absorber	upper	90 lb-in
	lower	150 lb-in

REAR SUSPENSION

Spring retainer		65 lb-ft
Universal joint companion flange		15 lb-ft
Transmission yoke		15 lb-ft
Axle drive shaft	to spindle	75 lb-ft
	to yoke	15 lb-ft
Stabilizer shaft bracket to frame		120 lb-ft
Drive spindle support to torque arm		30 lb-ft
Strut rod to spindle support		80 lb-ft
Bracket to carrier		45 lb-ft
Camber arm		65 lb-ft
Torque arm pivot		50 lb-ft

BRAKES

Master cylinder to dash	24 lb-ft
Master cylinder to booster	24 lb-ft
Vacuum booster to dash	24 lb-ft
Push rod to clevis	14 lb-ft
Brake line nuts (to master cyl. and valves)	150 lb-in
Brake line nuts (to front brake hose)	150 lb-in
Brake line nut (to rear brake hose)	150 lb-in
Brake line clip to frame	100 lb-in
Brake bleeder valves	65-100 lb-in
Caliper mounting bolt	70 lb-ft
Caliper housing bolt (front disc brakes)	130 lb-ft
Caliper housing bolt (rear disc brakes)	60 lb-ft
Flex hose to caliper	22 lb-ft
Brake pedal bracket to dash	20 lb ft
Distribution switch mounting	150 lb-in
Parking brake equalizer	70 lb-in
Parking brake assembly (to dash or floor)	100 lb-in
Parking brake cable pulley bolt	45 lb-ft
Stoplamp switch striker screw	120 lb-in
Rear cable bracket	100 lb-in

Lower control arm front bushing	50 lb-ft
Shock absorber	

Lower control arm front bushing		50 lb-ft
Shock absorber	upper	50 lb-ft
	lower	35 lb-ft
Leaf spring retainer		70 lb-ft
Universal joint companion flange		15 lb-ft
Wheel stud nuts		80 lb-ft
Axle drive shaft	to spindle	75 lb-ft
	to yoke	15 lb-ft

DIFFERENTIAL

Carrier cover	50 lb-ft
Ring gear	50 lb-ft
Bearing caps	55 lb-ft
Pinion lock screw	20 lb-ft
Filler plug	20 lb-ft

1980-1982 SUPPLEMENT 16

INDEX

THROTTLE BODY INJECTION

The Model 400 Throttle Body Injection is also called a 2 x 1 Throttle Body Injection as the system includes a pair of units, front and rear, each controlled by an Electronic Control Module. Each unit supplies the correct air/fuel mixture through long runners in the intake manifold to the bank of cylinders on the opposite side of the engine. Therefore the name Crossfire Injection.

The Electronic Control Module receives inputs from various sensors regarding operating conditions and referring to the stored program memory, performs calculations of fuel requirements and controls the fuel injector in each throttle body unit. The metering valve in each injector is pulsed or timed by the ECM.

The system includes an oxygen sensor in the exhaust to control the fuel delivery so the ideal air/fuel ratio of 14.7 is obtained. This in turn allows the catalytic converter to operate more efficiently. Other sensors take care of other operating conditions such as cranking, cold starting, altitude, acceleration and deceleration.

Both TBI units contain a throttle body with a single throttle valve, a fuel meter body assembly with an integral fuel pressure regulator in the rear unit and an integral fuel pressure compensator in the front unit and a fuel injector. Both units are interconnected by a common throttle rod and fuel line.

Fuel Metering

Filtered fuel from the tank mounted pump feeds the front TBI unit. When the ECM energizes the injector, fuel is injected into the throttle body above the throttle valve. The volume of flow is changed by varying the time the injector is open. Excess fuel passes by the fuel pressure compensator and fuel under pressure flows through the common fuel line to the rear unit. A similar function is performed here. The pressure regulator controls the pressure at each TBI unit and surplus fuel is returned to the fuel tank.

Idle Air Control

An Idle Air Control motor in each TBI unit controls fast and slow idle speeds. Inputs come from the ECM and are determined by the coolant temperature, driving modes, transmission gear, engagement of the A/C compressor, etc. A pintle valve in the air passage varies the amount of air bypassing the throttle valve for control of idle speeds.

Throttle Position Sensor

This is an electrical variable resistor mounted to the throttle body on the rear TBI unit and is electrically connected to the ECM. A lever on the end of the throttle shaft rotates a lever on the TPS. This mechanical movement is converted to an electrical signal to the ECM. This then becomes a reference point for the ECM to recalculate the correct air/fuel ratio.

16

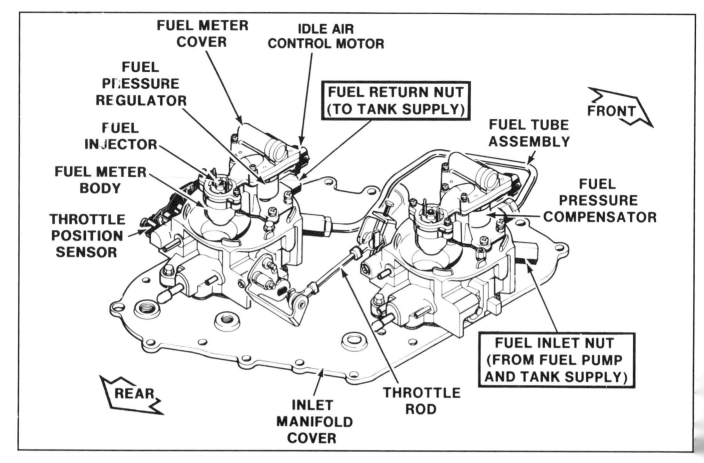

Throttle Body Injection System

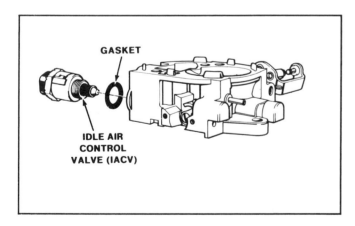

Idle Air Control Motor

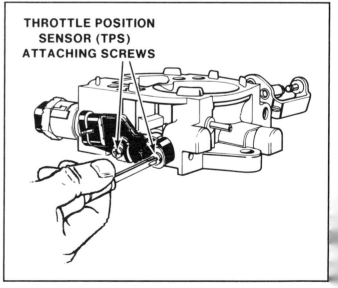

Throttle Position Sensor

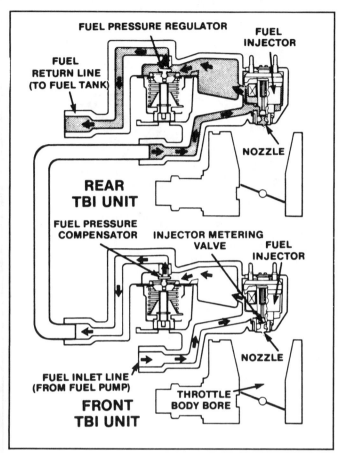

FUEL PRESSURE REGULATOR

FUEL INJECTOR

FUEL RETURN LINE (TO FUEL TANK)

NOZZLE

REAR TBI UNIT

FUEL PRESSURE COMPENSATOR

INJECTOR METERING VALVE

FUEL INJECTOR

NOZZLE

FUEL INLET LINE (FROM FUEL PUMP)

THROTTLE BODY BORE

FRONT TBI UNIT

Fuel Metering Diagram

Minimum Idle Speed

The throttle position of both throttle bodies must be balanced so the throttle plates open simultaneously. Make this adjustment only when a throttle body has been replaced or if there has been tampering with the minimum air adjustment screw or the idle balance screw.

Remove the air cleaner and gaskets and plug the vacuum port on the rear TBI unit for Thermac. Remove the tamper resistant plug on the minimum air adjustment screw. Block the wheels, apply the parking brake and connect a tachometer. Remove the IAC motors electrical connectors. Plug the idle air passages in each throttle body. Start the engine and stabilize the RPM at normal operating temperature.

Place the transmission in "D" — engine RPM should decrease below curb idle speed. If it does not, check for a vacuum leak. Remove the cap from the ported tube on the rear TBI unit and connect a water manometer. Adjust the minimum air adjustment screw to 6" of water on the manometer. Remove the manometer and install the cap.

Connect the manometer to the front TBI and is should read 6" of water. If it does not, locate the idle balance screw on the throttle linkage. If the screw is welded, break the weld and install a new screw with thread locking compound. Adjust the screw to obtain 6" of water on the manometer. Remove the manometer and install the cap.

Adjust the minimum air adjustment screw on the rear unit to obtain 475 RPM. Turn the ignition "OFF" and place the transmission in "N". Remove the idle air passage plugs and connect

the IAC motors. Start the engine — it may run at a high RPM which will decrease when the IAC motors close the air passages. Stop the engine when the RPM has decreased. Check the TPS voltage and adjust as necessary. Install the air cleaner gasket, connect the vacuum line to the TBI and install the air cleaner. Reset the IAC motors. To do this drive the vehicle at 30 MPH.

TBI Replacement

Remove the air cleaner and disconnect the electrical connectors at the injectors, IAC motors and TPS. Disconnect all vacuum lines, throttle cable, transmission detent cable and cruise control cable. Disconnect the fuel feed and return lines. Remove the bolts and nuts from the inlet manifold cover and remove the TBI unit. Protect the swirl plates located below each throttle plate. To install, reverse the above procedure and tighten the bolts and nuts to 120-168 in. lbs.

Rear TBI Replacement

Remove air cleaner, disconnect electrical connectors at the injector, IAC motor and TPS. Remove all vacuum lines. Remove throttle cable and cruise control cable. Disconnect fuel return line and fuel line to front TBI unit. Disconnect throttle rod between units. Remove bolts and then remove the rear TBI unit. To install, reverse the above procedure and tighten the bolts to 120-168 in. lbs. Check and adjust curb idle air rate RPM.

Front TBI Replacement

Remove the air cleaner and disconnect the electrical connectors at the injector and the IAC motor. Remove vacuum lines and transmission detent cable. Disconnect fuel feed line and fuel line to rear TBI unit. Disconnect throttle rod between units. Remove bolts and then remove the TBI unit. To install reverse the above procedure and tighten the bolts to 120-168 in. lbs. Check and adjust the curb idle air rate RPM.

TBI Unit Repair

Remove the complete Model 400 TBI unit and place it on a holding fixture so the swirl plates will not be damaged. The fuel meter cover with integral fuel pressure regulator or fuel pressure compensator are serviced as a complete assembly only. The fuel injector in each unit is also serviced as a complete assembly only, and these units must not be interchanged or excessive emissions will result and performance will suffer.

Fuel Meter Cover

Remove five fuel meter cover screws and lockwashers and lift off the fuel meter cover. Remove the cover gaskets. Do not remove the four screws holding the fuel pressure regulator or fuel pressure compensator to the fuel meter cover. This cover is serviced as a complete unit only. Do not immerse the fuel meter cover in any type of cleaner — damage to internal diaphragms and gaskets will result. Remove the foam sealing ring from the fuel meter body of the rear unit.

Fuel Injector

Gently grasp the center collar of the injector with a pair of small pliers and remove the injector with a twisting, lifting motion. Mark the injector for identification. This unit is serviced as a complete assembly only. Do not immerse in any cleaner. Remove the fuel filter from the base of the injector. Remove the "O" ring and steel back-up washer at the top of the fuel meter body. Remove the small "O" ring at the bottom of the injector cavity.

Fuel Meter Body

Remove the fuel inlet and outlet nuts and gaskets. Remove three screws and lockwashers and remove the fuel meter body. Remove the fuel body insulator gasket.

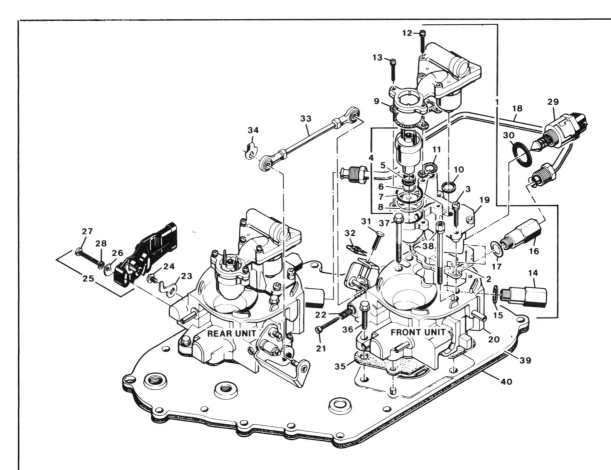

FUEL METERING PARTS

1 FUEL METER ASSY — F
  FUEL METER ASSY — R
2 GASKET — FUEL METER BODY
3 SCREW & WASHER ASSY — ATTACH. (3)
4 FUEL INJECTOR KIT — F
  FUEL INJECTOR KIT — R
5 FILTER — FUEL INJECTOR NOZZLE
6 SEAL — SMALL "O" RING
7 SEAL — LARGE "O" RING
8 BACK-UP WASHER — FUEL INJECTOR
9 GASKET — FUEL METER COVER
10 DUST SEAL — PRESSURE REG. — R
11 GASKET — FUEL METER OUTLET
12 SCREW & WASHER ASSY — LONG (3)
13 SCREW & WASHER ASSY — SHORT (2)
14 NUT — FUEL INLET
15 GASKET — FUEL INLET NUT
16 NUT — FUEL OUTLET
17 GASKET — FUEL OUTLET NUT
18 FUEL TUBE ASSEMBLY
19 FUEL METER BODY ASSEMBLY

THROTTLE BODY PARTS

20 THROTTLE BODY ASSEMBLY — F
   THROTTLE BODY ASSEMBLY — R

21 SCREW — IDLE STOP
22 SPRING — IDLE STOP SCREW
23 LEVER — TPS — R
24 SCREW — TPS LEVER ATTACHING — R
25 SENSOR — THROTTLE POSITION KIT — R
26 RETAINER — TPS (2)
27 SCREW — TPS ATTACHING (2)
28 WASHER — TPS SCREW (2)
29 IDLE AIR CONTROL VALVE
30 GASKET — CONTROL VALVE TO T.B.
31 SCREW — IDLE BALANCE
32 CLIP — IDLE BALANCE SCREW (SERVICE ONLY)
33 THROTTLE ROD & BEARING ASSEMBLY
34 CLIP — THROTTLE ROD (2)
35 GASKET — TBI MOUNTING
36 BOLT — TBI ATTACH. — SHORT (2)
37 BOLT — TBI ATTACH. — LONG (2)
38 STUD — TBI & AIR CLEANER ATTACH. (2)

INLET MANIFOLD PARTS

39 INLET MANIFOLD COVER
40 GASKET — MANIFOLD COVER

Throttle Body Injection

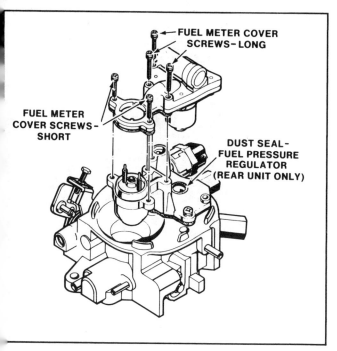

Fuel Meter Cover

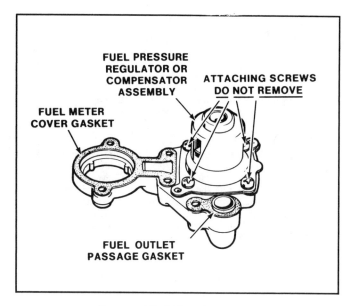

Bottom of Fuel Meter Cover

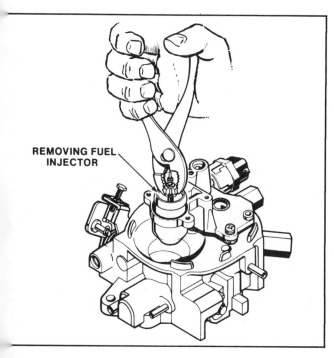

Fuel Injector

Throttle Body Assembly

Remove the Throttle Position Sensor from the rear unit. Do not immerse it in any cleaner. Remove the Idle Air Control motor and gasket. Do not immerse in any cleaner. Do not remove the throttle valve screws. The throttle body is serviced as a complete unit.

Thoroughly clean the throttle body injection parts in a cold immersion-type cleaner. Blow-dry the parts with air making sure all fuel passages are free of burrs and dirt.

Depending on the problem, check and repair or replace parts as required.

Flooding

Inspect both injector "O" rings for damage and check that the steel back-up washer is below the upper "O" ring. Use new "O" rings when installing the injector. Check the injector fuel filter for damage and replace as necessary. If the injector continued to supply fuel with the electrical connections removed, replace the injector.

Hesitation, Hard Starting or Poor Cold Operation

Check the injector fuel filter and clean or replace as necessary. Check for reversal of injectors. If improper fuel inlet and outlet pressure readings were obtained check for restricted passages or inoperative fuel pressure regulator. Repair or replace as necessary.

Rough Idle

Inspect both injector "O" rings for damage and check that the steel back-up washer is below the upper "O" ring. Use new "O" rings when installing the injector. Check the injector fuel filter for damage and replace as necessary. Check for reversal of injectors. If the injector continued to supply fuel with the electrical connections removed, replace the injector.

Throttle Body Assembly

Place the throttle body on a holding fixture. Using a new gasket, install the Idle Air Control Motor. Install the Throttle Position Sensor actuator lever by aligning the flats on the lever with the flats on the end of the shaft. Install the retaining screw and tighten securely. Use thread locking compound on the attaching screws.

Fuel Meter Body

Install new fuel meter body gasket on the throttle body assembly. Install the fuel meter body. Apply thread locking compound to the attaching screws and install the screws tightening them to 35 in. lbs. Install the fuel inlet and outlet nuts and new gaskets in the fuel meter body. Torque the nuts to 260 in. lbs.

Fuel Injectors

Install the fuel filter on the nozzle end of the injector. Lubricate the small "O" ring with lithium grease and install the "O" ring on the nozzle. Clean the injector cavity and install the steel back-up washer. Lubricate the large "O" ring with lithium grease and install it above the back-up washer. Install the injector aligning the raised lug on the injector with the notch in the fuel meter body cavity.

Fuel Meter Cover

Install a new dust seal for the fuel pressure regulator in the rear unit. Install a new gasket sealing the fuel passage on the fuel meter cover. Install the fuel meter cover with the fuel pressure regulator on the rear unit. Install the fuel meter cover with the fuel pressure compensator on the front unit. Apply thread locking compound to the five attaching screws for each unit and install the screws and lockwashers torquing them to 28 in. lbs.

Throttle Position Sensor

With the throttle valve in the normal closed idle position install the Throttle Position Sensor on the rear unit making sure the TPS pick-up lever is located above the tang on the throttle actuator lever. Install the retainers and two screws and lockwashers.

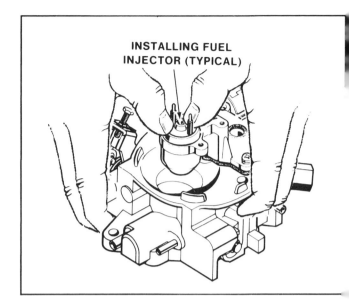

Installing Fuel Injector

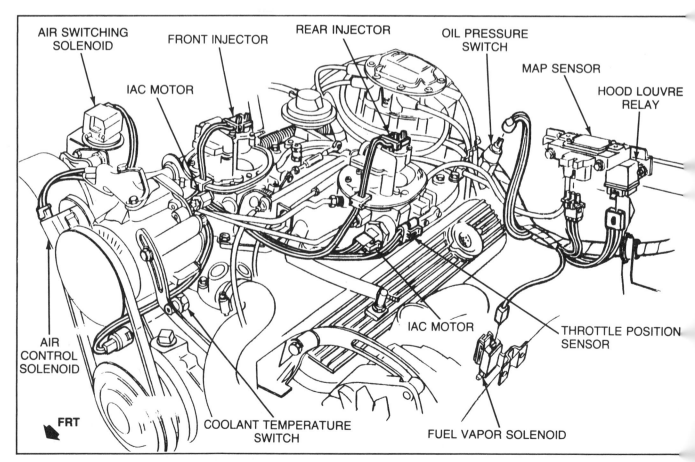

Electronic Controls – Left Side

ENGINE EMISSION CONTROLS

Emission controls fall into two broad categories: the more familiar positive crankcase ventilation (PCV), the thermostatic air cleaner and the new Computer Command Control system. The first two systems, PCV and thermostatic air cleaner systems are covered fully in the main part of this manual. This section relates primarily to the Computer Command Control system.

This is an electronically controlled exhaust emission system. An Electronic Control Module located in the passenger compartment behind the driver is the control center of the Computer Command Control system. It constantly monitors input information, processes it and generates output commands to the various systems.

The ECM has a removable Calibration Unit which customizes the ECM for the particular engine, transmission, vehicle weight, axle ratio, etc., in any one vehicle. These PROM units are not interchangeable.

The ECM also performs the diagnostic function. It recognizes operational problems, alerts the driver via the "CHECK ENGINE" light on the instrument panel and stores a code or codes which identify the problem(s). This light will remain "ON" with the engine running as long as the problem prevails. The trouble code will be retained in the memory for 50 starts of the engine and until the battery voltage falls to zero.

To perform the self diagnosis and to find out which circuit(s) do not operate properly, remove the ashtray and locate the 12-terminal connector. With the ignition turned "ON" and the engine stopped, the CHECK ENGINE lamp should be "ON". This is a bulb check to indicate that the lamp is working. Place a jumper between terminals A and B to activate the trouble code system and all the ECM controlled relays.

When the self-diagnosis is activated the CHECK ENGINE lamp will flash code "12". This consists of one flash, a short pause and two flashes in quick succession. After a longer pause, the code will repeat two more times. This code is not stored in memory and will only flash while the fault is present. This check indicates the self-diagnostic system is working.

If more than one fault is stored, the lowest number code will flash three times followed by the next code until all codes have been flashed. The faults will repeat as long as the diagnostic terminal is grounded.

A code indicates a problem in a particular circuit. A list of the codes together with service suggestions are listed below. A physical inspection of the malfunctioning circuit should be made. It should be checked for poor connectors snd frayed wires. After each check, the diagnostic circuit check should be made again to see if the fault has been corrected.

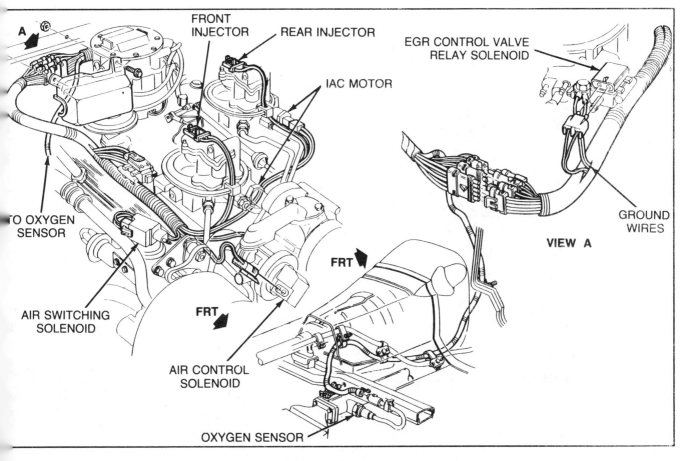

FRONT INJECTOR

REAR INJECTOR

EGR CONTROL VALVE RELAY SOLENOID

IAC MOTOR

TO OXYGEN SENSOR

GROUND WIRES

VIEW A

AIR SWITCHING SOLENOID

FRT

FRT

AIR CONTROL SOLENOID

OXYGEN SENSOR

Electronic Controls – Right Side

To prevent possible damage to the ECM, turn the ignition key off before clearing the memory. The trouble code memory is fed a continuous 12 volts even with the key off. To clear this memory after a fault has been corrected requires the removal of the 20-amp fuse located next to the battery for 10 seconds.

Diagnostic Circuits

Always block the vehicle drive wheels while checking the system. The oxygen sensor will cool off after a short period of operation at idle. This will put the system on "Open Loop". To restore "Closed Loop" operation, run the engine at part throttle for several minutes and accelerate from idle to part throttle several times.

Code 12: Distributor Pulses —This indicates the distributor spark tuning reference pulses generated by the pick-up coil are not received by the ECM.

Code 13: Oxygen Coolant Circuit — When the oxygen sensor is warm (above 390°F) the output voltage will be between 0 and 1.0 volts. The code 13 indicates that the oxygen sensor will not move above or below its cold voltage of 0.5 volts and therefore the system will not go to closed loop operation.

Code 14: Coolant Sensor Circuit — This indicates the coolant sensor signal voltage is too low because of shorted circuitry.

Code 15: Coolant Sensor Circuit — This indicates the coolant sensor signal voltage is too high because of an open circuit.

Code 21: Throttle Position Sensor Circuit — The ECM supplies a 5-volt signal to pin C of the Throttle Position Sensor and a ground path to pin A. The TPS signal from pin B is variable dependent on throttle position. This code indicates the voltage signal from the TPS is too high.

Code 22: Throttle Position Sensor Circuit — The ECM supplies a 5 volt signal to pin C of the Throttle Position Sensor and a ground path to pin A. The TPS signal from pin B is variable dependent on throttle position. This code indicates the voltage signal from the TPS is too low.

Code 24: Speed Sensor Circuit — The speed sensor generates an electrical signal dependent on vehicle speed. This signal is then amplified and inverted for the ECM. This code indicates that the speed signal is not being received by the ECM.

Code 33: Manifold Absolute Pressure Sensor Circuit — This indicates either an open circuit to ground or a short to the sensor circuit.

Code 34: Manifold Absolute Pressure Sensor Circuit — This indicates a plugged or leaking sensor vacuum line.

Code 42: Electronic Spark Timing Monitor — Bypass open or grounded. The ECM reduces the HEI bypass line to 0 volts when cranking. After the engine starts the ECM increases this voltage to approx. 5 volts. This 5-volt signal commands the HEI module to use the spark timing supplied by the ECM. The ECM monitors the HEI bypass feedback line to see if it is operating properly. This code indicates the ECM is not receiving bypass voltage.

Code 43: Electronic Spark Control Signal — This code indictates that the spark control is not in the retard mode as it should be.

Code 44: Lean Exhaust System — Under normal operation the oxygen sensor output voltage varies between 0 and 1.0 volts. This signal indicates that the voltage reading is below 0.3 volts.

Code 45: Rich Exhaust System — As in the Lean Exhaust System code above the voltage at the oxygen sensor output should be between 0 and 1.0 volts. This code indicates the voltage is above 0.6 volts.

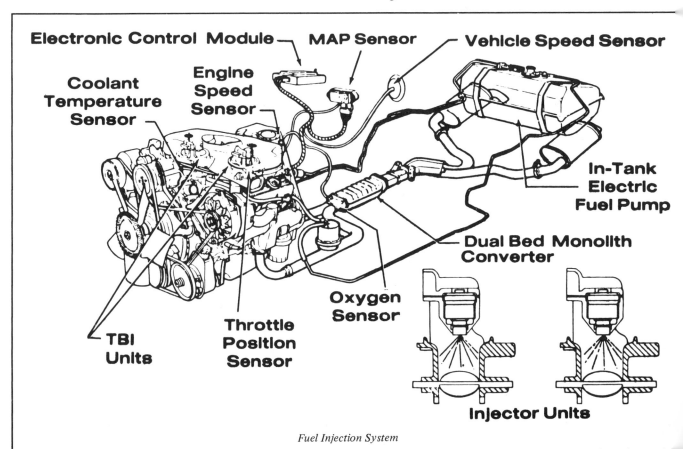

Fuel Injection System

Code 51: PROM Check — The PROM circuit is a calibration unit within the ECM that adjusts the ECM commands for basic vehicle variations such as engine size, transmission type, vehicle weight, etc. This code indicates that the PROM is not read properly by the ECM. Bent pins, missing pins or PROMS installed backwards could cause this code to set.

Code 55 ECM — This signal indicates a faulty ECM.

On-Car Service

When troubleshooting the Computer Command Control system check all vacuum and electrical connections. This should be done before testing or replacing components.

Electronic Control Module: This unit is located behind the driver's seat inside the battery compartment. Disconnect the negative battery terminal. The line drawing shows the mounting parts.

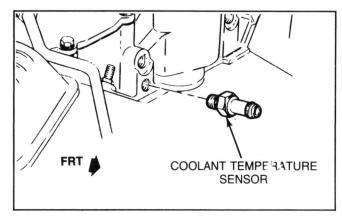

Coolant Sensor

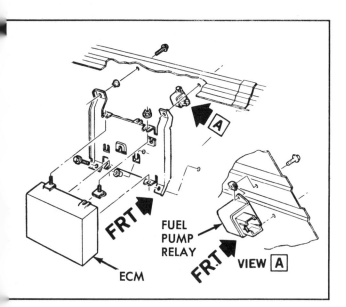

Electronic Control Module

PROM Replacement: The PROM plugs into a socket inside the ECM. If code 51 indicates the need for a PROM replacement be sure the correct new PROM is ordered — there are different units depending on vehicle specifications. If a new ECM is being installed, remove the PROM from the defective unit and use it in the new one.

Coolant Sensor: With the sensor tip at 59°F, the resistance across the terminals should be 4115-4750 ohms. Remove the electrical connection at the temperature switch. Replace the temperature switch and tighten it to 72 in. lbs. Connect the electrical terminal.

Manifold Absolute Pressure Sensor: This sensor is located on a bracket on the left cowl and dash assembly.

Oxygen Sensor: This sensor has a permanent pig-tail and connector. The electrical connector and louvered end must be kept clean. Remove this sensor when the engine temperature is above 120°F. If the old sensor is re-used coat the threads with anti-sieze compound.

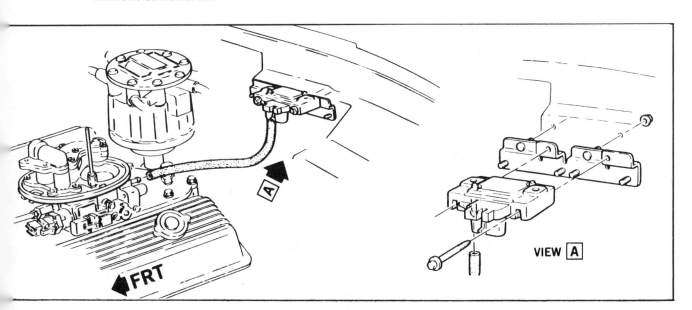

Manifold Absolute Pressure Sensor

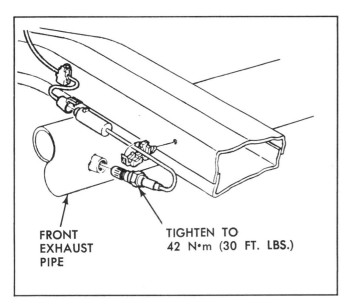

Oxygen Sensor

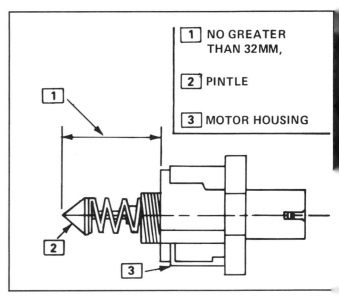

Idle Air Control Motor

Model 400 TBI

Idle Air Control Motor: Remove the air cleaner and remove the electrical connector from the motor. Remove the motor from the throttle body using a 1¼" wrench. Before installing a new idle air control motor measure the distance the valve is extended from the motor housing to the end of the pintle. It should be not more thatn 1¼". Install the new motor using a new gasket. Tighten the motor to 13 ft. lbs. Connect the electrical connector. Install the air cleaner.

Fuel Pressure Regulator or Compensator: Remove the air cleaner and disconnect the electrical connector at the injector. Remove five screws holding the fuel meter cover to the body. The fuel meter cover is serviced as a complete assembly only — do not remove the pressure regulator compensator from the cover. Do not immerse the cover in any type of cleaner.

Install new gaskets on the cover and a new dust seal, then install the cover. Reconnect the electrical connector. Install the air cleaner.

Injector Replacement: Remove the air cleaner and remove the elcetrical connector at the injector. Remove the fuel meter cover. Using small pliers, carefully remove the injector. Before installing the injector, remove all fuel from inside the fuel meter body. Lubricate the new small "O" ring with lithium grease and place it on the injector. Install the steel washer in the fuel meter body. Lubricate the large "O" ring with lithium grease and install it above the back-up washer. Install the injector aligning the raised lug on the injector base with the notch cast in the fuel meter body. Install the fuel meter cover. Install the electrical connector and the air cleaner.

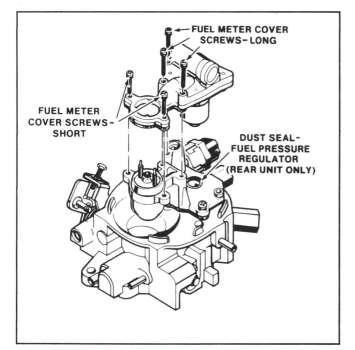

Fuel Regulator or Compensator

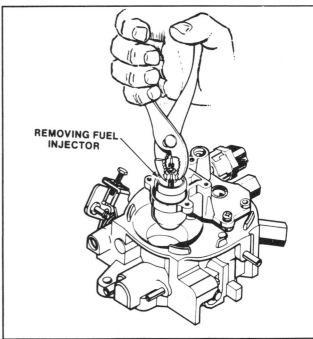

Fuel Injector

Throttle Position Sensor: Do not immerse this sensor in any cleaner. Remove the air cleaner and disconnect the wire harness. Connect the TPS harness to the TPS by means of three jumper wires. With the ignition ON measure the voltage between terminals A and B using a digital voltmeter. It should read 0.525 ± 0.075 volts. Adjust the TPS if required. With the ignition off, remove the jumpers and reconnect the harness. Install the air cleaner.

Throttle Position Sensor Adjustment: Remove the air cleaner and the TPS harness. Remove the TPS retaining screws and apply thread locking compound to the screws. Loosely reinstall the screws. Using jumper wires connect the harness to the TPS. With the ignition ON, measure the voltage between terminals A and B using a digital voltmeter. Rotate the TPS switch until 0.525 ± 0.075 volts is obtained. Tighten the screws. Turn the ignition OFF and remove the jumper wires. Reconnect the TPS harness and install the air cleaner.

Fuel Pump Relay: This unit is mounted in the passenger compartment behind the driver's seat. There are no adjustments to this unit — only complete replacement is made.

Electronic Spark Timing: This system consists of control circuitry within the ECM and the EST distributor with a seven-terminal HEI module. The distributor has neither a centrifugal nor a vacuum advance as the advance is compensated for by the ECM. Three terminals on the HRI module are called Reference, Bypass and EST. If the Ref or EST signals are interrupted by either open lines or a faulty ECM, the car will stop. The HEI/EST module will provide a timing signal

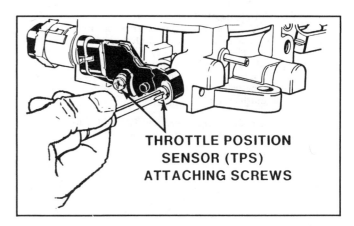

Throttle Position Sensor

based on engine RPM. If the bypass signal is lost, the ECM cannot control the HEI/EST module.

Electronic Spark Control: Trouble code 43 indicates a problem in the ESC system. Repair or replace the ESC.

Other emission control devices such as the air injection pump, exhaust gas recirculation, evaporative control system, etc., are covered elsewhere in this manual.

AUXILIARY COOLING FAN

1981 and later models are equipped with an auxiliary cooling fan. This fan is electrically powered through the circuit breaker in the fuse panel, when the switch is in the "RUN" position. It is controlled by an engine temperature switch which is screwed into the engine block. When the coolant temperature reaches 238°F, the switch closes and starts the fan. When the temperature decreases to 201°F, the switch opens and the fan is shut off.

Keep all objects away from this fan as it can start at any time if the ignition switch is in the "RUN" position.

To remove the fan, remove the negative battery cable and the fresh air scoop. Remove the engine fan and the upper radiator shroud screws. Remove the auxiliary fan and remove the wire connector at the motor.

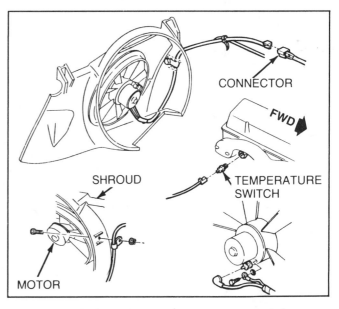

Cooling Fan Connector & Temperature Switch

AUTOMATIC TRANSMISSION

The Model 700-R4 automatic transmission consists of a three-element hydraulic torque converter with the addition of a converter clutch. The torque converter through the converter clutch provides a locked-up mechanical drive in fourth or top gear at the proper road speed and throttle position. The transmission provides four forward speeds with third speed direct and fourth overdrive.

With the exception of oil pressures and T.V. cable adjustments, refer to the Turbo Hydra-Matic 400 transmission for diagnosis and adjustments.

T.V. Oil Pressures

To check minimum T.V. line pressures set the T.V. cable to specifications and with the brakes applied, take the line pressure readings in the ranges and at the engine RPM's indicated in the chart.

To check the full T.V. line pressures hold the T.V. cable to the full extent of its travel and with the brakes applied, take the line pressure readings in the ranges and at the engine RPM's indicated in the chart.

Oil Pressures at 1000 RPM in PSI

Gear	Minimum T.V.	Full T.V.
P&N	55-65	130-170
R	90-105	210-285
D&3	55-65	130-170
2&L	100-120	100-120

T.V. Cable

The T.V. cable in the 700-R4 transmission does not operate as a downshift cable does in other transmissions. It controls line pressure, shift points, shift feel, part throttle downshifts and detent downshifts. It operates the throttle lever and bracket assembly. The function of this assembly is to transfer the carburetor throttle plate movement to the T.V. plunger in the control valve assembly as relayed by the T.V. cable and linkage. This causes T.V. pressure and line pressure to increase according to throttle opening and also controls part throttle and detent downshifts. Properly adjusted, the T.V. cable plunger when fully depressed will be flush with the T.V. bushing at wide open throttle.

It also prevents the transmission from operating at low pressures if the T.V. cable is broken or disconnected.

A sticking or binding T.V. linkage can result in delayed or full throttle shifts. Check the linkage for sticking or binding with the engine running at idle speed with the selector in N and the parking brake set. Pull the T.V. cable the full length of its travel through the cable terminal, then release the cable. It should return to the closed throttle position. If it remains ahead of the cable terminal, it may be caused by any one of the following malfunctions.

Sharp bends or a damaged cable housing. Straighten or replace the cable as necessary.

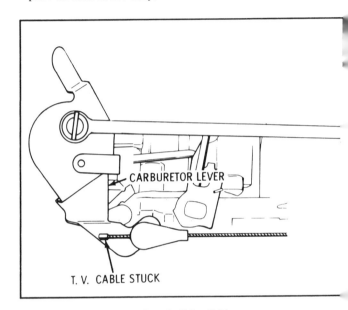

Sticking Throttle Valve Cable

A sharp burr on the T.V. link dragging in the cable housing. Smooth the end using a file or stone.

A bent T.V. link. Straighten or replace.

Misalignment of the throttle lever and bracket assembly. Correct or replace.

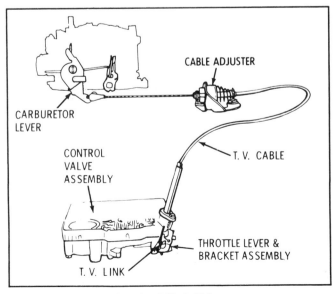

Throttle Valve Cable & Linkage

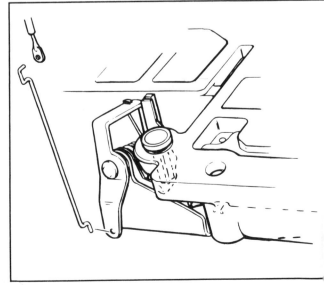

Alignment of Throttle Lever & Bracket

Throttle lever spring unhooked or damaged.

If the T.V. cable is adjusted too short it will raise the line pressure and shift points and it might limit the injector opening preventing full throttle operation.

T.V. Cable Adjustments
Check the transmission oil level and be sure the engine is operating properly and the brakes are not dragging. Be sure the vehicle has the correct cable and that it is connected at both ends.

With the engine stopped, depress the re-adjust tab. Move the slider through the fitting away from the throttle body until the slider stops against the fitting. Release the re-adjust tab. Open the throttle lever to full throttle stop position to automatically adjust the cable. Release the throttle lever. Road test the vehicle.

If delayed or only full throttle shifts still occur proceed as follows: Remove the oil pan and check that the exhaust valve lifter rod is not distorted and not binding in the control valve assembly or spacer plate. The exhaust check ball must move up and down as the lifter does. Be sure the lifter spring holds the lifter rod up against the bottom of the control valve assembly. Make sure the T.V. plunger is not stuck. Check for correct throttle lever to cable link.

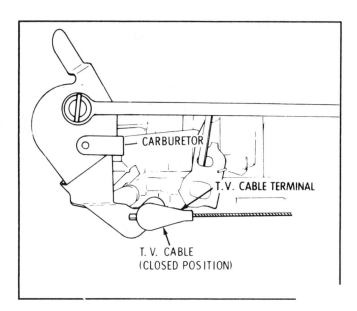

Throttle Valve Cable at Carburetor Lever

DRIVE AXLE AND DRIVE LINE

REAR SUSPENSION

Starting in 1981 some vehicles were produced with a new fiberglass-reinforced plastic leaf spring. Refer to the Drive Axle and Driveline section of this manual for spring removal. This procedure is the same and the same precautions should be used.

DIFFERENTIAL CARRIER AND COVER

1980 and later models have a new design of carrier and carrier support. To remove the carrier and cover assembly, first raise the vehicle and remove the spare tire. Remove the spare tire cover by removing the support hooks attached to the carrier cover.

Remove the exhaust system. Place jack stands under the front control arms. Remove the heat shield. Using a floor jack and C-clamp, raise the spring to relieve the load and disconnect the spring. Remove the spring at the cover plate. Mark the cam bolt and remove it from the bracket.

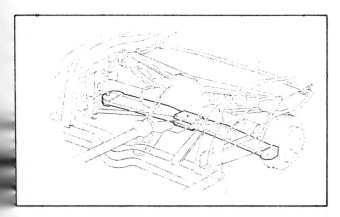

Fiberglass-Reinforced Plastic Leaf Spring

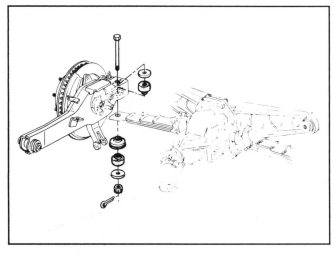

Spring To Spindle Support

Remove strut bracket to carrier bolts and lower the strut rods. Scratch mark the propeller shaft and disconnect it at the pinion companion flange to gain access to the carrier mounting bolts. Place a jack stand under the carrier and remove the carrier to body bolts. Lower the differential to gain access to all the cover bolts. Drain the differential and remove the cover.

If the carrier is to be removed, disconnect the drive shaft at the spindle companion flange. Lower and remove the differential assembly. Remove driveshafts from side yokes.

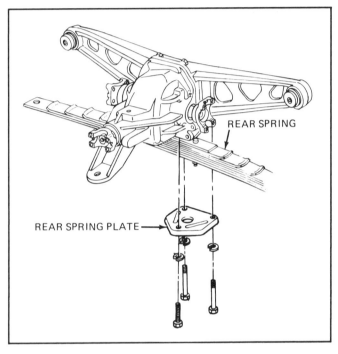

Spring To Carrier

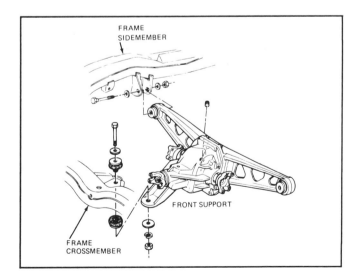

Differential Carrier Front Support

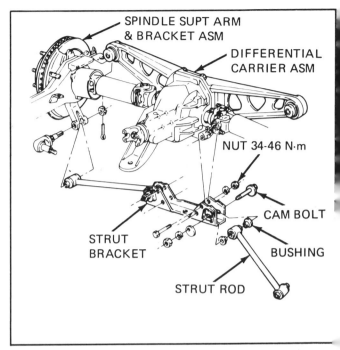

Strut To Carrier

BODY ELECTRICAL

WINDSHIELD WIPER SYSTEM

This system consists of a wiper motor, a pulse relay and electronic delay relay. There is also the washer and wiper switch, a controller assembly and the washer pump. Refer to the Electrical section of this manual for service. The wiper motor is located in the center engine bulkhead. The controller assembly is located in the jack storage box.

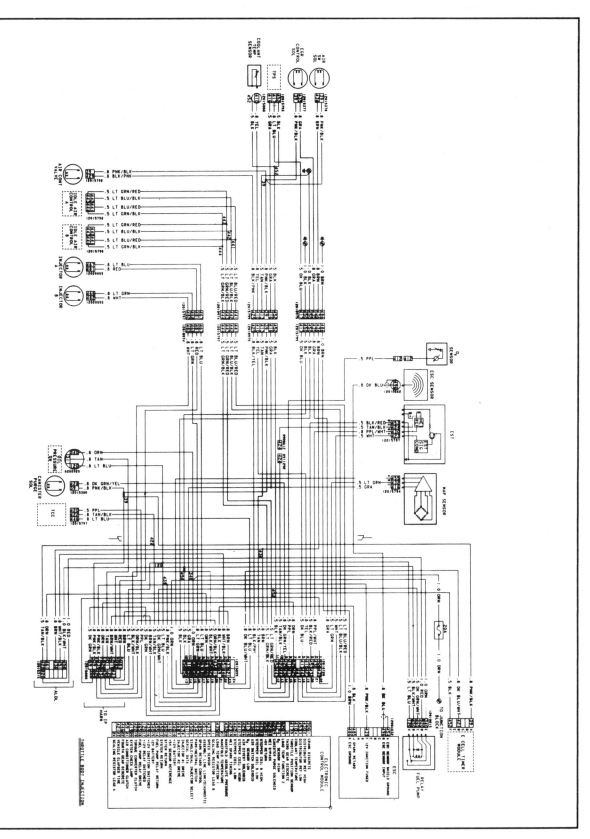

1982 Wiring Diagram

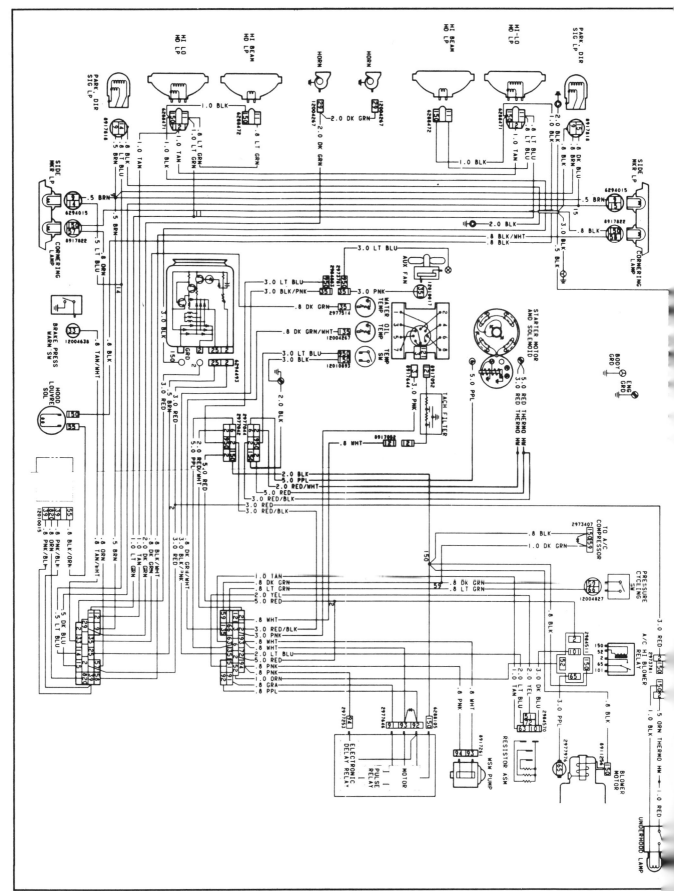

1982 Wiring Diagram

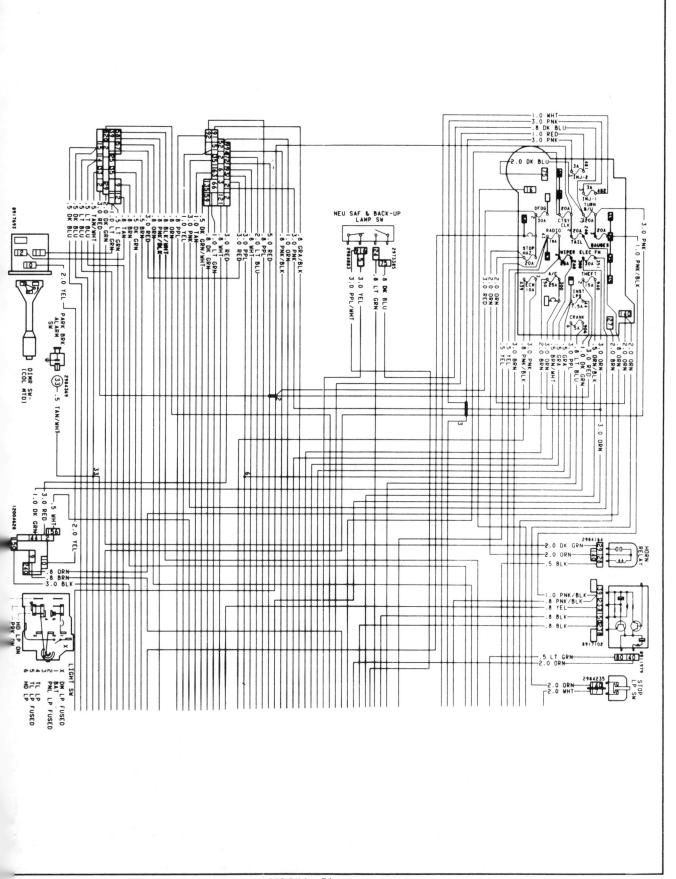

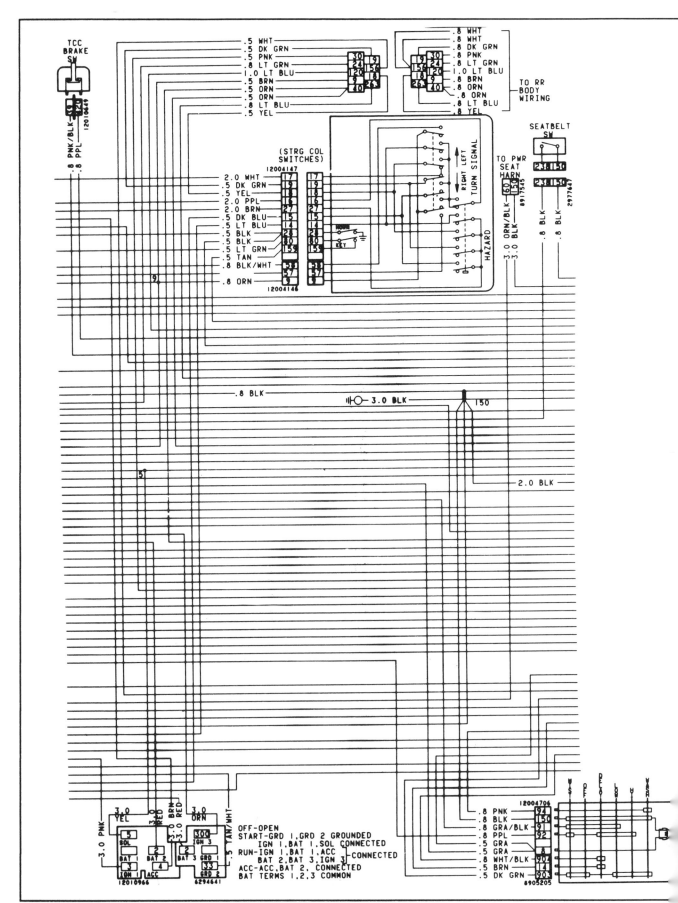

1982 Wiring Diagram

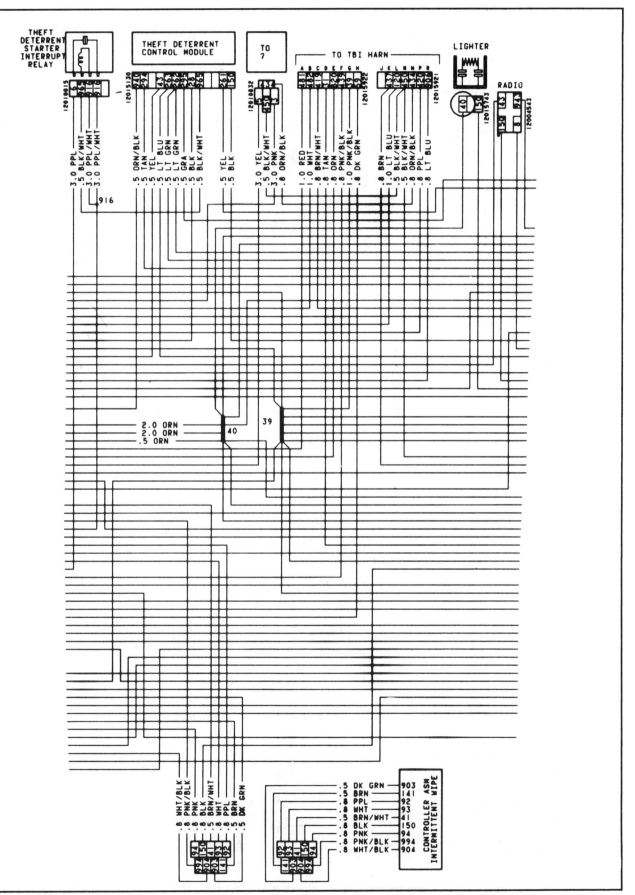

1982 Wiring Diagram

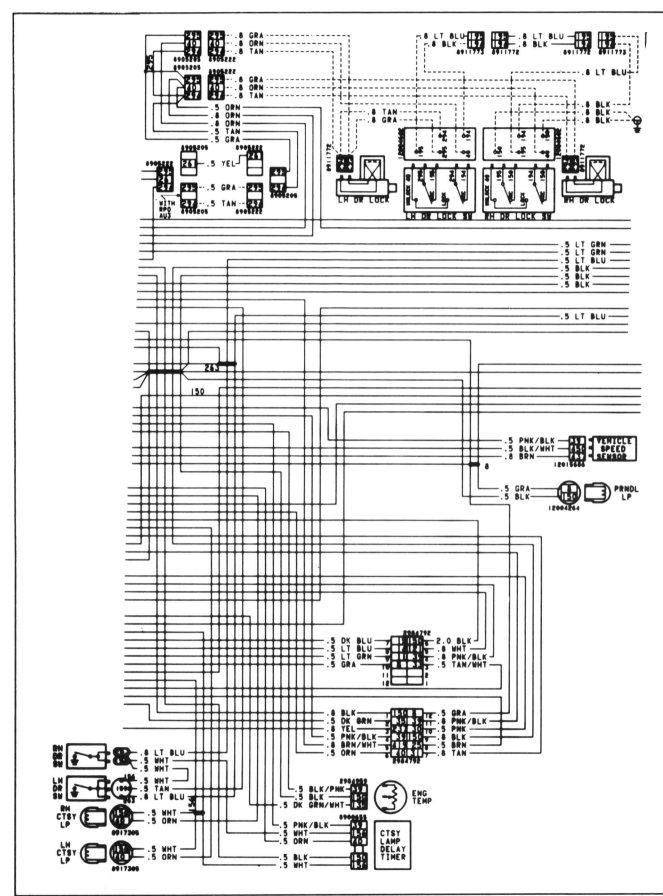

1982 Wiring Diagram

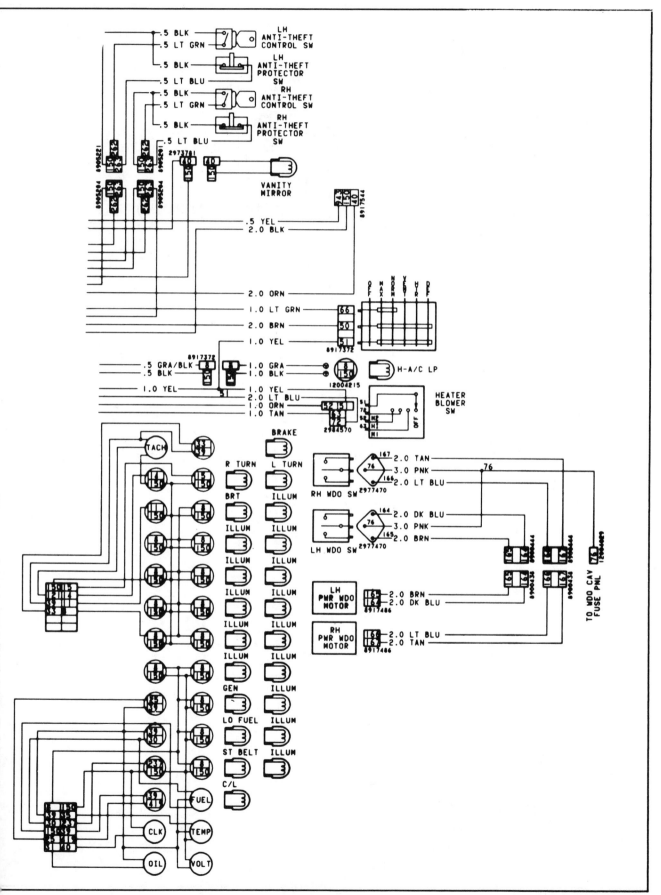

1982 Wiring Diagram

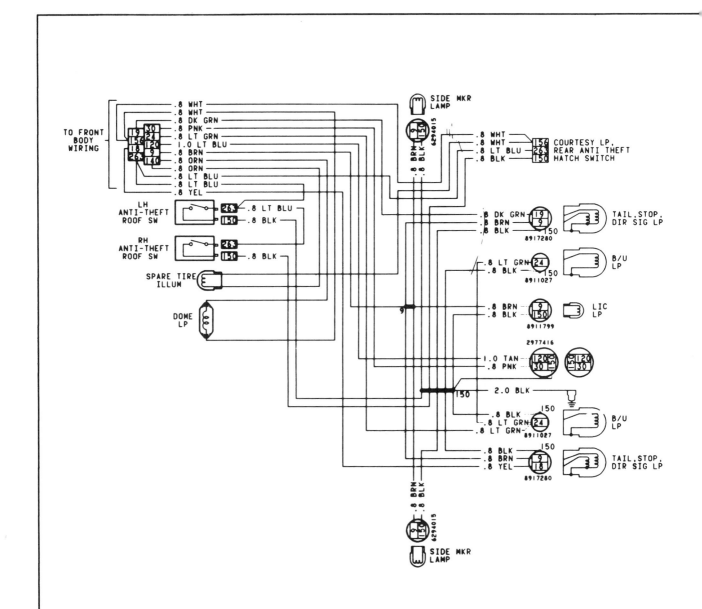

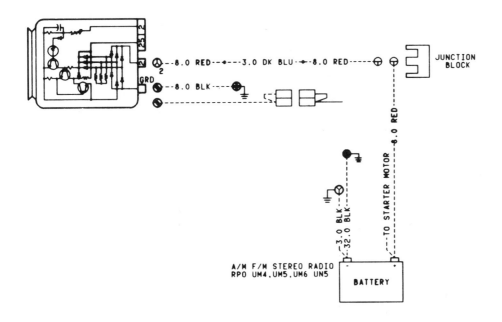

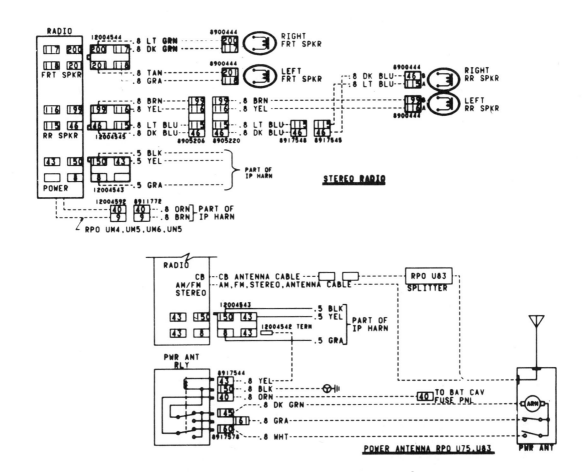

SPECIFICATIONS
CARBURETORS

Year	Model	Float Level	Air Valve Spring	Pump Rod	Vacuum Break Prim.	Vacuum Break Secon.	Choke Rod	Choke Unloader
1977	17057202	15/32	7/8	15/32	0.180	---	0.325	0.280
	17057204	15/32	3/4	9/32	0.160	---	0.325	0.280
	17057203	15/32	7/8	15/32	0.180	---	0.325	0.280
	17057502	15/32	7/8	15/32	0.165	---	0.325	0.280
	17057504	15/32	7/8	9/32	0.165	---	0.325	0.280
	17057210	15/32	1	15/32	0.180	---	0.325	0.280
	17057510	15/32	1	9/32	0.180	---	0.325	0.280
	17057528	15/32	1	9/32	0.180	---	0.325	0.280
	17057211	15/32	1	15/32	0.180	---	0.325	0.280
	17057228	13/32	1	15/32	0.180	---	0.325	0.280
	17057582	15/32	7/8	13/32	0.180	---	0.325	0.280
	17057584	15/32	1	9/32	0.180	---	0.325	0.280
1978	17058202	15/32	7/8	9/32	0.179	---	0.314	0.277
	17058203	15/32	7/8	9/32	0.179	---	0.314	0.277
	17058204	15/32	7/8	9/32	0.179	---	0.314	0.277
	17058210	15/32	1/2	9/32	0.203	---	0.314	0.277
	17058211	15/32	1/2	9/32	0.203	---	0.314	0.277
	17058228	15/32	7/8	9/32	0.203	---	0.314	0.277
	17058502	15/32	7/8	9/32	0.187	---	0.314	0.277
	17058504	15/32	7/8	9/32	0.187	---	0.314	0.277
	17058582	15/32	7/8	9/32	0.203	---	0.314	0.277
	17058584	15/32	7/8	9/32	0.203	---	0.314	0.277
1979	1705 9203	15/32	7/8	1/4	0.157	---	0.243	0.243
	17059207	15/32	7/8	1/4	0.157	---	0.243	0.243
	17059216	15/32	7/8	1/4	0.157	---	0.243	0.243
	17059217	15/32	7/8	1/4	0.157	---	0.243	0.243
	17059218	15/32	7/8	1/4	0.164	---	0.243	0.243
	17059222	15/32	7/8	1/4	0.164	---	0.243	0.243
	17059502	15/32	7/8	1/4	0.164	---	0.243	0.243
	17059504	15/32	7/8	1/4	0.164	---	0.243	0.243
	17059582	15/32	7/8	11/32	0.203	---	0.243	0.314
	17059584	15/32	7/8	11/32	0.203	---	0.243	0.314
	17059210	15/32	1	9/32	0.157	---	0.243	0.243
	17059211	15/32	1	9/32	0.157	---	0.243	0.243
	17029228	15/32	1	9/32	0.157	---	0.243	0.243
1980	17080202	7/16	7/8	1/4*	0.157	---	0.110	0.243
	17080204	7/16	7/8	1/4*	0.157	---	0.110	0.243
	17080207	7/16	7/8	1/4*	0.157	---	0.110	0.243
	17080228	7/16	7/8	9/32*	0.179	---	0.110	0.243
	17080243	3/16	9/16	9/32*	0.016	0.083	0.074	0.179
	17080274	15/32	5/8	5/16**	0.110	0.164	0.083	0.203
	17080282	7/16	7/8	11/32**	0.142	---	0.110	0.243
	17080284	7/16	7/8	11/32**	0.142	---	0.110	0.243
	17080502	1/2	7/8	Fixed	0.136	0.179	0.110	0.243
	17080504	1/2	7/8	Fixed	0.136	0.179	0.110	0.243
	17080542	3/8	9/16	Fixed	0.103	0.066	0.074	0.243
	17080543	3/8	9/16	Fixed	0.103	0.129	0.074	0.243
1981	17081202	11/32	7/8	Fixed	0.149	---	0.110	0.243
	17081203	11/32	7/8	Fixed	0.149	---	0.110	0.243
	17081204	11/32	7/8	Fixed	0.149	---	0.110	0.243
	17081207	11/32	7/8	Fixed	0.149	---	0.110	0.243
	17081216	11/32	7/8	Fixed	0.149	---	0.110	0.243
	17081217	11/32	7/8	Fixed	0.149	---	0.110	0.243
	17081218	11/32	7/8	Fixed	0.149	---	0.110	0.243
	17081242	5/16	9/16	Fixed	0.090	0.077	0.139	0.243
	17081243	1/4	9/16	Fixed	0.103	0.090	0.139	0.243

*Inner Hole   **Outer Hole

IDLE SPEEDS

Year	Engine/HP	Manual Neutral	Automatic Drive
1977	350/180	700	500/600
	350/210	800	500/600
1978	350/185	700	500
	350/220	900	700
1979	350/195	See Underhood Sticker	
	350/225	See Underhood Sticker	
1980	305/All	See Underhood Sticker	
	350/L48	See Underhood Sticker	
	350/L82	See Underhood Sticker	
1981	350/L81	See Underhood Sticker	
1982	350	See Underhood Sticker	

ENGINE SPECIFICATIONS

Year	Displ.	HP @ RPM	Comp. Ratio
1975	350	155@3800	8.5
	350	165@3800	8.5
	350	205@4800	9.0
1976	350	145@3800	8.5
	350	165@3800	8.5
	350	180@4000	8.5
	350	210@5200	9.0
1977	350	170@3800	8.5
	350	180@4000	8.5
	350	210@5200	9.0
1978	350	185@4000	8.4
	350	220@5200	8.9
1979	350	195@4000	8.2
	350	225@5200	8.9
1980	305	180@4200	8.6
	350	190@4400	8.2
	350	230@5200	9.0
1981	350	190@4400	8.2
1982	350	190@4400	8.2

The basic engine specifications which follow are for the 5.0 litre, 305 cu. in. displacement Model LG4 engine. For specifications on the 350 cu. in. engine, use the 1979 information together with the appropriate transmission.

Displacement, cu. in. 305
Bore & Stroke, in. . . . . 3.736 x 3.480
Firing Order 1-8-4-3-6-5-7-2
Compression Ratio 8.5:1
Tappet Clearance, hydraulic lifters . . . . One turn down from zero lash
Main Bearing Clearance, #1. . . . . .0008-.0020
　　　　　　　　　　　 #2, 3, 4 . . . . .0011-.0023
　　　　　　　　　　　 #5 . . . . .0017-.0032
Crankshaft Crankpin Dia. 2.0988-2.0998
　　　　　Rod Bearing Clearance 0013-.0035
　　　　　Rod Side Clearance 008-.014
Camshaft Journal Dia. 1.8682-1.8692
　　　　　Lobe Lift, Intake 2484
　　　　　　　　　 Exhaust 2667
Valve Lifters . . . . Hydraulic
　　　　　Face Angle, Intake & Exhaust 45º
　　　　　Seat Angle, Intake & Exhaust 46º
　　　　　Seat Width, Intake 1/32-1/16
　　　　　　　　　 Exhaust 1/16-3/32
　　　　　Stem Clearance, Intake 0010-.0027
　　　　　　　　　 Exhaust 0010-.0027
　　　　　Spring Free Length 2.03
Pistons, Clearance 0007-.0017
　　　　　Ring Groove Clearance, Top Comp. 0012-.0032
　　　　　　　　　　　　 2nd Comp. 0012-.0032
　　　　　　　　　　　　 Oil 002-.007
　　　　　Ring Gap, Top Comp. 010-.020
　　　　　　　　　 2nd Comp. 010-.025
　　　　　　　　　 Oil 015-.055
　　　　　Wrist Pin Dia. 9207-.9273
　　　　　　　　　 Clearance in Piston 00025-.00035
　　　　　　　　　 Fit in Rod 0008-.0016

SPARK PLUGS & TIMING

Year	Engine Cu.In./HP	Spark Plugs	Gap, In.	Timing, Deg. Manual	Auto.
1975	350/165	R44TX	.060	6B	6B
	350/205	R44TX	.060	12B	12B
1976	350/180	R45TS	.045	8B	8B
	350/210	R45TS	.045	12B	12B
1977	350/180	R45TS	.045	8B	8B
	350/210	R45TS	.045	12B	12B
1978	350/185	R45TS	.045	6B	6B
	350/220	R45TS	.045	12B	12B
1979	350/195	R45TS	.045	6B	6B
	350/225	R45TS	.045	12B	12B
1980	305/All	R43TS	.045	See Car Sticker	
	350/L48	R43TS	.045	See Car Sticker	
	350/L82	R43TS	.045	See Car Sticker	
1981	350/L81	R43TS	.045	See Car Sticker	
1982	350	- - - - -See Car Sticker - - - - - - - - -			

CAPACITIES

Year	Engine Cu. In.	Crankcase*	Transmission, Pts. Manual 3 Sp.	4 Sp.	Auto.	Rear Axle, Pts.	Gas Tank, Gals.	Cooling W/O A/C	With A/C
1975	350	4	- - -	3	8	4	18	17	17
1976	350	4	---	3	8	4	18	18	18
1977	350	4	---	3	8	4	17	21	21
1978	350	4	---	3	8	4	24	21	21
1979	350	4	---	3	8	3.75	24	21	21
1980	305	4	---	---	8	3.75	24	21	22
1981	350	4	---	3	8	3.75	24	21	21
1982	350	4	---	3	8	3.75	24	21	21

*Add one quart for filter.

Hydra-Matic 700-R4 Automatic Transmission

Ratios: First 3.06
Second 1.62
Third 1.00
Fourth .70
Rev. 2.29

WHEEL ALIGNMENT
Front
Caster . . . . 1-1/4P - 3-1/4P
Camber . . . . 0 - 1-1/2P
Toe-In . . . . .19-.31"
Steering Axis Inclination . 7.68°

Rear
Camber . . . . 0° ±
Toe-in (Per Wheel) . . . . .06"